Scientific Computing with Multicore and Accelerators

Chapman & Hall/CRC
Computational Science Series

SERIES EDITOR

Horst Simon
Associate Laboratory Director, Computing Sciences
Lawrence Berkeley National Laboratory
Berkeley, California, U.S.A.

AIMS AND SCOPE

This series aims to capture new developments and applications in the field of computational science through the publication of a broad range of textbooks, reference works, and handbooks. Books in this series will provide introductory as well as advanced material on mathematical, statistical, and computational methods and techniques, and will present researchers with the latest theories and experimentation. The scope of the series includes, but is not limited to, titles in the areas of scientific computing, parallel and distributed computing, high performance computing, grid computing, cluster computing, heterogeneous computing, quantum computing, and their applications in scientific disciplines such as astrophysics, aeronautics, biology, chemistry, climate modeling, combustion, cosmology, earthquake prediction, imaging, materials, neuroscience, oil exploration, and weather forecasting.

PUBLISHED TITLES

Chapman & Hall/CRC Computational Science Series

Scientific Computing with Multicore and Accelerators

Edited by

Jakub Kurzak
David A. Bader
Jack Dongarra

CRC Press
Taylor & Francis Group
Boca Raton   London   New York

CRC Press is an imprint of the
Taylor & Francis Group, an **informa** business

A CHAPMAN & HALL BOOK

CRC Press
Taylor & Francis Group
6000 Broken Sound Parkway NW, Suite 300
Boca Raton, FL 33487-2742

First issued in paperback 2017

© 2011 by Taylor and Francis Group, LLC
CRC Press is an imprint of Taylor & Francis Group, an Informa business

No claim to original U.S. Government works

ISBN-13: 978-1-4398-2536-5 (hbk)
ISBN-13: 978-1-138-11332-9 (pbk)

Library of Congress Cataloging-in-Publication Data

Scientific computing with multicore and accelerators / edited by Jakub Kurzak, David A. Bader, and Jack Dongarra.
    p. cm. -- (Chapman & Hall/CRC computational science ; 10)
    Summary: "The current trend in microprocessor architecture is toward powerful multicore designs in which a node contains several full-featured processing cores, private and shared caches, and memory. The IBM Cell Broadband Engine (B.E.) and Graphics Processing Units (GPUs) are two accelerators that are used for a variety of computations, including signal processing and quantum chemistry. This is the first reference on the use of Cell B.E. and GPUs as accelerators for numerical kernels, algorithms, and computational science and engineering applications. With contributions from leading experts, the book covers a broad range of topics on the increased role of these accelerators in scientific computing"-- Provided by publisher.
    Includes bibliographical references and index.
    ISBN 978-1-4398-2536-5 (hardback)
    1. Science--Data processing. 2. Engineering--Data processing. 3. High performance computing. 4. Multiprocessors. I. Kurzak, Jakub. II. Bader, David A. III. Dongarra, J. J. IV. Title. V. Series.

Q183.9.S325 2010
502.85--dc22   2010037123

Visit the Taylor & Francis Web site at
http://www.taylorandfrancis.com

and the CRC Press Web site at
http://www.crcpress.com

Contents

III   Multigrid Methods   111

6   Hardware-Oriented Multigrid Finite Element Solvers on GPU-Accelerated Clusters   113

Stefan Turek, Dominik Göddeke, Sven H.M. Buijssen, and Hilmar Wobker

V Combinatorial Algorithms   193

10 Combinatorial Algorithm Design on the Cell/B.E. Processor 195

David A. Bader, Virat Agarwal, Kamesh Madduri, and Fabrizio Petrini

VI Stencil Algorithms 217

VIII   Molecular Modeling   329

15 Drug Design on the Cell BE   331

*Cecilia González-Álvarez, Harald Servat, Daniel Cabrera-Benítez,
Xavier Aguilar, Carles Pons, Juan Fernández-Recio, and Daniel
Jiménez-González*

List of Figures

List of Tables

Preface

Recent activities of major chip manufacturers such as NVIDIA, Intel, AMD, and IBM make it more evident than ever that future designs of microprocessors and large high-performance computing (HPC) systems will be hybrid/heterogeneous in nature. These heterogeneous systems will rely on the integration of two major types of components in varying proportions:

Multi- and many-core CPU technology — The number of cores will continue to escalate because of the desire to pack more and more components on a chip while avoiding the power wall, instruction-level parallelism wall, and the memory wall.

Special purpose hardware and massively parallel accelerators — For example, graphics processing units (GPUs) from NVIDIA have outpaced standard central processing units (CPUs) in floating point performance in recent years. Furthermore, they have arguably become as easy, if not easier, to program than multicore CPUs.

The relative balance between these component types in future designs is not clear and will likely vary over time. There seems to be no doubt that future generations of computer systems, ranging from laptops to supercomputers, will consist of a composition of heterogeneous components. Indeed, the petaflop/s (10^{15} floating-point operations per second) performance barrier was breached by such a system accelerated with thousands of IBM Cell Broadband Engine processors, the chip originally developed for the Sony PlayStation3. Today high-performance computing is relying on commodity processors from gaming and entertainment (e.g. GPUs, Cell); not only is this book for supercomputing experts, but also for the masses who now have extreme performance in their laptop or game box.

And yet the problems and the challenges for developers in the new computational landscape of hybrid processors remain daunting. Critical parts of the software infrastructure are already having a very difficult time keeping up with the pace of change. In some cases, performance cannot scale with the number of cores because an increasingly large portion of time is spent on data movement rather than arithmetic. In other cases, software tuned for performance is delivered years after the hardware arrives and so is obsolete on delivery. And in some cases, as on some recent GPUs, software will not run at all because programming environments have changed too much.

This book gives an introduction to the area of hybrid computing by providing examples and insight into the process of constructing and effectively using the technology. It presents introductory concepts of parallel computing from simple examples to debugging (both logical and performance), as well as covering advanced topics and issues related to the use and building of many applications. Throughout the book, examples reinforce the concepts that have been presented.

Jakub Kurzak, David A. Bader, and Jack Dongarra

About the Editors

Jakub Kurzak is a research director in the Innovative Computing Laboratory in the Department of Electrical Engineering and Computer Science at the University of Tennessee. He received his Ph.D. in 2005 from the University of Houston, Texas. Currently, his research focuses on utilizing multicore systems and accelerators for scientific computing. Dr. Kurzak serves as program committee member for several international conferences and as a reviewer for a number of top ranking journals. He has written a number of journal papers and several book chapters about fast implementations of dense linear algebra computations on multicore processors and accelerators.

David A. Bader is a full professor in the School of Computational Science and Engineering, College of Computing, at Georgia Institute of Technology. Dr. Bader is the founding director of the NSF Center for Hybrid Multicore Productivity Research at Georgia Tech, a task leader in the Center for Adaptive Supercomputing Software for Multithreaded Architectures (CASS-MT), and has also served as director of the Sony-Toshiba-IBM Center of Competence for the Cell Broadband Engine Processor. He received his Ph.D. in 1996 from the University of Maryland and was awarded a National Science Foundation (NSF) postdoctoral research associateship in experimental computer science. He is an NSF CAREER Award recipient, is an investigator on several NSF and NIH awards, was a distinguished speaker in the IEEE Computer Society Distinguished Visitors Program, and is a member of the IBM PERCS team for the DARPA High Productivity Computing Systems program. Dr. Bader serves on the Research Advisory Council for Internet2 and the Steering Committees of the IPDPS and HiPC conferences. He is an associate editor for several high-impact publications including the *ACM Journal of Experimental Algorithmics (JEA), IEEE DSOnline*, and *Parallel Computing*; has been an associate editor for the *IEEE Transactions on Parallel and Distributed Systems (TPDS)*; and is an IEEE Fellow and a Member of the ACM. He is also the editor of the Chapman & Hall/CRC Press book, *Petascale Computing: Algorithms and Applications.* Dr. Bader's interests are at the intersection of high-performance computing and computational biology and genomics. He has co-chaired a series of meetings, the IEEE International Workshop on High-Performance Computational Biology (HiCOMB), co-organized the NSF Workshop on Petascale Computing in the Biological Sciences, written several book chapters, and co-edited special issues of the *Journal of Parallel and Distributed Computing*

(JPDC) and *IEEE TPDS* on *High-Performance Computational Biology.* His main areas of research are in parallel algorithms, combinatorial optimization, and computational biology and genomics.

Jack Dongarra received a Bachelor of Science in mathematics from Chicago State University in 1972 and a Master of Science in computer science from the Illinois Institute of Technology in 1973. He received his Ph.D. in applied mathematics from the University of New Mexico in 1980. He worked at the Argonne National Laboratory until 1989, becoming a senior scientist. He now holds an appointment as University Distinguished Professor of Electrical Engineering and Computer Science in the Computer Science Department at the University of Tennessee and holds the titles of Distinguished Research Staff in the Computer Science and Mathematics Division at Oak Ridge National Laboratory (ORNL), Turing Fellow at Manchester University, and an Adjunct Professor in the Computer Science Department at Rice University. He is the director of the Innovative Computing Laboratory at the University of Tennessee. He is also the director of the Center for Information Technology Research at the University of Tennessee.

He specializes in numerical algorithms in linear algebra, parallel computing, the use of advanced-computer architectures, programming methodology, and tools for parallel computers. His research includes the development, testing, and documentation of high-quality mathematical software. He has contributed to the design and implementation of the following open source software packages and systems: EISPACK, LINPACK, the BLAS, LAPACK, ScaLAPACK, Netlib, PVM, MPI, NetSolve, Top500, ATLAS, and PAPI. He has published approximately 300 articles, papers, reports, and technical memoranda and he is coauthor of several books. He was awarded the IEEE Sid Fernbach Award in 2004 for his contributions in the application of high-performance computers using innovative approaches; in 2008 he was the recipient of the first IEEE Medal of Excellence in Scalable Computing; in 2010 he was the first recipient of the SIAM Special Interest Group on Supercomputing's award for Career Achievement. He is a Fellow of the AAAS, ACM, the IEEE, and SIAM and a member of the National Academy of Engineering.

Contributor List

Virat Agarwal
IBM T.J. Watson Research Center
Yorktown Heights, New York, USA

Xavier Aguilar
PDC Center for High Performance
Computing
Stockholm, Sweden

Srinivas Aluru
Iowa State University
Ames, Iowa, USA

Wesley Alvaro
University of Tennessee
Knoxville, Tennessee, USA

Nathan Bell
NVIDIA Research
Santa Clara, California, USA

Daniel A. Brokenshire
IBM Corporation
Austin, Texas, USA

Sven H.M. Buijssen
TU Dortmund
Dortmund, Germany

Helmar Burkhart
University of Basel
Basel, Switzerland

Daniel Cabrera-Benítez
Barcelona Supercomputing
Center-CNS
Barcelona, Spain

Jonathan Carter
Lawrence Berkeley National
Laboratory
Berkeley, California, USA

Umit V. Catalyurek
The Ohio State University
Columbus, Ohio, USA

Jee Whan Choi
Georgia Institute of Technology
Atlanta, Georgia, USA

Matthias Christen
University of Basel
Basel, Switzerland

Alex Chunghen Chow
Dell Inc.
Round Rock, Texas, USA

Kaushik Datta
University of California, Berkeley
Berkeley, California, USA

Juan Fernández-Recio
Barcelona Supercomputing
Center-CNS
Barcelona, Spain

Renato Ferreira
Universidade Federal de Minas Gerais
Minas Gerais, Brazil

Gordon C. Fossum
IBM Corporation
Austin, Texas, USA

Michael Garland
NVIDIA Research
Santa Clara, California, USA

Dominik Göddeke
TU Dortmund
Dortmund, Germany

Cecilia González-Álvarez
Barcelona Supercomputing
Center-CNS
Barcelona, Spain

David J. Hardy
University of Illinois at
Urbana-Champaign
Urbana-Champaign, Illinois, USA

Mark Harris
NVIDIA Corporation
Santa Clara, California, USA

Timothy D. R. Hartley
The Ohio State University
Columbus, Ohio, USA

Barry Isralewitz
University of Illinois at
Urbana-Champaign
Urbana-Champaign, Illinois, USA

Daniel Jiménez-González
Barcelona Supercomputing
Center-CNS
Universitat Politècnica de Catalunya
Barcelona, Spain

Laxmikant V. Kalé
University of Illinois at
Urbana-Champaign
Urbana-Champaign, Illinois, USA

Michael Kistler
IBM Austin Research Laboratory
Austin, Texas, USA

David M. Kunzman
University of Illinois at
Urbana-Champaign
Urbana-Champaign, Illinois, USA

Kamesh Madduri
Lawrence Berkeley National
Laboratory Berkeley
California, USA

Rajib Nath
University of Tennessee
Knoxville, Tennessee, USA

Esra Neufeld
ETH Zurich
Zurich, Switzerland

Leonid Oliker
Lawrence Berkeley National
Laboratory
Berkeley, California, USA

John D. Owens
University of California, Davis
Davis, California, USA

Davide Pasetto
IBM Computational Science Center
Dublin, Ireland

Maarten Paulides
Erasmus MC-Daniel den Hoed
Cancer Center
Rotterdam, The Netherlands

Fabrizio Petrini
IBM T.J. Watson Research Center
Yorktown Heights, New York, USA

Carles Pons
Barcelona Supercomputing
Center-CNS
Spanish National Institute of
Bioinformatics
Barcelona, Spain

Vipin Sachdeva
IBM Systems and Technology Group
Indianapolis, Indiana, USA

Rafael Sachetto
Universidade Federal de Minas Gerais
Minas Gerais, Brazil

Abhinav Sarje
Iowa State University
Ames, Iowa, USA

Olaf Schenk
University of Basel
Basel, Switzerland

Klaus Schulten
University of Illinois at
Urbana-Champaign
Urbana-Champaign, Illinois, USA

Shubhabrata Sengupta
University of California, Davis
Davis, California, USA

Harald Servat
Barcelona Supercomputing
Center-CNS
Universitat Politècnica de Catalunya
Barcelona, Spain

John Shalf
Lawrence Berkeley National
Laboratory
Berkeley, California, USA

John E. Stone
University of Illinois at
Urbana-Champaign
Urbana-Champaign, Illinois, USA

Robert Strzodka
Max Planck Institut Informatik
Saarbrücken, Germany

Hari Subramoni
IBM T.J. Watson Research Center
Yorktown Heights, New York, USA

George Teodoro
Universidade Federal de Minas Gerais
Minas Gerais, Brazil

Stanimire Tomov
University of Tennessee
Knoxville, Tennessee, USA

Stefan Turek
TU Dortmund
Dortmund, Germany

Tzy-Hwa Kathy Tzeng
IBM Systems and Technology Group
Poughkeepsie, New York, USA

Vasily Volkov
University of California, Berkeley
Berkeley, California, USA

Richard Vuduc
Georgia Institute of Technology
Atlanta, Georgia, USA

Lukasz Wesolowski
University of Illinois at
Urbana-Champaign
Urbana-Champaign, Illinois, USA

Samuel Williams
Lawrence Berkeley National
Laboratory
Berkeley, California, USA

Hilmar Wobker
TU Dortmund
Dortmund, Germany

Katherine Yelick
Lawrence Berkeley National
Laboratory
Berkeley, California, USA

Jaroslaw Zola
Iowa State University
Ames, Iowa, USA

Part I

Dense Linear Algebra

Chapter 1

Implementing Matrix Multiplication on the Cell B. E.

Wesley Alvaro

Department of Electrical Engineering and Computer Science, University of Tennessee

Jakub Kurzak

Department of Electrical Engineering and Computer Science, University of Tennessee

Jack Dongarra

Department of Electrical Engineering and Computer Science, University of Tennessee
Computer Science and Mathematics Division, Oak Ridge National Laboratory
School of Mathematics & School of Computer Science, Manchester University

1.1   Introduction

Dense matrix multiplication is one of the most common numerical operations, especially in the area of dense linear algebra, where it forms the core of many important algorithms, including solvers of linear systems of equations, least square problems, and singular and eigenvalue problems. The Cell B. E. excels in its capabilities to process compute-intensive workloads, like matrix multiplication, in single precision, through its powerful SIMD capabilities. This chapter disects implementations of two single precision matrix

Reprinted from Parallel Computing, 35/3, J. Kurzak, W. Alvaro, J. Dongarra, Optimizing matrix multiplication for a short-vector SIMD architecture—CELL processor, 138–150, Copyright (2009), with permission from Elsevier.

multiplication kernels for the SIMD cores of the Cell B. E. (the SPEs), one implementing the $C = C - A \times B^T$ operation and the other implementing the $C = C - A \times B$ operation, for fixed size matrices of 64×64 elements. The unique dual-issue architecture of the SPEs provides for a great balance of the floating-point operations and the memory and permutation operations, leading to the utilization of the floating-point pipeline in excess of 99 % in both cases.

1.1.1   Performance Considerations

State of the art numerical linear algebra software utilizes *block algorithms* in order to exploit the memory hierarchy of traditional cache-based systems [8, 9]. Public domain libraries such as LAPACK [3] and ScaLAPACK [5] are good examples. These implementations work on square or rectangular submatrices in their inner loops, where operations are encapsulated in calls to *Basic Linear Algebra Subroutines* (BLAS) [4], with emphasis on expressing the computation as Level 3 BLAS, *matrix-matrix* type, operations. Frequently, the call is made directly to the matrix multiplication routine _GEMM. At the same time, all the other Level 3 BLAS can be defined in terms of _GEMM as well as a small amount of Level 1 and Level 2 BLAS [17]. Any improvement to the _GEMM routine immediately benefits the entire algorithm, which makes the optimization of the _GEMM routine yet more important. As a result, a lot of effort has been invested in optimized BLAS by hardware vendors as well as academic institutions through projects such as ATLAS [1] and GotoBLAS [2].

1.1.2   Code Size Considerations

In the current implementation of the Cell B. E. architecture, the SPEs are equipped with local memories (Local Stores) of 256 KB. It is a common practice to use tiles of 64×64 elements for dense matrix operations in single precision [6, 11, 12, 18, 19]. Such tiles occupy a 16 KB buffer in the Local Store. Between six and eight buffers are necessary to efficiently implement even such a simple operation as matrix multiplication [6, 11, 12]. Also, more complex operations, such as matrix factorizations, commonly allocate eight buffers [18, 19], which consume 128 KB of Local Store. In general, it is reasonable to assume that half of the Local Store is devoted to application data buffers. At the same time, the program may rely on library frameworks like ALF [14] or MCF [23], and utilize numerical libraries such as SAL [20], SIMD Math [15], or MASS [7], which consume extra space for the code. In the development stage, it may also be desirable to generate execution traces for analysis with tools like TATL^{TM} [21] or Paraver [10], which require additional storage for event buffers. Finally, the Local Store also houses the SPE stack, starting at the highest LS address and growing towards lower addresses with no overflow protection. As a result, the Local Store is a scarce resource and any *real-world* application is facing the problem of fitting tightly coupled components together in the limited space.

1.2 Implementation

1.2.1 Loop Construction

The main tool in loop construction is the technique of loop unrolling [13]. In general, the purpose of loop unrolling is to avoid pipeline stalls by separating dependent instructions by a distance in clock cycles equal to the corresponding pipeline latencies. It also decreases the overhead associated with advancing the loop index and branching. On the SPE it serves the additional purpose of balancing the ratio of instructions in the odd and even pipeline, owing to register reuse between iterations.

In the canonical form, matrix multiplication $C_{m \times n} = A_{m \times k} \times B_{k \times n}$ consists of three nested loops iterating over the three dimensions m, n, and k. Loop tiling [22] is applied to improve the locality of reference and to take advantage of the $O(n^3)/O(n^2)$ ratio of arithmetic operations to memory accesses. This way register reuse is maximized and the number of loads and stores is minimized.

Conceptually, tiling of the three loops creates three more inner loops, which calculate a product of a submatrix of A and a submatrix of B and updates a submatrix of C with the partial result. Practically, the body of these three inner loops is subject to complete unrolling to a single block of a straight-line code. The tile size is picked such that the cross-over point between arithmetic and memory operations is reached, which means that there is more FMA or FNMS operations to fill the even pipeline than there is load, store, and shuffle operations to fill the odd pipeline.

The resulting structure consists of three outer loops iterating over tiles of A, B, and C. Inevitably, nested loops induce mispredicted branches, which can be alleviated by further unrolling. Aggressive unrolling, however, leads quickly to undesired code bloat. Instead, the three-dimensional problem can be linearized by replacing the loops with a single loop performing the same traversal of the iteration space. This is accomplished by traversing tiles of A, B, and C in a predefined order derived as a function of the loop index. A straightforward row/column ordering can be used and tile pointers for each iteration can be constructed by simple transformations of the bits of the loop index.

At this point, the loop body still contains *auxiliary* operations that cannot be overlapped with arithmetic operations. These include initial loads, stores of final results, necessary data rearrangement with splats (copy of one element across a vector) and shuffles (permutations of elements within a vector), and

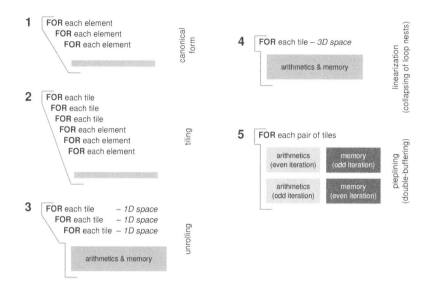

FIGURE 1.1: Basic steps of _GEMM loop optimization.

pointer advancing operations. This problem is addressed by *double-buffering*, on the register level, between two loop iterations. The existing loop body is duplicated and two separate blocks take care of the even and odd iteration, respectively. Auxiliary operations of the even iteration are hidden behind arithmetic instructions of the odd iteration and vice versa, and disjoint sets of registers are used where necessary. The resulting loop is preceded by a small body of *prologue* code loading data for the first iteration, and then followed by a small body of *epilogue* code, which stores results of the last iteration. Figure 1.1 shows the optimization steps leading to a high-performance implementation of the _GEMM inner kernel.

1.2.2   C = C − A × B trans

Before going into details, it should be noted that matrix storage follows C-style row-major format. It is not as much a careful design decision, as compliance with the common practice on the Cell B. E. It can be attributed to C compilers being the only ones allowed to exploit short-vector capabilities of the SPEs through C language SIMD extensions. If compliance with libraries relying on legacy FORTRAN API is required, a translation operation is necessary.

An easy way to picture the $C = C - A \times B^T$ operation is to represent it as the standard matrix vector product $C = C - A \times B$, where A is stored using

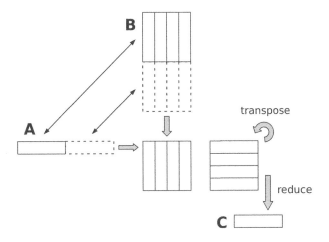

FIGURE 1.2: Basic operation of the $C = C - A \times B^T$ matrix multiplication micro-kernel.

row-major order and B is stored using column-major order. It can be observed that in this case a row of A can readily be multiplied with a column of B to yield a vector containing four partial results, which need to be summed up to produce one element of C. The vector reduction step introduces superfluous multiply-add operations. In order to minimize their number, four row-column products are computed, resulting in four vectors, which need to be internally reduced. The reduction is performed by first transposing the 4×4 element matrix represented by the four vectors and then applying four vector multiply-add operations to produce a result vector containing four elements of C. The basic scheme is depicted in Figure 1.2.

The crucial design choice to be made is the right amount of unrolling, which is equivalent to deciding the right tile size in terms of the triplet $\{m, n, k\}$ (here sizes express numbers of individual floating-point values, not vectors). Unrolling is mainly used to minimize the overhead of jumping and advancing the index variable and associated pointer arithmetic. However, both the jump and the jump hint instructions belong to the odd pipeline and, for compute intensive loops, can be completely hidden behind even pipeline instructions and thus introduce no overhead. In terms of the overhead of advancing the index variable and related pointer arithmetic, it will be shown in §1.2.4 that all of these operations can be placed in the odd pipeline as well. In this situation, the only concern is balancing even pipeline, arithmetic instructions with odd pipeline, data manipulation instructions.

Simple analysis can be done by looking at the number of floating-point operations versus the number of loads, stores, and shuffles, under the assumption that the size of the register file is not a constraint. The search space for the $\{m, n, k\}$ triplet is further truncated by the following criteria: only powers of two are considered in order to simplify the loop construction; the maximum possible number of 64 is chosen for k in order to minimize the number of extraneous floating-point instructions performing the reduction of partial results; only multiplies of four are selected for n to allow for efficient reduction of partial results with eight shuffles per one output vector of C. Under these constraints, the entire search space can be easily analyzed.

Table 1.1 shows how the number of each type of operation is calculated. Table 1.2 shows the number of even pipeline, floating-point instructions including the reductions of partial results. Table 1.3 shows the number of even pipeline instructions minus the number of odd pipeline instructions including loads, stores, and shuffles (not including jumps and pointer arithmetic). In other words, Table 1.3 shows the number of spare odd pipeline slots before

TABLE 1.1: Numbers of different types of operations in the computation of one tile of the $C = C - A \times B^T$ micro-kernel, as a function of tile size ($\{m, n, 64\}$ triplet).

Type of Operation	Pipeline Even	Pipeline Odd	Number of Operations
Floating point	✗		$(m \times n \times 64)/4 + m \times n$
Load A		✗	$m \times 64 / 4$
Load B		✗	$64 \times n / 4$
Load C		✗	$m \times n / 4$
Store C		✗	$m \times n / 4$
Shuffle		✗	$m \times n / 4 \times 8$

TABLE 1.2: Number of even pipeline, floating-point operations in the computation of one tile of the micro-kernel $C = C - A \times B^T$, as a function of tile size ($\{m, n, 64\}$ triplet).

M/N	4	8	16	32	64
1	68	136	272	544	1088
2	136	272	544	1088	2176
4	272	544	1088	2176	4352
8	544	1088	2176	4352	8704
16	1088	2176	4352	8704	17408
32	2176	4352	8704	17408	34816
64	4352	8704	17408	34816	69632

TABLE 1.3: Number of spare odd pipeline slots in the computation of one tile of the $C = C - A \times B^T$ micro-kernel, as a function of tile size ({m, n, 64} triplet).

M/N	4	8	16	32	64
1	-22	-28	-40	-64	-112
2	20	72	176	384	800
4	104	272	608	1280	2624
8	272	672	1472	3072	6272
16	608	1472	3200	6656	13568
32	1280	3072	6656	13824	28160
64	2624	6272	13568	28160	57344

TABLE 1.4: The size of code for the computation of one tile of the $C = C - A \times B^T$ micro-kernel, as a function of tile size ({m, n, 64} triplet).

M/N	4	8	16	32	64
1	1.2	1.2	2.3	4.5	8.9
2	1.0	1.8	3.6	7.0	13.9
4	1.7	3.2	6.1	12.0	23.8
8	3.2	5.9	11.3	22.0	43.5
16	6.1	11.3	21.5	42.0	83.0
32	12.0	22.0	42.0	82.0	162.0
64	23.8	43.5	83.0	162.0	320.0

jumps and pointer arithmetic are implemented. Finally, Table 1.4 shows the size of code involved in calculations for a single tile. It is important to note here that the double-buffered loop is twice the size.

It can be seen that the smallest unrolling with a positive number of spare odd pipeline slots is represented by the triplet {2, 4, 64} and produces a loop with 136 floating-point operations. However, this unrolling results in only 20 spare slots, which would barely fit pointer arithmetic and jump operations. Another aspect is that the odd pipeline is also used for instruction fetch, and near complete filling of the odd pipeline may cause instruction depletion, which in rare situations can even result in an indefinite stall [16].

The next larger candidates are triplets {4, 4, 64} and {2, 8, 64}, which produce loops with 272 floating-point operations, and 104 or 72 spare odd pipeline slots, respectively. The first one is an obvious choice, giving more room in the odd pipeline and smaller code. It turns out that the {4, 4, 64} unrolling is actually the most optimal of all, in terms of the overall routine footprint, when the implementation of pointer arithmetic is taken into account, as further explained in §1.2.4.

It can be observed that the maximum performance of the routine is ultimately limited by the extra floating-point operations, which introduce an overhead not accounted for in the formula for operation count in matrix mul-

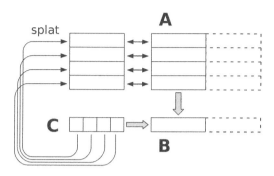

FIGURE 1.3: Basic operation of the $C = C - A \times B$ matrix multiplication micro-kernel.

tiplication: $2 \times m \times n \times k$. For matrices of size 64×64, every 64 multiply-add operations require four more operations to perform the intra-vector reduction. This sets a hard limit on the maximum achievable performance to $64/(64 + 4) \times 25.6 = 24.09$ $[Gflop/s]$, which is roughly 94 % of the peak.

1.2.3 $\mathbf{C = C - A \times B}$

Here, same as before, row-major storage is assumed. The key observation is that multiplication of one element of A with one row of B contributes to one row of C. As a result, the elementary operation splats an element of A over a vector, multiplies this vector with a vector of B, and accumulates the result in a vector of C (Figure 1.3). Unlike for the other kernel, in this case no extra floating-point operations are involved.

Same as before, the size of unrolling has to be decided in terms of the triplet $\{m, n, k\}$. This time, however, there is no reason to fix any dimension. Nevertheless, similar constraints to the search space apply: all dimensions have to be powers of two, and additionally only multiples of four are allowed for n and k to facilitate efficient vectorization and simple loop construction. Table 1.5 shows how the number of each type of operation is calculated. Table 1.6 shows the number of even pipeline, floating-point instructions. Table 1.7 shows the number of even pipeline instructions minus the number of odd pipeline instructions including loads, stores, and splats (not including jumps and pointer arithmetic). In other words, Table 1.7 shows the number of spare odd pipeline slots before jumps and pointer arithmetic are implemented. Finally, Table 1.8 shows the size of code involved in calculations for a single tile. It should be noted again that the double-buffered loop is twice the size.

It can be seen that the smallest unrolling with a positive number of spare odd pipeline slots produces a loop with 128 floating-point operations. Five

TABLE 1.5: Numbers of different types of operations in the computation of one tile of the $C = C - A \times B$ micro-kernel, as a function of tile size ({m, n, k}).

Type of Operation	Pipeline		Number of Operations
	Even	Odd	
Floating point	✘		$(m \times n \times k)/4$
Load A		✘	$m \times k / 4$
Load B		✘	$k \times n / 4$
Load C		✘	$m \times n / 4$
Store C		✘	$m \times n / 4$
Splat		✘	$m \times k$

TABLE 1.6: Number of even pipeline operations in the computation of one tile of the micro-kernel $C = C - A \times B$, as a function of tile size ({m, n, k}).

K	M/N	4	8	16	32	64
4	1	4	8	16	32	64
4	2	8	16	32	64	128
4	4	16	32	64	128	256
4	8	32	64	128	256	512
4	16	64	128	256	512	1024
4	32	128	256	512	1024	2048
4	64	256	512	1024	2048	4096
8	1	8	16	32	64	128
8	2	16	32	64	128	256
8	4	32	64	128	256	512
8	8	64	128	256	512	1024
8	16	128	256	512	1024	2048
8	32	256	512	1024	2048	4096
8	64	512	1024	2048	4096	8192
16	1	16	32	64	128	256
16	2	32	64	128	256	512
16	4	64	128	256	512	1024
16	8	128	256	512	1024	2048
16	16	256	512	1024	2048	4096
16	32	512	1024	2048	4096	8192
16	64	1024	2048	4096	8192	16384

TABLE 1.7: Number of spare odd pipeline slots in the computation of one tile of the $C = C - A \times B$ micro-kernel, as a function of tile size ($\{m, n, k\}$).

K	M/N	4	8	16	32	64
4	1	-7	-9	-13	-21	-37
4	2	-10	-10	-10	-10	-10
4	4	-16	-12	-4	12	44
4	8	-28	-16	8	56	152
4	16	-52	-24	32	144	368
4	32	-100	-40	80	320	800
4	64	-196	-72	176	672	1664
8	1	-12	-14	-18	-26	-42
8	2	-16	-12	-4	12	44
8	4	-24	-8	24	88	216
8	8	-40	0	80	240	560
8	16	-72	16	192	544	1248
4	32	-136	48	416	1152	2624
4	64	-264	112	864	2368	5376
16	1	-22	-24	-28	-36	-52
16	2	-28	-16	8	56	152
16	4	-40	0	80	240	560
16	8	-64	32	224	608	1376
16	16	-112	96	512	1344	3008
16	32	-208	224	1088	2816	6272
16	64	-400	480	2240	5760	12800

possibilities exist, with the triplet $\{4, 16, 8\}$ providing the highest number of 24 spare odd pipeline slots. Again, such unrolling would both barely fit pointer arithmetic and jump operations and be a likely cause of instruction depletion.

The next larger candidates are unrollings that produce loops with 256 floating-point operations. There are 10 such cases, with the triplet $\{4, 32, 8\}$ being the obvious choice for the highest number of 88 spare odd pipeline slots and the smallest code size. It also turns out that this unrolling is actually the most optimal in terms of the overall routine footprint, when the implementation of pointer arithmetic is taken into account, as further explained in §1.2.4.

Unlike for the other routine, the maximum performance is not limited by any extra floating-point operations, and performance close to the peak of 25.6 $Gflop/s$ should be expected.

1.2.4   Advancing Tile Pointers

The remaining issue is the one of implementing the arithmetic calculating the tile pointers for each loop iteration. Due to the size of the input matrices and the tile sizes being powers of two, this is a straightforward task. The tile offsets can be calculated from the tile index and the base addresses of the input matrices using integer arithmetic and bit manipulation instructions (bitwise logical instructions and shifts). Figure 1.4 shows a sample implementation of

```
int tile;

vector float *Abase;
vector float *Bbase;
vector float *Cbase;

vector float *Aoffs;
vector float *Boffs;
vector float *Coffs;

Aoffs = Abase + ((tile & ~0x0F) << 2);
Boffs = Bbase + ((tile &  0x0F) << 6);
Coffs = Cbase +  (tile &  0x0F)
               + ((tile & ~0x0F) << 2);
```

FIGURE 1.4: Sample C language implementation of pointer arithmetic for the kernel $C = C - A \times B^T$ with unrolling corresponding to the triplet $\{4, 4, 64\}$.

TABLE 1.8: The size of code for the computation of one tile of the $C = C - A \times B$ micro-kernel, as a function of tile size ($\{m, n, k\}$).

K	M/N	4	8	16	32	64
4	1	0.1	0.1	0.2	0.3	0.6
4	2	0.1	0.2	0.3	0.5	1.0
4	4	0.2	0.3	0.5	1.0	1.8
4	8	0.4	0.6	1.0	1.8	3.4
4	16	0.7	1.1	1.9	3.4	6.6
4	32	1.4	2.2	3.7	6.8	12.9
4	64	2.8	4.3	7.3	13.4	25.5
8	1	0.1	0.2	0.3	0.6	1.2
8	2	0.2	0.3	0.5	1.0	1.8
8	4	0.3	0.5	0.9	1.7	3.2
8	8	0.7	1.0	1.7	3.1	5.8
8	16	1.3	1.9	3.3	5.9	11.1
4	32	2.5	3.8	6.4	11.5	21.8
4	64	5.0	7.6	12.6	22.8	43.0
16	1	0.2	0.3	0.6	1.1	2.2
16	2	0.4	0.6	1.0	1.8	3.4
16	4	0.7	1.0	1.7	3.1	5.8
16	8	1.3	1.9	3.1	5.6	10.6
16	16	2.4	3.6	6.0	10.8	20.3
16	32	4.8	7.1	11.8	21.0	39.5
16	64	9.6	14.1	23.3	41.5	78.0

```
lqa   $2,tile
lqa   $3,Abase
andi  $4,$2,-16
andi  $2,$2,15
shli  $6,$4,2
shli  $4,$4,6
shli  $5,$2,10
a     $2,$2,$6
a     $4,$4,$3
shli  $2,$2,4
lqa   $3,Bbase
stqa  $4,Aoffs
a     $5,$5,$3
lqa   $3,Cbase
stqa  $5,Boffs
a     $2,$2,$3
stqa  $2,Coffs
```

FIGURE 1.5: The result of compiling the code from Figure 1.4 to assembly language, with even pipeline instructions in bold.

pointer arithmetic for the kernel $C = C - A \times B^T$ with unrolling corresponding to the triplet $\{4, 4, 64\}$. *Abase*, *Bbase*, and *Cbase* are base addresses of the input matrices, and the variable *tile* is the tile index running from 0 to 255; *Aoffs*, *Boffs*, and *Coffs* are the calculated tile offsets.

Figure 1.5 shows the result of compiling the sample C code from Figure 1.4 to assembly code. Although a few variations are possible, the resulting assembly code will always involve a similar combined number of integer and bit manipulation operations. Unfortunately, all these instructions belong to the even pipeline and will introduce an overhead, which cannot be hidden behind floating point operations, like it is done with loads, stores, splats, and shuffles.

One way of minimizing this overhead is extensive unrolling, which creates a loop big enough to make the pointer arithmetic negligible. An alternative is to eliminate the pointer arithmetic operations from the even pipeline and replace them with odd pipeline operations. With the unrolling chosen in §1.2.2 and §1.2.3, the odd pipeline offers empty slots in abundance. It can be observed that, since the loop boundaries are fixed, all tile offsets can be calculated in advance. At the same time, the operations available in the odd pipeline include loads, which makes it a logical solution to precalculate and tabulate tile offsets for all iterations. It still remains necessary to combine the offsets with the base addresses, which are not known beforehand. However, under additional alignment constraints, offsets can be combined with bases using shuffle instructions, which are also available in the odd pipeline. As will be further shown, all instructions that are not floating point arithmetic can be removed from the even pipeline.

The precalculated offsets have to be compactly packed in order to preserve

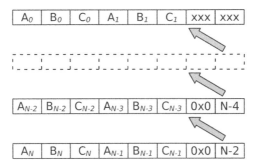

FIGURE 1.6: Organization of the tile offset lookup table. N is the number of tiles.

space consumed by the lookup table. Since tiles are 16 KB in size, offsets consume 14 bits and can be stored in a 16-bit halfword. Three offsets are required for each loop iteration. With eight halfwords in a quadword, each quadword can store offsets for two loop iterations or a single interation of the pipelined, double-buffered loop. Figure 1.6 shows the organization of the offset lookup table.

The last arithmetic operation remaining is the advancement of the iteration variable. It is typical to decrement the iteration variable instead of incrementing it, and branch on non-zero, in order to eliminate the comparison operation, which is also the case here. This still leaves the decrement operation, which would have to occupy the even pipeline. In order to annihilate the decrement, each quadword containing six offsets for one iteration of the double-buffered loop also contains a seventh entry, which stores the index of the quadword to be processed next (preceding in memory). In other words, the iteration variable, which also serves as the index to the lookup table, is tabulated along with the offsets and loaded instead of being decremented.

Normally, the tile pointers would have to be calculated as a sum of an 18-bit base address and a 14-bit offset, which would require the use of integer addition residing in the even pipeline. With the additional constraint of 16 KB alignment of the base addresses, 14 less significant bits of the base are zero and can be simply replaced with the bits of the offset. The replacement could be implemented with the logical *AND* operation. This would however, again, involve an even pipeline instruction. Instead, both the base addresses and the offsets are initially shifted left by two bits, which puts the borderline between offsets and bases on a byte boundary. At this point the odd pipeline shuffle instruction operating at byte granularity can be used to combine the base with the offset. Finally, the result has to be shifted right by two bits, which can be accomplished by a combination of bit and byte quadword rotations,

TABLE 1.9: The overall footprint of the micro-kernel $C = C - A \times B^T$, including the code and the offset lookup table, as a function of tile size ({m, n, 64} triplet).

M/N	4	8	16	32	64
1	9.2	6.3	6.6	10.0	18.4
2	6.0	5.7	8.1	14.5	28.0
4	5.4	7.4	12.8	24.3	47.6
8	7.4	12.3	22.8	44.1	87.1
16	12.8	22.8	43.1	84.1	166.0
32	24.3	44.1	84.1	164.0	324.0
64	47.6	87.1	166.0	324.0	640.0

which also belong to the odd pipeline. Overall, all the operations involved in advancing the double-buffered loop consume 29 extra odd pipeline slots, which is small, given that 208 are available in the case of the first kernel and 176 in the case of the second.

This way, all operations involved in advancing from tile to tile are implemented in the odd pipeline. At the same time, both the branch instruction and the branch hint belong to the odd pipeline. Also, a correctly hinted branch does not cause any stall. As a result, such an implementation produces a continuous stream of floating-point operations in the even pipeline, without a single cycle devoted to any other activity.

The last issue to be discussed is the storage overhead of the lookup table. This size is proportional to the number of iterations of the unrolled loop and reciprocal to the size of the loop body. Using the presented scheme (Figure 1.6), the size of the lookup table in bytes equals $N^3/(m \times n \times k) \times 8$. Table 1.9 presents the overall footprint of the $C = C - A \times B^T$ micro-kernel as a function of the tile size. Table 1.10 presents the overall footprint of the $C = C - A \times B$ micro-kernel as a function of the tile size. As can be clearly seen, the chosen tile sizes result in the lowest possible storage requirements for the routines.

1.3   Results

Both presented SGEMM kernel implementations produce a continuous stream of floating-point instructions for the duration of the pipelined loop. In both cases, the loop iterates 128 times, processing two tiles in each iteration. The $C = C - A \times B^T$ kernel contains 544 floating-point operations in the loop body and, on a 3.2 GHz processor, delivers 25.54 Gflop/s (99.77 % of peak) if actual operations are counted, and 24.04 Gflop/s (93.90 % of peak) if the

TABLE 1.10: The overall footprint of the micro-kernel $C = C - A \times B$, including the code and the offset lookup table, as a function of tile size ({m, n, 64} triplet).

K	M/N	4	8	16	32	64
4	1	128.1	64.2	32.4	16.7	9.3
4	2	64.2	32.3	16.6	9.1	6.1
4	4	32.4	16.6	9.0	5.9	5.7
4	8	16.7	9.1	5.9	5.6	7.8
4	16	9.4	6.2	5.8	7.9	13.6
4	32	6.8	6.3	8.4	14.0	26.0
4	64	7.5	9.6	15.1	27.0	51.1
8	1	64.2	32.4	16.6	9.2	6.3
8	2	32.4	16.6	9.0	5.9	5.7
8	4	16.7	9.1	5.8	5.3	7.3
8	8	9.3	6.0	5.4	7.1	12.1
8	16	6.6	5.9	7.5	12.3	22.5
4	32	9.1	9.6	13.8	23.5	43.8
4	64	12.1	16.1	25.8	45.8	86.1
16	1	32.4	16.7	9.2	6.3	6.4
16	2	16.7	9.1	5.9	5.6	7.8
16	4	9.3	6.0	5.4	7.1	12.1
16	8	6.5	5.8	7.3	11.8	21.5
16	16	6.9	8.3	12.5	21.8	40.6
16	32	10.6	14.8	23.8	42.1	79.1

standard formula, $2N^3$, is used for operation count. The $C = C - A \times B$ kernel contains 512 floating-point operations in the loop body and delivers 25.55 Gflop/s (99.80 % of peak). Here, the actual operation count equals $2N^3$. At the same time, neither implementation overfills the odd pipeline, which is 31 % empty for the first case and 17 % empty for the second case. This guarantees no contention between loads and stores and DMA operations, and no danger of instruction fetch starvation. Table 1.11 shows the summary of the kernels' properties.

1.4 Summary

Computational micro-kernels are architecture specific codes, where no portability is sought. It has been shown that systematic analysis of the problem combined with exploitation of low-level features of the Synergistic Processing Unit of the Cell B. E. leads to dense matrix multiplication kernels achieving peak performance without code bloat.

This proves that great performance can be achieved on SIMD architecture by optimizing code manually. The question remains, whether similar results

TABLE 1.11: Summary of the properties of the SPE SIMD SGEMM micro-kernels.

Characteristic	C=C-A×B^T	C=C-A×B
Performance	24.04 Gflop/s	25.55 Gflop/s
Execution time	21.80 μs	20.52 μs
Fraction of peak USING THE 2xMxNxK FORMULA	93.90 %	99.80 %
Fraction of peak USING ACTUAL NUMBER OF FLOATING–POINT INSTRUCTIONS	99.77 %	99.80%
Dual issue rate ODD PIPELINE WORKLOAD	68.75 %	82.81 %
Register usage	69	69
Code segment size	4008	3992
Data segment size	2192	2048
Total memory footprint	6200	6040

can be accomplished by automatic vectorization techniques or a combination of auto-vectorization with heuristic techniques based on searching the parameter space. It is likely that good results could be achieved by a combination of the *Superworld Level Parallelism* technique for auto-vectorization [24] with heuristic search similar to the ATLAS [1] methodology.

1.5   Code

The code is freely available, under the BSD license and can be downloaded from the author's web site `http://icl.cs.utk.edu/~alvaro/`. A few comments can be useful here. In absence of better tools, the code has been developed with the help of a spreadsheet, mainly for easy manipulation of two columns of instructions for the two pipelines of the SPE. Other useful features were taken advantage of as well. Specifically, color coding of blocks of instructions greatly improves the readability of the code. It is the hope of the authors that such visual representation of code considerably helps the reader's understanding of the techniques involved in construction of optimized SIMD assembly code. Also, the authors put forth considerable effort in making the software self-contained, in the sense that all tools involved in construction of the code are distributed alongside. This includes the lookup table generation code and the scripts facilitating translation from spreadsheet format to SPE assembly language.

Bibliography

[1] ATLAS. http://math-atlas.sourceforge.net/.

[2] GotoBLAS. http://www.tacc.utexas.edu/resources/software/.

[3] E. Anderson, Z. Bai, C. Bischof, L. S. Blackford, J. W. Demmel, J. J. Dongarra, J. Du Croz, A. Greenbaum, S. Hammarling, A. McKenney, and D. Sorensen. *LAPACK Users' Guide.* SIAM, Philadelphia, PA, 1992. http://www.netlib.org/lapack/lug/.

[4] Basic Linear Algebra Technical Forum. *Basic Linear Algebra Technical Forum Standard,* August 2001. http://www.netlib.org/blas/blast-forum/blas-report.pdf.

[5] L. S. Blackford, J. Choi, A. Cleary, E. D'Azevedo, J. Demmel, I. Dhillon, J. J. Dongarra, S. Hammarling, G. Henry, A. Petitet, K. Stanley, D. Walker, and R. C. Whaley. *ScaLAPACK Users' Guide.* SIAM, Philadelphia, PA, 1997. http://www.netlib.org/scalapack/slug/.

[6] T. Chen, R. Raghavan, J. N. Dale, and E. Iwata. Cell Broadband Engine architecture and its first implementation — A performance view. *IBM J. Res. & Dev.,* 51(5):559–572, 2007. DOI: 10.1147/rd.515.0559.

[7] IBM Corporation. Mathematical Acceleration Subsystem — Product overview. http://www-306.ibm.com/software/awdtools/mass/, March 2007.

[8] J. W. Demmel. *Applied Numerical Linear Algebra.* SIAM, 1997. ISBN: 0898713897.

[9] J. J. Dongarra, I. S. Duff, D. C. Sorensen, and H. A. van der Vorst. *Numerical Linear Algebra for High-Performance Computers.* SIAM, 1998. ISBN: 0898714281.

[10] European Center for Parallelism of Barcelona, Technical University of Catalonia. *Paraver, Parallel Program Visualization and Analysis Tool Reference Manual, Version 3.1,* October 2001.

[11] D. Hackenberg. Einsatz und Leistungsanalyse der Cell Broadband Engine. Institut für Technische Informatik, Fakultät Informatik, Technische Universität Dresden, February 2007. Großer Beleg.

[12] D. Hackenberg. Fast matrix multiplication on CELL systems. http://tu-dresden.de/die_tu_dresden/zentrale_einrichtungen/ zih/forschung/architektur_und_leistungsanalyse_von_ hochleistungsrechnern/cell/matmul/, July 2007.

[13] J. L. Hennessy and D. A. Patterson. *Computer Architecture, Fourth Edition: A Quantitative Approach.* Morgan Kaufmann, 2006.

[14] IBM Corporation. *ALF for Cell BE Programmer's Guide and API Reference*, November 2007.

[15] IBM Corporation. *SIMD Math Library API Reference Manual*, November 2007.

[16] IBM Corporation. *Preventing Synergistic Processor Element Indefinite Stalls Resulting from Instruction Depletion in the Cell Broadband Engine Processor for CMOS SOI 90 nm, Applications Note, Version 1.0*, November 2007.

[17] B. Kågström, P. Ling, and C. van Loan. GEMM-Based Level 3 BLAS: High-performance model implementations and performance evaluation benchmark. *ACM Trans. Math. Soft.*, 24(3):268–302, 1998.

[18] J. Kurzak, A. Buttari, and J. J. Dongarra. Solving systems of linear equation on the CELL processor using Cholesky factorization. *Trans. Parallel Distrib. Syst.*, 19(9):1175–1186, 2008. DOI: TPDS.2007.70813.

[19] J. Kurzak and J. J. Dongarra. Implementation of mixed precision in solving systems of linear equations on the CELL processor. *Concurrency Computat.: Pract. Exper.*, 19(10):1371–1385, 2007. DOI: 10.1002/cpe.1164.

[20] Mercury Computer Systems, Inc. *Scientific Algorithm Library (SAL) Data Sheet*, 2006. http://www.mc.com/uploadedfiles/SAL-ds.pdf.

[21] Mercury Computer Systems, Inc. *Trace Analysis Tool and Library (TATL^{TM}) Data Sheet*, 2006. http://www.mc.com/uploadedfiles/ tatl-ds.pdf.

[22] S. Muchnick. *Advanced Compiler Design and Implementation.* Morgan Kaufmann, 1997.

[23] M. Pepe. *Multi-Core Framework (MCF), Version 0.4.4.* Mercury Computer Systems, October 2006.

[24] J. Shin, J. Chame, and M. W. Hall. Exploiting superword-level locality in multimedia extension architectures. *J. Instr. Level Parallel.*, 5:1–28, 2003.

Chapter 2

Implementing Matrix Factorizations on the Cell B. E.

Jakub Kurzak

Department of Electrical Engineering and Computer Science, University of Tennessee

Jack Dongarra

Department of Electrical Engineering and Computer Science, University of Tennessee
Computer Science and Mathematics Division, Oak Ridge National Laboratory
School of Mathematics & School of Computer Science, Manchester University

2.1 Introduction

It is clear that the impact of the multicore processors and accelerators will be ubiquitous. There are obvious advantages, however, to look at linear algebra in general and dense linear algebra in particular. This type of software is critically important to computational science across an enormous spectrum of disciplines and applications. Yet more importantly, dense linear algebra has strategic advantages as a research vehicle, because the methods and algorithms that underlie it have been so thoroughly studied and are so well understood [5, 6, 10, 17]. This chapter dissects highly optimized Cell B. E. implementations of two classic dense linear algebra computations, the Cholesky factorization and the QR factorization.

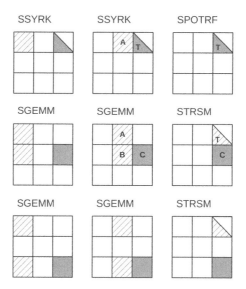

FIGURE 2.1: Tile operations in the Cholesky factorization. The sequence is left-to-right and top-down. Hatching indicates input data, shade of gray indicates in/out data.

2.2   Cholesky Factorization

The Cholesky factorization (or Cholesky decomposition) is mainly used for the numerical solution of linear equations $Ax = b$, where A is symmetric and positive definite. Such systems arise often in physics applications, where A is positive definite due to the nature of the modeled physical phenomenon. This happens frequently in numerical solutions of partial differential equations. The Cholesky factorization of an $n \times n$ real symmetric positive definite matrix A has the form

$$A = LL^T,$$

where L is an $n \times n$ real lower triangular matrix with positive diagonal elements.

The algorithm can be expressed using either the top-looking version, the left-looking version or the right-looking version. The first one follows depth-first exploration of the task graph and the last one follows the breadth-first exploration of the task graph. The left-looking variant is used here. The algorithm relies on four basic operations implemented by four computational kernels (Figure 2.1). Figure 2.2 shows the generic pseudocode of the left-looking Cholesky factorization.

```
FOR k = 0..TILES-1
  FOR n = 0..k-1
    A[k][k] ← SSYRK(A[k][n], A[k][k])
  A[k][k] ← SPOTRF(A[k][k])
  FOR m = k+1..TILES-1
    FOR n = 0..k-1
      A[m][k] ← SGEMM(A[k][n], A[m][n], A[m][k])
    A[m][k] ← STRSM(A[k][k], A[m][k])
```

FIGURE 2.2: Pseudocode of the (left-looking) Cholesky factorization.

SSYRK: The kernel applies updates to a diagonal (lower triangular) tile T of the input matrix, resulting from factorization of the tiles A to the left of it. The operation is a symmetric rank-k update.

SPOTRF: The kernel performs the Cholesky factorization of a diagonal (lower triangular) tile T of the input matrix and overrides it with the final elements of the output matrix.

SGEMM: The operation applies updates to an off-diagonal tile C of the input matrix, resulting from factorization of the tiles to the left of it. The operation is a matrix multiplication.

STRSM: The operation applies an update to an off-diagonal tile C of the input matrix, resulting from factorization of the diagonal tile above it and overrides it with the final elements of the output matrix. The operation is a triangular solve.

2.3 Tile QR Factorization

The QR factorization (or QR decomposition) offers a numerically stable way of solving underdetermined and overdetermined systems of linear equations (least squares problems) and is also the basis for the *QR algorithm* for solving the eigenvalue problem.

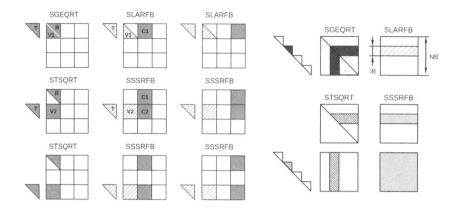

FIGURE 2.3: *Left:* Tile operations in the tile QR factorization. The sequence is left-to-right and top-down. Hatching indicates input data, shade of gray indicates in/out data. *Right:* Inner blocking in the tile QR factorization.

The QR factorization of an $m \times n$ real matrix A has the form

$$A = QR,$$

where Q is an $m \times m$ real orthogonal matrix and R is an $m \times n$ real upper triangular matrix. The traditional algorithm for QR factorization applies a series of elementary Householder matrices of the general form

$$H = I - \tau v v^{T},$$

where v is a column reflector and τ is a scaling factor. In the block form of the algorithm a product of nb elementary Householder matrices is represented in the form

$$H_1 H_2 \ldots H_{nb} = I - V T V^{T},$$

where V is an $N \times nb$ real matrix whose columns are the individual vectors v, and T is an $nb \times nb$ real upper triangular matrix [2, 16].

Here a derivative of the block algorithm is used called the *tile QR* factorization. The ideas behind the tile QR factorization are very well known. The tile QR factorization was initially developed to produce a high-performance "out-of-memory" implementation (typically referred to as "out-of-core") [11] and, more recently, to produce a high-performance implementation on "standard" (x86 and alike) multicore processors [3, 4]. The tile QR algorithm relies on four basic operations implemented by four computational kernels (Figure 2.3). Figure 2.4 shows the pseudocode of the tile QR factorization.

```
FOR k = 0..TILES-1
    A[k][k], T[k][k] ← SGEQRT(A[k][k])
    FOR m = k+1..TILES-1
        A[k][k], A[m][k], T[m][k] ← STSQRT(A[k][k], A[m][k], T[m][k])
    FOR n = k+1..TILES-1
        A[k][n] ← SLARFB(A[k][k], T[k][k], A[k][n])
        FOR m = k+1..TILES-1
            A[k][n], A[m][n] ← SSSRFB(A[m][k], T[m][k], A[k][n], A[m][n])
```

FIGURE 2.4: Pseudocode of the tile QR factorization.

SGEQRT: The kernel performs the QR factorization of a diagonal tile of the input matrix and produces an upper triangular matrix R and a unit lower triangular matrix V containing the Householder reflectors. The kernel also produces the upper triangular matrix T as defined by the compact WY technique for accumulating Householder reflectors [2, 16]. The R factor overrides the upper triangular portion of the input and the reflectors override the lower triangular portion of the input. The T matrix is stored separately.

STSQRT: The kernel performs the QR factorization of a matrix built by coupling the R factor, produced by SGEQRT or a previous call to STSQRT, with a tile below the diagonal tile. The kernel produces an updated R factor, a square matrix V containing the Householder reflectors and the matrix T resulting from accumulating the reflectors V. The new R factor overrides the old R factor. The block of reflectors overrides the square tile of the input matrix. The T matrix is stored separately.

SLARFB: The kernel applies the reflectors calculated by SGEQRT to a tile to the right of the diagonal tile, using the reflectors V along with the matrix T.

SSSRFB: The kernel applies the reflectors calculated by STSQRT to two tiles to the right of the tiles factorized by STSQRT, using the reflectors V and the matrix T produced by STSQRT.

A naive implementation, where the full T matrix is built, results in 25 % more floating point operations than the standard algorithm. In order to minimize this overhead, the idea of *inner-blocking* is used, where the T matrix has sparse (block-diagonal) structure (Figure 2.3) [7–9].

2.4 SIMD Vectorization

The keys to maximum utilization of the synergistic processing elements (SPEs) are highly optimized implementations of the computational kernels, which rely on efficient use of the short-vector single instruction multiple data (SIMD) architecture. For the most part, the kernels are developed by applying standard loop optimization techniques, including tiling, unrolling, reordering, fusion, fission, and sometimes also collapsing of loop nests into one loop spanning the same iteration space with appropriate pointer arithmetics. Tiling and unrolling are mostly dictated by Local Store latency and the size of the register file, and aim at hiding memory references and reordering of vector elements, while balancing the load of the two execution pipelines. Due to the huge size of the SPEs' register file, unrolling is usually quite aggressive.

Implementation of the tile kernels assumes a fixed size of the tiles. Smaller tiles (finer granularity) have a positive effect on scheduling for parallel execution and facilitate better load balance and higher parallel efficiency. Bigger tiles provide better performance in sequential execution on a single SPE. In the case of the CELL chip, the crossover point is rather simple to find for problems in dense linear algebra. From the standpoint of this work, the most important operation is matrix multiplication in single precision. It turns out that this operation can achieve the peak performance of the SPE for matrices of size 64×64 (see the preceeding chapter). The fact that the peak performance can be achieved for a tile of such a small size has to be attributed to the large size of the register file and fast access to the Local Store, undisturbed with any intermediate levels of memory. Also, such a matrix occupies a 16 KB block of memory, which is the maximum size of a single DMA transfer. Eight such matrices fit in half of the Local Store providing enough flexibility for multibuffering while, at the same time, leaving enough room for the code.

Table 2.1 shows characteristics of the Cholesky kernels and the tile QR kernels. It can be observed that the Cholesky kernels required moderate effort. Initially, all the kernels were coded using C language SIMD extensions (intrinsics) and required roughly 300 lines of code per kernel. However, preprocessor macros were used and the resulting assembly code is significantly longer. Nevertheless, the effort associated with development and maintenance of this code is rather small. At the same time, the delivered performance is more than satisfactory. Specifically, the SGEMM and SSYRK kernels deliver 90 and 79% of the peak respectively, which has to be considered quite good for SIMD code developed in a higher level language. The STRSM kernel delivers poorer performance due to a lower level of SIMD parallelism available and the SPOTRF kernel performs the poorest for the same reason. The SPOTRF kernel simply performs the Cholesky factorization within a tile and is the the most complex operation to SIMD'ize with the lowest level of available SIMD parallelism.

TABLE 2.1: Complexity and performance characteristics of SPE micro-kernels for the Cholesky factorization (top) and the tile QR factorization (bottom). Bold font highlights the largest codes and the highest performance. All operations are for matrices of size 64×64 (n=64).

Kernel Name	Lines of Code in C	Lines of Code in ASM	Object Size [KB]	Exec. Time [μs]	Flop Count Formula	Exec. Rate [Gflop/s]	Fraction of Peak [%]
SGEMM$_C$	330	2000	7.8	23	$2n^3$	23.03	90
SGEMM$_{ASM}$	-	**3900**	6.2	22	$2n^3$	**24.04**	94
SSYRK	160	1000	3.6	13	n^3	20.11	79
STRSM	310	1600	6.2	16	n^3	16.26	64
SPOTRF	340	800	3.1	14	$^1/_3n^3$	6.57	26
SSSRFB	1600	2200	8.8	47	$4n^3$	**22.20**	87
STSQRT	1900	**3600**	14.2	46	$2n^3$	11.40	45
SLARFB	600	600	2.2	41	$2n^3$	12.70	50
SGEQRT	1600	2400	9.0	57	$^4/_3n^3$	6.15	24

Table 2.1 also includes the SGEMM kernel developed in the SPE assembly language, which was described in the previous chapter. In this case the effort was rather huge and involved development of 3900 lines of hand-tuned assembly code. At the same time, the performance gain is less than 5 %. Such an effort is justified in reasearch circles, but would be questionable in commercial environments. Nevertheless, the performance for parallel runs, presented further in the text, relies on the fast assembly kernel.

It should be pointed out that the performance of the kernels developed in the C language is very sensitive to the version of the compiler used and the compilation flags. The authors exhaustively tried all the combinations and the table reports the best results achieved. Many times high performance was only achievable while using the spu-gcc, version 3.4.1, released in SDK 1.1, toolchain 2.3. Most of the time either the flag -O3 or the flag -Os delivered the best performance. Since the follow up versions of the compiler delivered poorer performance for the kernels, the code posted online by the authors (2.9) includes the kernels precompiled to assembly using the old compiler.

The development of the Cholesky kernels was moderately difficult. Three of them implement simple Level 3 BLAS operations, while the fourth one implements the Cholesky factorization on a tile, which is not overly complicated. The same cannot be claimed about the tile QR factorization kernels. None of the kernel operations is a simple BLAS operation, and the technique of inner-blocking further complicates matters.

Inner-blocking in the tile QR algorithm is required to minimize the number of extraneous floating-point operations (beond the $^4/_3n^3$ formula) coming from the accumulation of Householder reflectors. As necessary as the

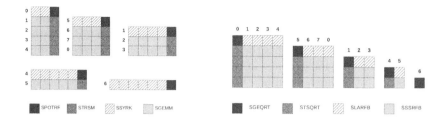

FIGURE 2.5: Assignment of work to SPEs for the Cholesky factorization (left) and the tile QR factorization (right).

inner-blocking is, it also restricts the level of available SIMD parallelism. Ideally, the size of the inner block would be chosen in the process of autotuning. However, such an approach would require some means of automatic code generation. Since such capabilities were not available here, the size was picked arbitrarily. For productivity reasons, the size of four elements was picked to match the size of the SIMD vector length in single precision.

As Table 2.1 shows, it was also possible to achieve good performance for the tile QR kernels coded in the C language using intrinsics. Most importantly, good performance was achieved for the SSSRFB kernel, which is as performance-critical to the tile QR factorization as the SGEMM kernel performance is critical to the Cholesky factorization. At the same time, much heavier coding effort was involved, resulting in three kernels larger that 1500 lines of source code (SGEQRT, SSSRFB, STSQRT) and the STSQRT kernel ultimately translating to 3600 lines of assembly code.

2.5   Parallelization—Single Cell B. E.

Matrix factorizations represent computations with a very clear structure and regular data access pattern. This motivates the use of static partitioning of work to the SPEs, shown in Figure 2.5. For the Cholesky factorization, in each step of the factorization, each SPE goes through one row of tiles. The assignment of rows is cyclic, from step to step, and the SPE which "runs out of work" in a given step immediately follows to the consecutive step, a behavior resembling the popular technique of *lookahead*. The scheme is followed for the tile QR factorization, except here each SPE goes through one column of tiles.

Due to the regular nature of these workloads, static scheduling is extemely straightforward to implement. Using a simple formula on tiles' idexes, each SPE can traverse its own path through the iteration space. Additionally, at each step, a check for data dependencies is required. The SPE does that by looking

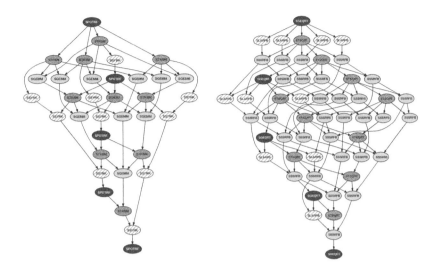

FIGURE 2.6: Direct Acyclic Graphs of the Cholesky factorization (left) and the tile QR factorization (right) for a matrix of size 5×5 tiles.

up a progress table in its Local Store. The progress table contains the global progress information and is replicated on all SPEs. The progress table holds one entry (a byte) for each tile of the input matrix, indicating progress of the computation associated with that tile. At the completion of each operation, the SPE broadcasts the progress information to progress tables of all SPEs with an SPE-to-SPE DMA.

Alternatively to the static scheduling, a dynamic scheduling could be used, based on representing the computation as a task graph or *Direct Acyclic Graph* (DAG). The task is rather non-trivial due to the complexity of the DAGs of dense matrix factorizations (Figure 2.6). One framework capable of such scheduling on the Cell B. E. is the *Cell Superscalar* (CellSs) project from the Barcelona Supercomputer Center [1,15]. Unfortunately, due to the overheads of dynamically scheduling complex DAGs, the software is still not competitive, in terms of performance, with the approach presented here.

An important aspect of the algorithm is overlapping of communication and computation by double-buffering of data. At each step, the tiles of the input matrix are exchanged between the main memory and Local Store. Since scheduling is static, upcoming operations can be anticipated and the necessary data prefetched. In fact, all data buffers are duplicated and, at each operation, a prefetch of data is initiated for the upcoming operation (again, subject to a dependency check). If the prefetch fails for dependency reasons, data are fetched in a blocking mode right before the operation. Algorithm 1 shows the mechanism of double-buffering in matrix factorizations.

The pipelined scheduling scheme along with double-buffering of data trans-

---

Algorithm 1 General scheme of double-buffering in the Cholesky and tile QR factorizations.

 1: **while** more work to do **do**
 2:     **if** data not prefetched **then**
 3:         wait for dependencies
 4:         fetch data
 5:     **end if**
 6:     **if** more work to follow **then**
 7:         **if** dependencies met **then**
 8:             prefetch data
 9:         **end if**
10:     **end if**
11:     compute
12:     swap buffers
13: **end while**

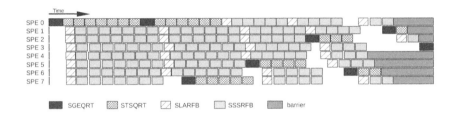

FIGURE 2.7: Execution trace of the tile QR factorization of a 512×512 matrix. (total time: 1645 μs, execution rate: 109 Gflop/s).

fers provide for smooth execution with minimal idle time caused by dependency stalls and almost no time lost to data transfers. This is clearly visible on a trace of the tile QR factorization presented in Figure 2.7.

2.6   Parallelization—Dual Cell B. E.

Given the single-Cell B. E. implementation, extension to dual-Cell B. E. implementation (e.g., IBM QS20, IBM QS22) is relatively straightforward. A single PPE process can launch 16 SPE threads, eight on each Cell B. E. The single-Cell B. E. code is going to run correctly on a dual-Cell B. E. system by simply increasing the number of SPEs to 16.

The problem is a one of performance of the memory system. The dual-Cell blades are *Non-Uniform Memory Access* (NUMA) architectures. Each Cell

B. E. is associated with a separate memory node. Peak bandwidth to the local node is 25.6 GB/s. Cross-traffic, however, is handled at a much lower bandwidth (roughly half of that number). It is important, then, that each SPE satisfies its data needs mostly from the local memory node.

This situation is addressed by duplicating the input matrix in both memory nodes (*libnuma* is used for correct memory placement). Each SPE reads data only from the local node, but writes data to both nodes. From the perspective of the shared memory model, it can be viewed as a manual implementation of the write-back memory consistency protocol. From the perspective of a distributed memory model, it can be viewed as non-blocking collective communication (broadcast) or as one-sided communication. The obvious limitation is that the approach would not be scalable to larger NUMA systems. As of today, however, larger Cell-based NUMA systems do not exist.

One technical detail to be mentioned here is the acknowledgment DMAs implementing the synchronization protocol between SPEs. When 16 SPEs are used, each SPE needs to send 16 acknowledgment messages following a write of data to the system memory. The acknowledgment DMA is fenced with the data DMA and the SPE also sends such a message to its own progress table (hence 16 messages are sent and not 15). The DMA request queue is, however, only 16 entries deep, and issuing 16 acknowledgment requests at the same time stalls data transfers until some requests clear the queue. A simple remedy is the use of a DMA list with 16 elements, where the elements point to appropriate Local Store locations of the other SPEs. The code alternates between two such lists in the double-buffered communication cycle.

2.7   Results

Results presented here are produced by the two 3.2 GHz Cell B. E. chips of the QS20 dual-socket blade running Fedora Core 7 Linux. The code is cross-compiled using x86 SDK 3.1, although the kernels are cross-compiled with an old x86 SPU GCC 3.4.1 cross-compiler, since this compiler yields the highest performance. It also needs to be mentioned that the implementation utilizes *Block Data Layout* (BDL) [13, 14], where each tile is stored in a continuous 16 KB portion of the main memory, which can be transferred in a single DMA, which puts an equal load on all 16 memory banks. Tiles are stored in the row-major order, and also data within tiles are arranged in the row-major order, a common practice on the Cell B.E. Translation from standard (FORTRAN) layout to BDL can be implemented very efficiently on the Cell B.E. [12]. Here the translation is not included in timing results. Also, in order to avoid the problem of TLB misses, all the memory is allocated in huge TLB pages and "faulted in" at initialization. As a result, an SPE never incurs

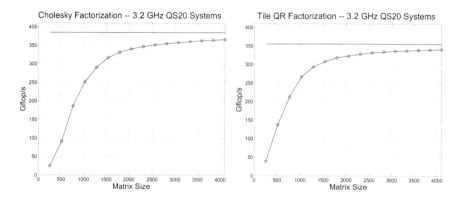

FIGURE 2.8: Performance of the Cholesky factorization (left) and the tile QR factorization (right) in single precision on an IBM QS20 blade using two Cell B. E. chips (16 SPEs). Square matrices were used. The solid horizontal line marks the performance of the SGEMM kernel for the Cholesky factorization, and the SSSRFB kernel for the tile QR factorization, multiplied by the number of SPEs (16).

a TLB miss during the run. Correct NUMA memory placement is enforced using the *libnuma* library.

Figure 2.8 and Tables 2.2 and 2.3 show the performance. Not only do the factorizations get close to the peak performance of the hardware, but also the performance curves raise very quickly with the sizes of the matrices, i.e., the code delivers very good performance even for relatively small problem sizes. Ultimately the algorithm's performance is limited by the performance of the critical SPE kernels, SGEMM for Cholesky and SSSRFB for tile QR.

2.8   Summary

It has been shown here that a silicon chip can provide an outstanding performance for compute-intensive scientific workloads by combining short-vector SIMD capabilities with multicore architecture and also providing for explicit control over caches (Local Stores). It is also an important factor that the SPEs allow for implementation of complex synchronization mechanisms and thus for efficiently exploiting task-level parallelism in workloads with complex data dependencies, such as dense matrix factorizations. One point to be made here is that successful implementation relies on addressing all aspects of performance optimization: exploiting data-level parallelism through short-vector SIMD vectorization, exploiting task-level parallelism through SPE-parallelization and

TABLE 2.2: Performance of the Cholesky factorization in single precision on two 3.2 GHz Cell B. E. chips of the IBM QS20 dual-socket blade (16 SPEs).

Matrix Size	Execution Rate [Gflop/s]	Fraction of Peak [%]	Fraction of SGEMM Peak [%]
256	24	6	6
512	91	22	24
768	186	46	49
1024	251	61	65
1280	290	71	75
1536	316	77	82
1792	330	81	86
2048	340	83	88
3072	357	87	93
4096	365	89	95

TABLE 2.3: Performance of the tile QR factorization in single precision on two 3.2 GHz Cell B. E. chips of the IBM QS20 dual-socket blade (16 SPEs).

Matrix Size	Execution Rate [Gflop/s]	Fraction of Peak [%]	Fraction of SSSRFB Peak [%]
256	38	9	11
512	137	34	39
768	212	52	60
1024	266	65	75
1280	293	72	83
1536	307	75	87
1792	317	78	90
2048	322	79	91
3072	335	82	95
4096	340	83	96

exploiting the memory hierarchy through explicit control of the local memories.

2.9   Code

The code is freely available under the BSD license and can be downloaded from the author's web site http://icl.cs.utk.edu/~kurzak/. Although the authors put a lot of effort into making the code both robust and readable, it is a proof-of-concept prototype and not a production-quality code.

Bibliography

[1] P. Bellens, J. M. Perez, R. M. Badia, and J. Labarta. CellSs: A programming model for the Cell BE architecture. In *Proceedings of the 2006 ACM/IEEE Conference on Supercomputing*, Tampa, FL, November 11-17 2006. ACM. DOI: 10.1145/1188455.1188546.

[2] C. Bischof and C. van Loan. The WY representation for products of Householder matrices. *J. Sci. Stat. Comput.*, 8:2–13, 1987.

[3] A. Buttari, J. Langou, J. Kurzak, and J. J. Dongarra. Parallel tiled QR factorization for multicore architectures. *Concurrency Computat.: Pract. Exper.*, 20(13):1573–1590, 2008. DOI: 10.1002/cpe.1301.

[4] A. Buttari, J. Langou, J. Kurzak, and J. J. Dongarra. A class of parallel tiled linear algebra algorithms for multicore architectures. *Parallel Comput. Syst. Appl.*, 35:38–53, 2009. DOI: 10.1016/j.parco.2008.10.002.

[5] J. W. Demmel. *Applied Numerical Linear Algebra*. SIAM, 1997. ISBN: 0898713897.

[6] J. J. Dongarra, I. S. Duff, D. C. Sorensen, and H. A. van der Vorst. *Numerical Linear Algebra for High-Performance Computers*. SIAM, 1998. ISBN: 0898714281.

[7] E. Elmroth and F. G. Gustavson. New serial and parallel recursive QR factorization algorithms for SMP systems. In *Applied Parallel Computing, Large Scale Scientific and Industrial Problems, 4th International Workshop, PARA'98*, Umeå, Sweden, June 14-17 1998. *Lecture Notes in Computer Science* 1541:120-128. DOI: 10.1007/BFb0095328.

[8] E. Elmroth and F. G. Gustavson. Applying recursion to serial and parallel QR factorization leads to better performance. *IBM J. Res. & Dev.*, 44(4):605–624, 2000.

[9] E. Elmroth and F. G. Gustavson. High-performance library software for QR factorization. In *Applied Parallel Computing, New Paradigms for HPC in Industry and Academia, 5th International Workshop, PARA 2000*, Bergen, Norway, June 18-20 2000. *Lecture Notes in Computer Science* 1947:53–63. DOI: 10.1007/3-540-70734-4_9.

[10] G. H. Golub and C. F. van Loan. *Matrix Computations*. The Johns Hopkins University Press, 1996. ISBN: 0801854148.

[11] B. C. Gunter and R. A. van de Geijn. Parallel out-of-core computation and updating the QR factorization. *ACM Transactions on Mathematical Software*, 31(1):60–78, 2005. DOI: 10.1145/1055531.1055534.

[12] J. Kurzak and J. J. Dongarra. Implementation of mixed precision in solving systems of linear equations on the CELL processor. *Concurrency Computat.: Pract. Exper.*, 19(10):1371–1385, 2007. DOI: 10.1002/cpe.1164.

[13] N. Park, B. Hong, and V. K. Prasanna. Analysis of memory hierarchy performance of block data layout. In *Proceedings of the 2002 International Conference on Parallel Processing, ICPP'02*, pages 35–44, Vancouver, Canada, August 18-21 2002. IEEE Computer Society. DOI: 10.1109/ICPP.2002.1040857.

[14] N. Park, B. Hong, and V. K. Prasanna. Tiling, block data layout, and memory hierarchy performance. *IEEE Trans. Parallel Distrib. Syst.*, 14(7):640–654, 2003. DOI: 10.1109/TPDS.2003.1214317.

[15] J. M. Perez, P. Bellens, R. M. Badia, and J. Labarta. CellSs: Making it easier to program the Cell Broadband Engine processor. *IBM J. Res. & Dev.*, 51(5):593–604, 2007. DOI: 10.1147/rd.515.0593.

[16] R. Schreiber and C. van Loan. A storage-efficient WY representation for products of Householder transformations. *J. Sci. Stat. Comput.*, 10:53–57, 1991.

[17] L. N. Trefethen and D. Bau. *Numerical Linear Algebra.* SIAM, 1997. ISBN: 0898713617.

Chapter 3

Dense Linear Algebra for Hybrid GPU-Based Systems

Stanimire Tomov

Department of Electrical Engineering and Computer Science, University of Tennessee

Jack Dongarra

Department of Electrical Engineering and Computer Science, University of Tennessee
Computer Science and Mathematics Division, Oak Ridge National Laboratory
School of Mathematics & School of Computer Science, Manchester University

3.1 Introduction

Since the introduction of multicore architectures, hardware designs are going through a renaissance due to the need for new approaches to manage the exponentially increasing:

1. Appetite for power, and

2. Gap between compute and communication speeds.

Hybrid graphics processing unit (GPU)-based multicore platforms, composed of both homogeneous multicores and GPUs, stand out among a confluence of current hardware trends as they provide an effective solution to these two challenges. Indeed, as power consumption is typically proportional to the cube of the frequency, GPUs have a clear advantage against current homogeneous multicores, as GPUs' compute power is derived from many cores that are of low frequency. Furthermore, initial GPU experiences across academia, industry, and national research laboratories have provided a long list of success stories for specific applications and algorithms, often reporting speedups of order 10 to 100× compared to current x86-based homogeneous multicore systems [2].

3.1.1 Linear Algebra (LA)—Enabling New Architectures

Despite the current success stories involving hybrid GPU-based systems, the large-scale enabling of those architectures for computational science would still depend on the successful development of fundamental numerical libraries for them. Major issues in terms of developing new algorithms, programmability, reliability, and user productivity must be addressed. This chapter describes some of the current efforts toward the development of these fundamental libraries, and in particular, libraries in the area of dense linear algebra (DLA).

Historically, linear algebra has been in the vanguard of efforts to enable new architectures for computational science for good strategic reasons. First, a very wide range of science and engineering applications depend on linear algebra; these applications will not perform well unless linear algebra libraries perform well. Second, linear algebra has a rich and well understood structure for software developers to exploit algorithmically, so these libraries represent an advantageous starting point for the effort to bridge the yawning software gap that has opened up within the HPC community today.

3.1.2 MAGMA—LA Libraries for Hybrid Architectures

The Matrix Algebra on GPU and Multicore Architectures (MAGMA) project, and the MAGMA and MAGMA BLAS libraries [3] stemming from it, are used to demonstrate the techniques and their effect on performance. Designed to be similar to LAPACK in functionality, data storage, and interface, the MAGMA libraries will allow scientists to effortlessly port their LAPACK-relying software components and to take advantage of the new hybrid architectures. Current work targets GPU-based systems, and the efforts are supported by both government and industry, including NVIDIA, who recently recognized the University of Tennessee, Knoxville's (UTKs) Innovative Computing Laboratory (ICL) as a CUDA Center of Excellence. This is to further promote, expand, and support ICL's commitment toward developing **LA Libraries for Hybrid Architectures**.

Against this background, the main focus of this chapter will be to provide some high-level insight on **how to code/develop DLA for GPUs**. The approach described here is based on the idea that, in order to deal with the complex challenges stemming from the heterogeneity of the current GPU-based systems, optimal software solutions will themselves have to hybridize, combining the strengths of the system's hybrid components. That is, *hybrid algorithms* that match algorithmic requirements to the architectural strengths of the system's hybrid components must be developed. It has been shown that properly designed hybrid algorithms for GPU-based multicore platforms work for the core DLA routines, namely the one- and two-sided matrix factorizations.

3.2 Hybrid DLA Algorithms

The development of high-performance DLA for homogeneous multicores has been successful in some cases, like the one-sided factorizations, and difficult for others, like the two-sided factorizations. The situation is similar for GPUs—some algorithms map well, others do not. Developing algorithms for a combination of these two architectures (to use both multicore and GPUs) though can be beneficial and should be exploited, especially since in many situations, the computational bottlenecks for one of the components (of this hybrid system) may not be for the other. Thus, developing **hybrid algorithms** that properly split and schedule the computation over different hardware components may lead to very efficient algorithms. The goal is to develop these new, hybrid algorithms for the area of DLA, and moreover, show that the new developments:

- Leverage prior DLA developments, and

- Achieve what has not been possible so far, e.g., using just homogeneous multicores.

3.2.1 How to Code DLA for GPUs?

The question of how to code for any architecture, including GPUs, is complex, e.g., involving issues in terms of choosing a language, programming model, developing new (GPU specific) algorithms, programmability, reliability, and user productivity. Nevertheless, it is possible to give some major considerations and directions that have already shown promising results:

Use CUDA / OpenCL CUDA is currently the language of choice for programming GPUs. It facilitates a data-based parallel programming model that has turned out to be a remarkable fit for many applications. Moreover, current results show its programming model allows applications to scale on many cores. DLA is no exception as performance relies on the performance of Level 2/3 BLAS—essentially a data parallel set of subroutines that are scaling on current many-core GPUs (see also Chapter 4). The approach described here also shows how the BLAS scalability is in fact translated into scalability on higher level routines (LAPACK).

Similarly to CUDA, OpenCL also has its roots in the data-based parallelism (now both moving to support task-based parallelism). OpenCL is still yet to be established but the fact that it is based on a programming model with already established potential and the idea of providing portability—across heterogeneous platforms consisting of CPUs, GPUs, and other processors—makes it an excellent candidate for coding hybrid algorithms.

Use GPU BLAS Performance of DLA critically depends on the availability of fast BLAS, especially on the Level 3 BLAS matrix-matrix multiplication. Older generation GPUs did not have memory hierarchy and their performance exclusively relied on high bandwidth. Therefore, although there has been some work in the field, the use of older GPUs has not led to significantly accelerated DLA. For example, Fatahalian et al. studied SGEMM and their conclusion was that CPU implementations outperform most GPU implementations. Similar results were produced by Galoppo et al. on LU factorization. The introduction of **memory hierarchy** in current GPUs though has drastically changed the situation. Indeed, by having memory hierarchy, GPUs can be programmed for memory reuse and hence not rely exclusively on their high bandwidth. An illustration of this fact is given in Figure 3.1, showing the performance of correspondingly a compute-bound (matrix-matrix multiplication on the left) and a memory-bound routine (matrix-vector multiplication on the right). Having fast BLAS is significant because algorithms for GPUs can now leverage prior DLA developments, which have traditionally relied on fast BLAS. Of course there are GPU-specific optimizations, as will be shown, like performing extra-operations, BLAS fusion, etc, but the important fact is, high-performance algorithms can be coded at a high level, just using BLAS, often abstracting the developer from the need of low-level GPU-specific coding.

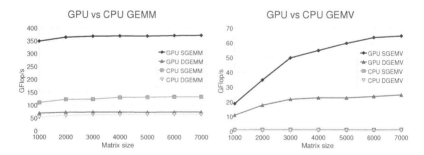

FIGURE 3.1: BLAS on GPU (GTX 280) *vs* CPU (8× Intel Xeon 2.33GHz).

Use Hybrid Algorithms Current GPUs feature massive parallelism but serial kernel execution. For example NVIDIA's GTX280 has 30 multiprocessors, each multiprocessor having eight SIMD functional units, each unit capable of executing up to three (single floating point) operations per cycle. At the same time, kernels are executed serially; only one kernel is allowed to run at a time using the entire GPU. This means that only large, highly parallelizable kernels can run efficiently on GPUs. The idea of using hybrid algorithms presents an opportunity to remedy this situation and therefore enable the efficient use of GPUs well beyond the case of data-parallel applications. Namely, the solution and advice to developers is to use a hybrid coding approach, where small, non-parallelizable kernels would be executed on the CPU, and only large, parallelizable kernels on the GPU. Although GPUs move towards supporting task-based parallelism as well (e.g., advertised for the next generation NVIDIA GPUs, code named "Fermi" [4]), small tasks that arise in DLA would still make sense to be done on the CPU for various reasons, e.g., to use the x-86 software infrastructure. Moreover, efficient execution would still require parallelism and small tasks still may be difficult or impossible to parallelize.

3.2.2   The Approach—*Hybridization of DLA Algorithms*

The above considerations are incorporated in the following *Hybridization of DLA Algorithms* approach:

- Represent DLA algorithms as a collection of BLAS-based tasks and dependencies among them (see Figure 3.2):

  - Use parametrized task granularity to facilitate auto-tuning;
  - Use performance models to facilitate the task splitting/mapping.

- Schedule the execution of the BLAS-based tasks over the multicore and the GPU:

  - Schedule small, non-parallelizable tasks on the CPU and large, parallelizable on the GPU;
  - Define the algorithm's *critical path* and prioritize its execution/scheduling.

The splitting of the algorithms into tasks is in general easy, as it is based on the splitting of large BLAS into smaller ones. More challenging is choosing the granularity and shape of the splitting and the subsequent scheduling of the sub-tasks. There are two main guiding directions on how to design the splitting and scheduling of tasks. First, the splitting and scheduling should allow for asynchronous execution and load balance among the hybrid components. Second, it should harness the strengths of the components of a hybrid architecture by properly matching them to algorithmic/task requirements. Examples demonstrating these general directions are given in the next two sections.

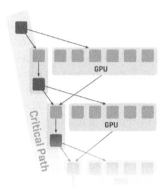

FIGURE 3.2: Algorithms as a collection of BLAS-based tasks and dependencies among them (DAGs) for hybrid GPU-based computing.

Next, choosing the task granularity can be done by parametrizing the tasks' sizes in the implementations and tuning them empirically [7]. The process can be automated [13], often refered to as *auto-tuning*. Auto-tuning is crucial for the performance and the maintenance of modern numerical libraries, especially for hybrid architectures and algorithms for them. Figuratively speeking, it can be regarded as both *the Beauty and the Beast* behind hybrid DLA libraries (e.g., MAGMA) as it is an elegant and very practical solution for easy maintenance and performance portability, while often being a brute force, empirically based exhaustive search that would find and set automatically the best performing algorithms/kernels for a specific hardware configuration. The "exhaustive" search is often relaxed by applying various performance models.

Finally, the problem of scheduling is of crucial importance for the efficient execution of an algorithm. In general, the execution of the critical path of an algorithm should be scheduled as soon as possible. This often remedies the problem of synchronizations introduced by small, non-parallelizable tasks (often on the critical path; scheduled on the CPU) by overlapping their execution with the execution of larger more parallelizable ones (often Level 3 BLAS; scheduled on the GPU).

These principles are general enough to be applied in areas well beyond DLA. Usually they come with specifics, induced by the architecture and the algorithms considered. The following two sections present some of these specifics for, correspondingly, the classes of one-sided and two-sided dense matrix factorizations.

3.2.3 One-Sided Factorizations

This section describes the hybridization of LAPACK's one-sided factorizations—LU, QR, and Cholesky—on dense matrices. These factorizations are important because they are the first of two steps in solving dense linear systems of equations. The factorization represents the bulk of the computation—$O(N^3)$ floating point operations in the first step *vs* $O(N^2)$ in the second step—and therefore has to be highly optimized. LAPACK uses block-partitioned algorithms, and the corresponding hybrid algorithms are based on them.

The opportunity for acceleration using hybrid approaches (CPU and GPU) has been noticed before in the context of one-sided factorizations. In particular, while developing algorithms for GPUs, several groups observed that panel factorizations are often faster on the CPU than on the GPU, which led to the development of highly efficient, one-sided hybrid factorizations for a single CPU core and a GPU [6,7,9], multiple GPUs [6,10], and multicore+GPU systems [11]. M. Fatica [12] developed hybrid DGEMM and DTRSM for GPU-enhanced clusters and used them to accelerate the Linpack benchmark. This approach, mostly based on BLAS level parallelism, results only in minor or no modifications to the original source code.

MAGMA provides two interfaces to the hybrid factorizations. These are the CPU interface, where the input matrix is given in the CPU memory and the result is expected also in the CPU memory, and the GPU interface, where the input and the output are on the GPU memory. In both cases the bulk of the computation is done on the GPU and along the computation; the sub-matrices that remain to be factored always reside on the GPU memory. Panels are copied to the CPU memory, processed on the CPU using LAPACK, and copied back to the GPU memory (see Figure 3.3). Matrix updates are done on the GPU. Thus, for any panel transfer of size $n \times nb$ elements, a sub-matrix of size $n \times (n - nb)$ is updated on the GPU (e.g., this is the case for the right-looking versions of LU and QR; for the left-looking Cholesky the ratios are $nb \times nb$ elements transfered *vs* $n \times nb$ elements updated), where nb is a blocking size and n is the size of the sub-matrix that remains to be factored. These ratios of communications *vs* computations allow for mostly overlapping the panel factorizations on the CPU with updates on the GPU (see below the specifics). Figure 3.3 illustrates this typical pattern (as just described) of hybridization for the one-sided factorizations in the GPU interface.

The CPU interface can reuse the GPU interface implementation by wrapping it around two matrix copies—one at the beginning copying the input matrix from the CPU to the GPU memory, and one at the end copying the final result from the GPU back to the CPU memory. This overhead can be partially avoided, though. The input matrix can be copied to the GPU by starting an asynchronous copy of the entire matrix except the first panel and overlapping this copy with the factorization of the first panel on the CPU. The GPU-to-CPU copy at the end of the factorizations can be replaced by a

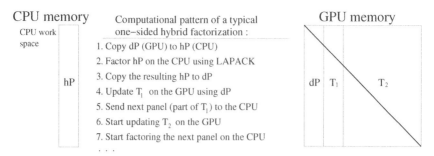

FIGURE 3.3: A typical hybrid pattern of computation and communication for the one-sided matrix factorizations in MAGMA's GPU interface.

continuous accumulation of the output directly on the CPU. One way to do it is by sending the entire block of columns containing the current panel (*vs* just the panel) and keeping copies of the factored panels on the CPU. This way, as the computation progresses, the final result is accumulated on the CPU as a byproduct of the computation, avoiding the overhead of an entire matrix copy.

Below are given some of the specifics in the development of the hybrid Cholesky, QR, and LU factorizations, respectively:

Cholesky Factorization MAGMA uses the left-looking version of the block Cholesky factorization. Figure 3.4 shows an iteration of the standard block Cholesky algorithm coded in correspondingly MATLAB and LA-PACK style, and how these standard implementations can easily be translated into a hybrid implementation. Note the simplicity and the similarity of the hybrid code with the LAPACK code. The only difference is the two CUDA calls needed to offload data back and forth from the CPU to the GPU. Also, note that steps (2) and (3) are independent and can be overlapped—step (2), a Cholesky factorization task on a small diagonal block, is scheduled on the CPU using a call to LAPACK and step (3), a large matrix-matrix multiplication, on the GPU. This is yet another illustration of the general guidelines mentioned in the previous two sections. The performance of this algorithm is given in Section 3.3.

QR Factorization Currently, MAGMA uses static scheduling and a right-looking version of the block QR factorization. The panel factorizations are scheduled on the CPU using calls to LAPACK, and the Level 3 BLAS updates on the trailing sub-matrices are scheduled on the GPU. The trailing matrix updates are split into 2 parts—one that updates just the next panel and a second one updating the rest, i.e., correspondingly sub-matrices T_1 and T_2 as given in Figure 3.3. The next panel update (i.e., T_1) is done first, sent to the CPU, and the panel factorization on the CPU is overlapped with the second part of the trailing matrix

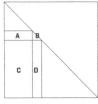

MATLAB code	LAPACK code	Hybrid code
[1] B = B – A*A'	ssyrk_("L", "N", &nb, &j, &mone, hA(j,0), ...)	cublasSsyrk('L', 'N', nb, j. mone, dA(j,0), ...)
		cublasGetMatrix(nb, nb, 4, dA(j, j), *lda, hwork, nb)
[2] B = chol(B, 'lower')	spotrf_("L", &nb, hA(j, j), lda, info)	cublasSgemm('N', 'T', j, ...)
[3] D = D – C*A'	sgemm_("N", "T", &j, ...)	spotrf_("L", &nb, hwork, &nb, info)
		cublasSetMatrix(nb, nb, 4, hwork, nb, dA(j, j), *lda)
[4] D = D\B	strsm_("R", "L", "T", "N", &j, ...)	cublasStrsm('R', 'L', 'T', 'N', j, ...)

FIGURE 3.4: Pseudo-code implementation of the hybrid Cholesky. hA and dA are pointers to the matrix to be factored correspondingly on the host (CPU) and the device (GPU).

(i.e., T_2). This technique is known as *look-ahead*, e.g., used before in the Linpack benchmark [13]. Its use enables the overlap of CPU and GPU work (and some communications). Figure 3.5 illustrates this by quantifying the CPU-GPU overlap for the case of QR in single precision arithmetic. Note that, in this case for matrices of size above 6,000, the

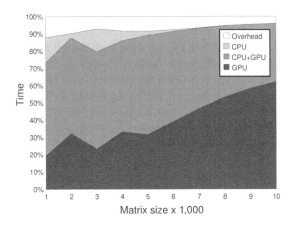

FIGURE 3.5: Time breakdown for the hybrid QR from MAGMA in single precision arithmetic on GTX280 GPU and Intel Xeon 2.33GHz CPU.

CPU work on the panels is entirely overlapped by work on the GPU.

It is important to note that block factorizations (both one and two sided) inherently lead to multiplications with triangular matrices. To execute them in data-parallel fashion on GPUs, it is often more efficient to put zeroes in the unused triangular parts and perform the multiplication as having general dense matrices. To give an example, the block QR algorithm accumulates orthogonal Level 2 BLAS transformations during the panel factorization. The accumulated transformation has the form

$$Q_i = I - V_i T_i V_i^T$$

where V_i is of size $k \times nb$, $k \geq nb$ and T_i is $nb \times nb$. This transformation is than applied as a Level 3 BLAS to the trailing sub-matrix. The LA-PACK implementation splits the multiplication with V_i into two Level 3 BLAS calls, as the top $nb \times nb$ sub-matrix of V_i is triangular. A GPU implementation is more efficient if the multiplication with V_i is performed as one BLAS call, and therefore is enabled, e.g., by providing an efficient mechanism to put zeroes in the unused triangular part of V_i (before the multiplication) and restoring the original (after the multiplication).

This is just one example of a GPU-specific optimization technique where to get higher performance extra flops must be done or separate kernels must be fused into one.

LU Factorization Similarly to QR, MAGMA uses a right-looking version of the LU factorization. The scheduling is static using the look-ahead technique. Interchanging rows of a matrix stored in column major format, needed in the pivoting process, cannot be done efficiently on current GPUs. We use the LU factorization algorithm by V. Volkov and J. Demmel [6] that removes the above mentioned bottleneck. The idea behind it is to transpose the matrix in the GPU memory. This is done once at the beginning of the factorization so that row elements are contiguous in memory, i.e., equivalent to changing the storage format to row major. Row interchanges now can be done efficiently using coalescent memory accesses on the GPU (*vs* strided memory accesses for a matrix in column major format). The panels are being transposed before being sent to the CPU for factorization, i.e., moved back to the standard for LAPACK column major format. Compared to the non-transposed version, this algorithm runs approximately 50% faster on current NVIDIA GPUs, e.g., GTX 280.

3.2.4   Two-Sided Factorizations

If the importance of the one-sided factorizations stems from their role in solving linear systems of equations, the importance of the two-sided factorizations stems from their role in solving eigenvalue problems. The two-sided factorizations are an important first step in solving eigenvalue problems. Similarly to the one-sided, the two-sided factorizations are compute intensive ($O(N^3)$ flops) and therefore also have to be highly optimized.

The two-sided factorizations can be organized as block-partitioned algorithms. This is used in LAPACK to develop efficient signle CPU/core algorithms and can be used as the basis for developing hybrid algorithms. The development follows the main approach already described. This section will describe the specifics involved and will actually demonstrate much higher speedups of accelerating two-sided factorizations *vs* the speedups in accelerating the one-sided factorizations.

The hybridization of the two-sided factorizations can be best explained with the reduction to upper Hessenberg form, denoted further by HR, which is stressed in this section. The operation count for the reduction of an $N \times N$ matrix is approximately $\frac{10}{3}N^3$, which, in addition to not running efficiently on current architectures, makes the reduction a very desirable target for acceleration.

The bottleneck: The problem of accelerating the two-sided factorization algorithms stems from the fact that these algorithms are rich in Level 2 BLAS operations, which do not scale on multicore architectures and actually run only at a fraction of the machine's peak performance. This is shown in Figure 3.1, right. There are dense linear algebra (DLA) techniques that can replace Level 2 BLAS operations with Level 3 BLAS. These are the block algorithms, already mentioned several times, where the application of consecutive Level 2 BLAS operations that occur in the algorithms can be delayed and accumulated so that at a later moment the accumulated transformation is applied at once as a Level 3 BLAS (see LAPACK [14] and the specifics related to QR above). This approach totally removes Level 2 BLAS flops from Cholesky and reduces its amount to $O(n^2)$ flops in LU and QR thus making it asymptotically insignificant compared to the total $O(n^3)$ amount of operations for these factorizations. The same technique can be applied to HR [15], but in contrast to the one-sided factorizations, it still leaves about 20% of the total number of operations as Level 2 BLAS. Note that 20% of Level 2 BLAS is significant because, in practice, using a single core of a multicore machine, this 20% can take about 70% of the total execution time, thus leaving the grim perspective that multicore use—no matter how many cores would be available—can ideally reduce only the 30% of the execution time that is spent on Level 3 BLAS.

Bottleneck identification: For large applications, tools like TAU [16] can help in locating performance bottlenecks. TAU can generate execution profiles and traces that can be analyzed with other tools like ParaProf [17], Jumpshot [18], and Kojak [19]. Profiles of the execution time (or of PAPI [20] hardware counters) on runs using various numbers of cores of a multicore processor can be compared using ParaProf to easily see which functions scale, what is the performance for various parts of the code, what percentage of the execution time is spent in the various functions, etc. In the case of HR, using a dual socket quad-core Intel Xeon at 2.33GHz, this analysis easily identifies a call to a single matrix-vector product that runs at about 70% of the total time, does not scale with multicore use, and has only 20% of the total amount of flops. Algorithmically speaking, this matrix-vector product is part of the panel factorization, where the entire trailing matrix (denoted by A_j in Figure 3.6) has to multiply a currently computed Householder reflector vector v_j (see [10] for further datail on the algorithm). These findings are also illustrated in Figure 3.6.

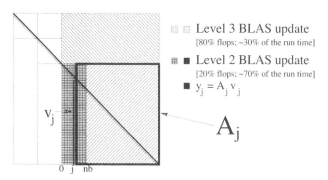

FIGURE 3.6: HR computational bottleneck: Level 2 BLAS $y_j = A_j v_j$.

Since fast Level 2 BLAS are available for GPUs (Figure 3.1, right), it is clear that the operation has to be scheduled for execution on the GPU. In other words, having fast implementations for all kernels, the development of the hybrid HR algorithms is now just a matter of splitting the computation into tasks and properly scheduling the tasks' execution on the available hybrid hardware components. Below are the details of the main steps in this process of *hybridization* of the HR algorithm:

Task Spliting Studying the execution and data flow of an algorithm is important in order to properly split the computation into tasks and dependencies. It is important to study the memory footprint of the routines, e.g., what data are accessed and what data are modified. Moreover, it is important to identify the algorithm's critical path and to decouple from it as much work as possible, as the tasks outside the critical path in general would be the ones trivial to parallelize. Applied to HR, this analysis identifies that the computation must be split into three main tasks, further denoted by P_i, M_i, and G_i, where i is iteration index for the block HR and is described as follows. The splitting is motivated by and associated with operations updating three corresponding matrices, for convenience denoted by the name of the task updating them. The splitting is illustrated in Figure 3.7, left, and described as follows:

- The panel factorization task P_i (20% of the flops) updates the current panel and accumulates matrices V_i, T_i, and Y_i needed for the trailing matrix update;

- Task G_i (60% of the flops) updates the trailing sub-matrix

$$G_i = (I - V_i \ T_i \ V_i^T) \ G_i \ (I - V_i \ T_i \ V_i(nb + 1 : \ , \ : \ )^T)$$

- Task M_i (20% of the flops) updates the sub-matrix, determined to fall **outside of the critical path** of the algorithm

$$M_i = M_i \ (I - V_i \ T_i \ V_i^T).$$

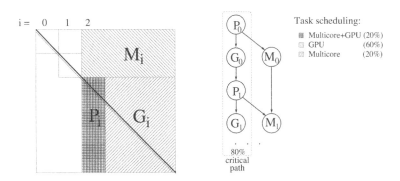

FIGURE 3.7: Main tasks and their scheduling.

We note that this splitting is motivated by the memory footprint analysis. Using this particular splitting, one can see that task M_i gets independent of G_i and falls outside of the critical path of the algorithm (illustrated in Figure 3.7, right). This is important for scheduling the tasks' execution over the components of the hybrid system. Note that the critical path is still 80% of the total amount of flops.

Task Scheduling The scheduling is given also in Figure 3.7, right. The tasks on the critical path must be scheduled as fast as possible—and the scheduling must be hybrid, using both the Multicore and the GPU. The memory footprint of task P_i, with 'P' standing for panel, is both P_i and G_i, but G_i is accessed only for the time consuming computation of $y_j = A_j v_j$ (see Figure 3.6). Therefore, the part of P_i that is constrained to the panel (not rich in parallelism, with flow control statements) is scheduled on the multicore using LAPACK, and the time consuming $y_j = A_j v_j$ (highly parallel but requiring high bandwidth) is scheduled on the GPU. G_i, with 'G' standing for GPU, is scheduled on the GPU. This is a Level 3 BLAS update and can be done very efficiently on the GPU. Moreover, note that G_{i-1} contains the matrix A_j needed for task P_i, so for the computation of $A_j v_j$ we have to only send v_j to the GPU and the resulting y_j back from the GPU to the multicore. The scheduling so far heavily uses the GPU, so in order to simultaneously make the critical path execution faster and to make a better use of the multicore, task M_i, with 'M' standing for multicore, is scheduled on the multicore.

Hybrid HR Figure 3.8 illustrates the communications between the multicore and GPU for i^{th} iteration (outer) of the hybrid block HR algorithm. Shown is the jth inner iteration. Note that, as in the case of the one-sided factorizations, the matrix to be factored resides originally on the GPU memory, and as the computation progresses the result is continuously accumulated on the CPU. At the end, the factored matrix is available on the CPU memory in a format identical to LAPACK's. The tasks scheduling is as given above.

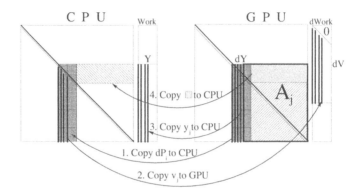

FIGURE 3.8: CPU/GPU communications for inner/outer iteration j/i.

This approach can be applied to the rest of the two-sided matrix factorizations. The key component—having fast GPU implementations of the various Level 2 BLAS matrix-vector products that may be needed in the different algorithms—is now available through the MAGMA BLAS library (see Chapter 4). Having the kernels needed, as stated above, it is now simply a matter of organizing the computation in a hybrid fashion over the available hardware components.

3.3   Performance Results

The performance results provided in this section use NVIDIA's GeForce GTX 280 GPU and its multicore host, a dual socket quad-core Intel Xeon running at 2.33GHz. Kernels executed on the multicore use LAPACK and BLAS from MKL 10.1, and BLAS kernels executed on the GPU use a combination of CUBLAS 2.1 and MAGMA BLAS 0.2, unless otherwise noted.

The performance of the hybrid Cholesky factorization is given in Figure

3.9. It runs asymptotically at 300 Gflop/s in single and at almost 70 Gflop/s in double precision arithmetic.

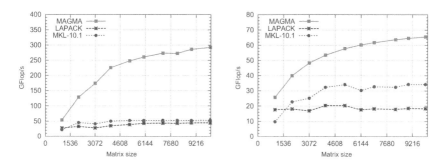

FIGURE 3.9: Performance of MAGMA's hybrid Cholesky in single (left) and double precision (right) arithmetic on GTX 280 *vs* MKL 10.1 and LAPACK (with multi-threaded BLAS) on Intel Xeon dual socket quad-core 2.33GHz.

The performance of the hybrid QR factorization is given in Figure 3.10. It runs asymptotically almost at 290 Gflop/s in single and at almost 68 Gflop/s in double precision arithmetic.

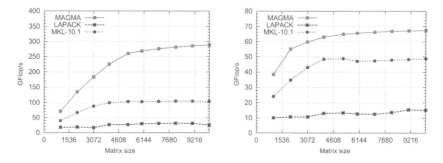

FIGURE 3.10: Performance of MAGMA's hybrid QR in single (left) and double precision (right) arithmetic on GTX 280 *vs* MKL 10.1 and LAPACK (with multi-threaded BLAS) on Intel Xeon dual socket quad-core 2.33GHz.

The performance of the hybrid LU factorization is given in Figure 3.11. It runs asymptotically almost at 320 Gflop/s in single and at almost 70 Gflop/s in double precision arithmetic.

Figure 3.12 shows the performance of two versions of the hybrid HR algorithms in double precision arithmetic. The performance is also compared to the block HR on single core and multicore. The basic hybrid HR is for one core accelerated with one GPU. The "Multicore+GPU" hybrid algorithm is the one described in this chapter where tasks M_i are executed on the available CPU cores. The result shows an enormous speedup of 16× when compared to the current block HR running on just homogeneous multicore. The techniques

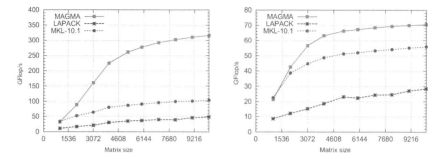

FIGURE 3.11: Performance of MAGMA's hybrid LU in single (left) and double precision (right) arithmetic on GTX 280 *vs* MKL 10.1 and LAPACK (with multi-threaded BLAS) on Intel Xeon dual socket quad-core 2.33GHz.

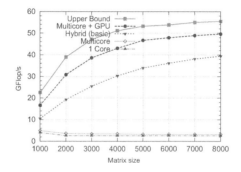

FIGURE 3.12: Performance of the hybrid HR in double precision (bottom) arithmetic on GTX 280 *vs* MKL 10.1 on Intel Xeon dual socket quad-core 2.33GHz.

in the basic implementation account for most of the acceleration, which is due to the use of hybrid architectures and the proper algorithmic design—splitting the computation into tasks and their scheduling so that we match algorithmic requirements to architectural strengths of the hybrid components.

This performance gets to be asymptotically within 90% of the "upper bound" performance, as shown in Figure 3.12. Here upper bound denotes the performance of just the *critical path* of the algorithm when no synchronizations and data transfer times are taken into account, i.e., this is the performance of tasks P_i and G_i (without counting M_i) just based on the performance of the BLAS used.

3.4 Summary

This work is on the development of numerical linear algebra libraries and is motivated by hardware changes that have rendered legacy numerical libraries inadequate for the new architectures. GPUs, homogeneous multicore architectures, and hybrid systems based on them adequately address the main requirements of new architectures—to keep power consumption and the gap between compute and communication speeds low. There is every reason to believe that future systems will continue the general trend taken by GPUs and hybrid combinations of GPUs with homogeneous multicores—freeze the frequency and keep escalating the number of cores; stress on data-parallelism to provide high bandwidth to the escalating number of cores—and thus render this work relevant for future system designs.

This chapter presented a concept on how to code/develop dense linear algebra algorithms for GPUs. The approach is general enough to be applicable to areas well beyond dense linear algebra and is implemented on a high-enough level to guarantee easy portability and relevance for future hybrid systems. In particular, the approach uses CUDA to develop low-level kernels when needed, but mostly relies on high-level libraries like LAPACK for CPUs and BLAS for CPUs and GPUs. Moreover, the approach is based on the development of hybrid algorithms, where in general small, non-parallelizable tasks are executed on the CPU, reusing the existing software infrastructure for standard architectures, and large, data-parallel tasks are executed on the GPU.

It was shown that using this approach in the area of dense linear algebra, one can leverage prior DLA developments and achieve what has not been possible so far, e.g., using just homogeneous multicore architectures. In particular, the approach uses LAPACK to execute small, non-parallelizable tasks on the CPU. Tasks that are bottlenecks for the CPU, like Level 2 BLAS, are properly designed in the new algorithms and scheduled for execution on the GPU.

Specific examples were given on fundamental dense linear algebra algorithms. Namely, it was demonstrated how to develop hybrid one-sided and two-sided matrix factorizations. Although the one-sided factorizations can be represented as Level 3 BLAS and ran efficiently on current multicore architectures, GPUs still can accelerate them significantly, depending on the hardware configuration, e.g., $O(1)\times$ for the systems used in the numerical experiments presented. In the case of two-sided factorizations, the speedups are even higher, e.g., $O(10)\times$, as implementations using homogeneous multicores, no matter how many cores are available, currently run them at the speed of a single core.

The implementations are now freely available through the MAGMA library site http://icl.eecs.utk.edu/magma/.

Bibliography

[1] NVIDIA CUDA Compute Unified Device Architecture - Programming Guide. http://developer.download.nvidia.com, 2007.

[2] General-purpose computation using graphics hardware. http://www.gpgpu.org.

[3] S. Tomov, R. Nath, P. Du, and J. Dongarra. MAGMA version 0.2 Users' Guide. http://icl.cs.utk.edu/magma, November 2009.

[4] NVIDIA. NVIDIA's Next Generation CUDA Compute Architecture: Fermi. http://www.nvidia.com/object/fermi_architecture.html, 2009.

[5] Y. Li, J. Dongarra, and S. Tomov. A Note on Auto-tuning GEMM for GPUs. In *ICCS '09: Proceedings of the 9th International Conference on Computational Science*, pages 884–892, Berlin, Heidelberg, 2009. Springer-Verlag.

[6] R. Whaley, A. Petitet, and J. Dongarra. Automated Empirical Optimizations of Software and the ATLAS Project. *Parallel Computing* 27(1-2):2001, 3–35.

[7] J. Dongarra, S. Moore, G. Peterson, S. Tomov, J. Allred, V. Natoli, and D. Richie. Exploring new architectures in accelerating CFD for Air Force applications. In *Proceedings of HPCMP Users Group Conference 2008*, July 14-17 2008, http://www.cs.utk.edu/~tomov/ugc2008_final.pdf.

[8] V. Volkov and J. Demmel. Benchmarking GPUs to tune dense linear algebra. In *SC '08: Proceedings of the 2008 ACM/IEEE Conference on Supercomputing*, pages 1–11, Piscataway, NJ, 2008. IEEE Press.

[9] M. Baboulin, J. Dongarra, and S. Tomov. Some issues in dense linear algebra for multicore and special purpose architectures. LAPACK Working Note 200, May 2008.

[10] H. Ltaief, S. Tomov, R. Nath, P. Du, and J. Dongarra. A scalable high performant cholesky factorization for Multicore with GPU Accelerators. LAPACK Working Note 223, November 2009.

[11] S. Tomov, J. Dongarra, and M. Baboulin. Towards dense linear algebra for hybrid GPU accelerated manycore systems. LAPACK Working Note 210, October 2008.

[12] M. Fatica Accelerating Linpack with CUDA on heterogenous clusters. GPGPU-2, pages 46–51, Washington, DC, 2009.

[13] J. Dongarra, P. Luszczek, and A. Petitet. The LINPACK benchmark: Past, present, and future. *Concurrency and Computation: Practice and Experience*, 15:820, 2003, pp. 1–18.

[14] E. Anderson, Z. Bai, C. Bischof, L. S. Blackford, J. W. Demmel, J. J. Dongarra, J. Du Croz, A. Greenbaum, S. Hammarling, A. McKenney, and D. Sorensen. *LAPACK Users' Guide*. SIAM, Philadelphia, PA, 1992. http://www.netlib.org/lapack/lug/.

[15] S. Hammarling, D. Sorensen, and J. Dongarra. Block reduction of matrices to condensed forms for eigenvalue computations. *J. Comput. Appl. Math* 27:1987, 215–227.

[16] S. Shende and A. Malony. The TAU parallel performance system. *Int. J. High Perform. Comput. Appl.* 20(2):2006, 287–311.

[17] R. Bell, A. Malony, and S. Shende. ParaProf: A portable, extensible, and scalable tool for parallel performance profile analysis Euro-Par, 2003, pp. 17–26.

[18] O. Zaki, E. Lusk, W. Gropp, and D. Swider. Toward scalable performance visualization with Jumpshot. *HPC Applications* 13(2):1999, 277–288.

[19] F. Wolf and B. Mohr. Kojak—A tool set for automatic performance analysis of parallel applications. Proceedings of Euro-Par 2003 Klagenfurt, Austria, August 26–29, 2003, pp. 1301–1304.

[20] S. Browne, C. Deane, G. Ho, and P. Mucci. PAPI: A portable interface to hardware performance counters. Proc. of DoD HPCMP u&c pp. 7–10.

[21] S. Tomov and J. Dongarra. Accelerating the reduction to upper Hessenberg form through hybrid GPU-based computing. Technical Report 219, LAPACK Working Note 219, May 2009.

Chapter 4

BLAS for GPUs

Rajib Nath

Department of Electrical Engineering and Computer Science, University of Tennessee

Stanimire Tomov

Department of Electrical Engineering and Computer Science, University of Tennessee

Jack Dongarra

Department of Electrical Engineering and Computer Science, University of Tennessee
Computer Science and Mathematics Division, Oak Ridge National Laboratory
School of Mathematics & School of Computer Science, Manchester University

4.1 Introduction

Recent activities of major chip manufacturers, such as Intel, AMD, IBM and NVIDIA, make it more evident than ever that future designs of microprocessors and large HPC systems will be hybrid/heterogeneous in nature, relying on the integration (in varying proportions) of two major types of components:

1. Multi-/many-cores CPU technology, where the number of cores will con-

tinue to escalate while avoiding the power wall, instruction level parallelism wall, and the memory wall [1]; and

2. Special purpose hardware and accelerators, especially GPUs, which are in commodity production, have outpaced standard CPUs in performance, and have become as easy—if not easier—to program than multicore CPUs.

The relative balance between these component types in future designs is not clear, and will likely vary over time, but there seems to be no doubt that future generations of computer systems, ranging from laptops to supercomputers, will consist of a composition of heterogeneous components.

These hardware trends have inevitably brought up the need for updates on existing legacy software packages, such as the sequential LAPACK [2], from the area of dense linear algebra (DLA). To take advantage of the new computational environment, successors of LAPACK must incorporate algorithms of three main characteristics: high parallelism, reduced communication, and heterogeneity-awareness. In all cases though, the development can be streamlined if the new algorithms are designed at a high level (see Chapter 3), using just a few, highly optimized low-level kernels. Chapter 3 demonstrated a hybridization approach that indeed streamlined the development of high-performance DLA for multicores with GPU accelerators. The new algorithms, covering core DLA routines, are now part of the MAGMA library [3], a successor to LAPACK for the new heterogeneous/hybrid architectures. Similarly to LAPACK, MAGMA relies on the efficient implementation of a set of low-level linear algebra kernels. In the context of GPU-based hybrid computing, this is a subset of BLAS [4] for GPUs.

The goal of this chapter is to provide guidance on **how to develop high performance BLAS for GPUs**—a key prerequisite to enabling GPU-based hybrid approaches in the area of DLA. Section 4.2 describes some of the basic principles on how to write high-performance BLAS kernels for GPUs. Section 4.3 gives GPU-specific, generic kernel optimization techniques—pointer redirecting, padding, and auto-tuning—and their application in developing high-performance BLAS for GPUs. Finally, Section 4.4 gives a summary.

4.2   BLAS Kernels Development

Implementations of the BLAS interface are a major building block of DLA libraries, and therefore must be highly optimized. This is true for GPU computing as well, especially after the introduction of shared memory in modern GPUs. This is important because it enabled fast Level 3 BLAS implementations [5–7], which in turn made possible the development of DLA for GPUs to be based on BLAS for GPUs (see Chapter 3 and references [3,6]).

Despite the current success in developing highly optimized BLAS for GPUs [5–7], the area is still new and presents numerous opportunities for improvements. Addressed are several very important kernels, including the matrix-matrix multiplication, crucial for the performance throughout DLA, and matrix-vector multiplication, crucial for the performance of linear solvers and two-sided matrix factorizations (and hence eigen-solvers). The new implementations are included in the MAGMA version 0.2 BLAS Library [3].

This section describes some of the basic principles on how to write high-performance kernels for GPUs. Along with the specifics on developing each of the BLAS considered, the stress is on two important issues for achieving high performance. Namely, these are:

Blocking Blocking is a DLA optimization technique where a computation is organized to operate on blocks/submatrices of the original matrix. The idea is that blocks are of small enough size to fit into a particular level of the CPU's memory hierarchy so that once loaded to reuse the blocks' data to perform the arithmetic operations that they are involved in. This idea can be applied for GPUs, using GPUs' shared memory. As demonstrated below, the application of blocking is crucial for the performance of numerous GPU kernels.

Coalesced Memory Access GPU global memory accesses are costly and not cached, making it crucial for the performance to have the right access pattern to get maximum memory bandwidth. There are two access requirements [8]. The first is to organize global memory accesses in terms of parallel consecutive memory accesses—16 consecutive elements at a time by the threads of a half-warp (16 threads)—so that memory accesses (to 16 elements at a time) are coalesced into a single memory access. This is demonstrated in the kernels' design throughout the section. Second, the data should be properly aligned. In particular, the data to be accessed by half-warp should be aligned at $16 * \texttt{sizeof(element)}$, e.g., 64 for single precision elements.

Clearly, fulfilling the above requirements will involve partitioning the computation into blocks of fixed sizes (e.g., multiple of 16) and designing memory accesses that are coalescent (properly aligned and multiple of 16 consecutive elements). This is demonstrated in the kernels' design throughout the section. The problems of selecting best performing partitioning sizes/parameters for the various algorithms as well as the cases where (1) the input data are not aligned to fulfill coalescent memory accesses and (2) the problem sizes are not divisible by the partitioning sizes required for achieving high performance need special treatment and are considered in Section 4.3. The main ideas in this section are demonstrated on general and symmetric matrices, in both the transpose and non-transpose cases.

The BLAS considered are not exhaustive; only subroutines that are critical for the performance of MAGMA are discussed. Moreover, these would often

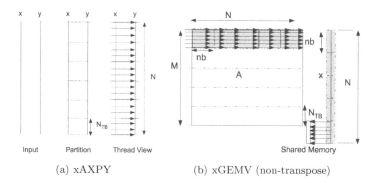

(a) xAXPY   (b) xGEMV (non-transpose)

FIGURE 4.1: Algorithmic view of Level 1 and Level 2 BLAS.

be DLA-specific cases that can be accelerated compared to CUBLAS [5], an implementation of the BLAS standard provided by NVIDIA.

Further down a *thread block* will be denoted by TB, its size by N_{TB} (or $N_{TBX} \times N_{TBY}$ in 2D), the number of threads in a TB by N_T (or $N_{TX} \times N_{TY}$ in 2D), and the size associated with blocking (as described above) by nb.

4.2.1 Level 1 BLAS

Implementing Level 1 BLAS, especially reduce-type operations like dot-product, isamax, etc., is of general interest for parallel computing, but not in the area of DLA. The reason is that Level 1 BLAS are of very low computational intensity (flops *vs* data required) and are avoided at first place (at algorithm design level) in DLA. Even when they cannot be avoided algorithmically, e.g., the use of isamax in LU for pivoting, their computation on the GPU is avoided by scheduling their execution on the CPU (see the hybrid approach described in Chapter 3). One operation that fits very well with the GPU architecture, and therefore can be efficiently executed on GPUs, is xAXPY:

$$y := \alpha x + y,$$

where x and y are vectors of size N, and α is a scalar. An example of its use is the mixed-precision iterative refinement solvers in MAGMA [9].

The implementation is straightforward—one dimensional TB of size N_{TB} computes N_{TB} consecutive elements of the resulting vector y (a thread per element; also illustrated in Figure 4.1(a)). Important for achieving high performance in this case, as discussed at the beggining of this section, is coalesced memory accesses, tuning N_{TB} and properly handling the case when N is not divisible by N_{TB} (i.e., $N \% N_{TB} \neq 0$). These are recurring issues for obtaining high-performance BLAS and will be further discussed in the context of other BLAS kernels and GPU optimization techniques like auto-tuning (in Section 4.3.3) and pointer redirecting (in Section 4.3.1).

Note that the algorithm described satisfies the first requirement for co-alescent memory access—to organize global GPU memory accesses in terms of parallel consecutive memory accesses. The pointer redirecting technique in Section 4.3.1 deals with the second requirement for coalescent memory access, namely cases where the starting address of x is not a multiple of $16 * \texttt{sizeof(element)}$ and/or $N \% N_{TB} \neq 0$. The same applies for the other BLAS kernels in the section and will not be explicitly mentioned again.

4.2.2 Level 2 BLAS

Level 2 BLAS routines, similar to Level 1 BLAS, are of low computational intensity and, ideally, DLA algorithms must be designed to avoid them. An example from the area of DLA is the *delayed update* approach where the appli-cation of a sequence of Level 2 BLAS is delayed and accumulated in order to be applied at once as a more efficient single matrix-matrix multiplication [2]. In many cases, like MAGMA's mixed-precision iterative refinement solvers [9] or two-sided matrix factorizations [10], this is not possible, and efficient imple-mentations are crutial for the performance. This section considers the GPU implementations of two fundamental Level 2 BLAS operations, namely the matrix-vector multiplication routines for correspondingly general (xGEMV) and symmetric matrices (xSYMV).

4.2.2.1 xGEMV

The xGEMV matrix-vector multiplication routine performs one of:

$$y := \alpha A x + \beta y \quad \text{or} \quad y := \alpha A^T x + \beta y,$$

where A is an M by N matrix, x and y are vectors, and α and β are scalars. The two cases are considered separately as follows:

Non-Transposed Matrix: The computation in this case can be organized in one-dimensional grid of TBs of size N_{TB} where each block has $N_T = N_{TB}$ threads, as shown in Figure 4.1(b). Thus, each thread computes one element of the resulting vector y.

GEMV is the first of the kernels considered to which *blocking* can be ap-plied. Although matrix A cannot be reused in any blocking, vector x can be reused by the threads in a TB. Specifically, the computation is blocked by loading nb consecutive elements of x at a time into the shared memory (using all N_T threads). This part of x is then used by all T_N threads in a TB to multiply it by the corresponging $N_{TB} \times nb$ submatrix of A. The process is repeated $\frac{N}{N_{TB}}$ times.

Note that the algorithm as described depends on two parameters: N_{TB} and nb. Figures 4.2(a), 4.2(b) compare the performance for cases $N_{TB} = nb = 16, 32, 64$ with that of CUBLAS 2.3. The performances are for matrix sizes $M = N$ that are divisible by the corresponding blocking sizes. Also, the start-ing addresses of A, x, and y are taken to be divisible by $16 * \texttt{sizeof(element)}$

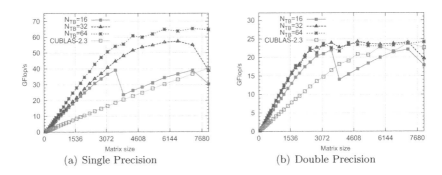

(a) Single Precision   (b) Double Precision

FIGURE 4.2: Performance of xGEMV (non-transpose) on a GTX 280.

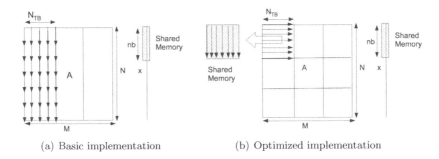

(a) Basic implementation   (b) Optimized implementation

FIGURE 4.3: Two memory access implementations of xGEMV (transpose).

and the leading dimension of A is divisible by 16. This guarantees that all memory accesses in the algorithm are coalescent.

Transposed Matrix: Following the non-transposed version approach leads to poor performance because the memory acceses are not going to be coalesced (see Figure 4.3(a)). To improve the speed on accessing the data, blocks of the matrix A can be first loaded into the shared memory using coalesced memory accesses, and second, data only from the shared memory can be used to do all the necessary computations (see Figure 4.3(b)).

Although the new version significantly improves the performance, experiments that increase the design space of the algorithm show that further improvements are possible. In particular, one exploration direction is the use of higher numbers of threads in a TB, e.g., 64, as high-performance DLA kernels are associated with the use of 64 threads (and occasionally more). Using 64 threads directly does not improve performance though because the amount of shared memory used (a 64×64 matrix) gets to be excessive, prohibiting the effective scheduling of that amount of threads [8]. Decreasing the use of shared memory, e.g., to a 32×32 matrix, while having a higher level of thread parallelism, e.g., a grid of 32×2 threads, is possible in the following way:

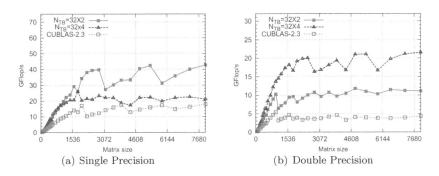

(a) Single Precision   (b) Double Precision

FIGURE 4.4: Performance of xGEMV (transpose) on a GTX 280.

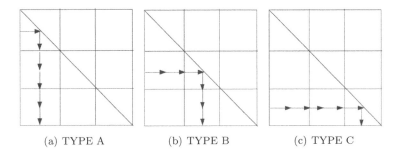

(a) TYPE A   (b) TYPE B   (c) TYPE C

FIGURE 4.5: Three cases of TB computations in xSYMV.

(1) two groups of 32×1 threads, e.g., denoted by 32_j where $j = 0/1$, load correspondingly the two 32×16 submatrices of the shared memory matrix using coalesced memory accesses, (2) each group performs the computation from the second GEMV version but constrained to the 16×32 submatrix of the shared memory matrix, accumulating their independent y_j results. The final result $y := y_0 + y_1$ can be accumulated by one of the $j = 0/1$ threads.

The same idea can be used with more threads, e.g., 32×4, while using the same amount of shared memory. Performance results are shown in Figure 4.4 along with a comparison to the performance from CUBLAS 2.3.

4.2.2.2 xSYMV

The xSYMV matrix-vector multiplication routine performs:

$$y := \alpha Ax + \beta y,$$

where α and β are scalars, x and y are vectors of size N, and A is an N by N symmetric matrix, stored in the upper or lower triangular part of a two-dimensional array of size $N \times N$. The difficulty of designing a high-performance SYMV kernel stems from the triangular data storage, which is more challenging to organize a data parallel computation with coalescent memory accesses.

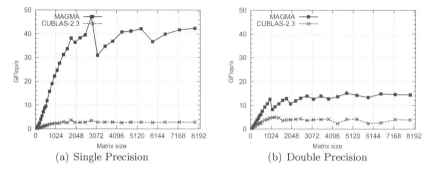

(a) Single Precision   (b) Double Precision

FIGURE 4.6: Performance of xSYMV on a GTX 280.

Indeed, if A is given as an $N \times N$ array, storing both the upper and lower triangular parts of the symmetric matrix A, the SYMV kernel can be implemented using GEMV. Similar to GEMV, the computation is organized in one-dimensional grid of TBs of size N_{TB}, where each block has $N_T = N_{TB}$ threads. A TB computation can be classified as one of three cases (see the illustration in Figure 4.5):

- Type A—TB threads do SYMV followed by GEMV (transpose);

- Type B—threads do GEMV (non-transpose) followed by SYMV and GEMV (transpose);

- Type C—threads do GEMV (non-transpose) followed by SYMV.

This way the computation within a TB is converted into one/two GEMVs (to reuse the GEMV kernels) and an SYMV involving a matrix of size $N_{TB} \times N_{TB}$. The remaining SYMV is also converted into a GEMV by loading the $N_{TB} \times N_{TB}$ matrix into the GPU's shared memory and generating the missing symmetric part in the shared memory (a process defined as *mirroring*). Figure 4.6 compares the performance for kernel with parameters $N_{TB} = nb = 32$, $N_T = 32 \times 4$ with that of CUBLAS 2.3.

4.2.3   Level 3 BLAS

Level 3 BLAS routines are of high computational intensity, enabling their implementations (and that of high-level DLA algorithms based on Level 3 BLAS) to get close within the computational peak of ever evolving architectures, despite that architectures are evolving with an exponentially growing gap between their compute and communication speeds. The shared memory of GPUs, similar to memory hierarchy in standard CPUs, can be used to develop highly efficient Level 3 BLAS kernels. This section describes the GPU implementations of three primary Level 3 BLAS operations: the matrix-matrix

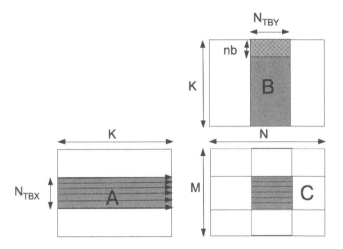

FIGURE 4.7: The GPU GEMM ($C = AB$) of a single TB.

multiplication (xGEMM), the symmetric rank-k update (xSYRK), and the triangular matrix solver (xTRSM).

4.2.3.1 xGEMM

The xGEMM matrix-matrix multiplication routine performs one of:

$$C := \alpha \ op(A)op(B) + \beta C,$$

where $op(X)$ is X, or X^T, α and β are scalars; and A, B, and C are matrices. Crutial for the performance is the application of blocking—schematicly represented in Figure 4.7 for the case of $C := \alpha AB + \beta C$ and described as follows [6]. The computation is done on a two-dimensional grid of TBs of size $N_{TBX} \times N_{TBY}$ and each TB is assigned to $N_T = N_{TX} \times N_{TY}$ threads. For simplicity, take $N_T = N_{TBX}$. Then, each thread is coded to compute a row of the submatrix assigned to the TB. Each thread accesses its corresponding row of A, as shown by an arrow, and uses the $K \times N_{TBY}$ submatrix of B for computing the final result. This TB computation can be blocked, which is crucial for obtaining high performance. In particular, submatrices of B of size $nb \times N_{TBY}$ are loaded into shared memory and multiplied nb times by the corresponding $N_{TBX} \times 1$ submatrices of A. The $N_{TBX} \times 1$ elements are loaded and kept in registers while multiplying them with the $nb \times N_{TBY}$ part of B. The result is accumulated to the resulting $N_{TBX} \times N_{TBY}$ submatrix of C, which is kept in registers throughout the TB computation (a row per thread, as already mentioned). This process is repeated until the computation is over. All memory accesses are coalesced. Kernels for various $N_{TBX}, N_{TBY}, N_{TX}, N_{TY}$, and nb can be automatically generated (see Section 4.3.3) to select the best performing kernel for particular architecture and GEMM parameters. A sam-

Kernel	N_{TBX}	N_{TBY}	nb	N_{TX}	N_{TY}
K1	32	8	4	8	4
K2	64	16	4	16	4
K3	128	16	8	16	8
K4	256	16	16	16	16

TABLE 4.1: Key parameters of a sample of GPU GEMM kernels.

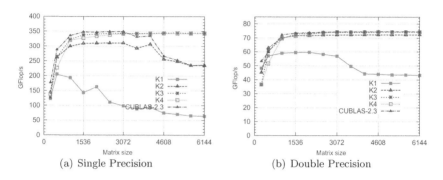

(a) Single Precision   (b) Double Precision

FIGURE 4.8: Performance of GEMM ($C = \alpha AB^T + \beta C$) on a GTX 280.

ple choice of these kernels is shown in Table 4.1. Figure 4.8 compares their performances with that of CUBLAS 2.3 on square matrices. K1 performs well for small matrices (e.g., of dimension ≤ 512) as it provides more parallelism compared to the other kernels in Table 4.1. The performance deteriorations experienced by some of the kernels are due to the GPUs global memory layout and memory access patterns of hitting a particular memory module (a phenomenon referred to by NVIDIA as *partition camping*).

This particular configuration works well when $Op(A) = A$, $Op(B) = B$. The $Op(A) = A^T$, $Op(B) = B^T$ case is similar—only the argument order and the update location of C at the end of the kernel have to be changed, as:

$$C := \alpha \, A^T B^T + \beta C \quad \text{or} \quad C^T := \alpha \, BA + \beta C^T.$$

The $Op(A) = A^T$, $Op(B) = B$ kernel can be analogously developed except that both A and B must be stored into shared memory.

4.2.3.2   xSYRK

The xSRYK routine performs one of the symmetric rank-k updates:

$$C := \alpha AA^T + \beta C \quad \text{or} \quad C := \alpha A^T A + \beta C,$$

where α and β are scalars, C is an $N \times N$ symmetric matrix, and A is an $N \times K$ matrix in the first case and a $K \times N$ matrix in the second case. A TB

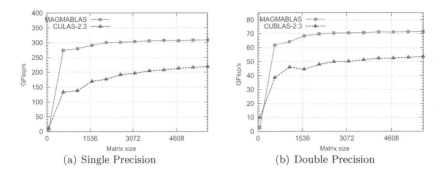

(a) Single Precision   (b) Double Precision

FIGURE 4.9: Performance of xSYRK on a GTX 280.

index reordering technique can be used to initiate and limit the computation only to TBs that are on the diagonal or in the lower (correspondingly upper) triangular part of the matrix. In addition, all the threads in a diagonal TB compute redundantly half of the block in a data parallel fashion in order to avoid expensive conditional statements that would have been necessary otherwise. Some threads also load unnecessary data to ensure coalescent global memory accesses. At the end, the results from the redundant computations (in the diagonal TBs) are discarded and the data tile is correctly updated. Figure 4.9 illustrates the performance gains in applying this technique.

4.2.3.3   xTRSM

The xTRSM routine solves one of the matrix equations:

$$op(A)X = \alpha B \quad \text{or} \quad Xop(A) = \alpha B,$$

where α is a scalar, X and B are $M \times N$ matrices, A is upper/lower triangular matrix and $op(A)$ is A or A^T. Matrix B is overwritten by X.

Trading off parallelism and numerical stability, especially in algorithms related to triangular solvers, has been known and studied before [11,12]. Some of these TRSM algorithms are getting extremely relevant with the emerging highly parallel architectures, especially GPUs. In particular, the MAGMA library includes implementations that explicitly invert blocks of size 32×32 on the diagonal of the matrix and use them in blocked xTRSM algorithms. The inverses are computed simultaneously, using one GPU kernel, so that the critical path of the blocked xTRSM can be greatly reduced by doing it in parallel (as a matrix-matrix multiplication). Variations are possible, e.g., the inverses to be computed on the CPU, to use various block sizes, including recursively increasing it from 32, etc. Similarly to xSYRK, extra flops can be performed to reach better performance—the empty halves of the diagonal triangular matrices can be set to zeros and the multiplications with them done with GEMMs instead of with TRMMs. This avoids diverting warp threads and

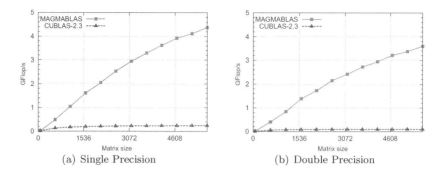

(a) Single Precision   (b) Double Precision

FIGURE 4.10: Performance of xTRSM on a GTX 280.

ensures efficient parallel execution. Figure 4.10 illustrates the performance gains in applying this technique.

4.3   Generic Kernel Optimizations

This section addresses three optimization techniques that are crutial for developing high performance GPU kernels. The first two techniques—pointer redirecting (Section 4.3.1) and padding (Section 4.3.2)—are used in cases where the input data are not aligned to directly allow coalescent memory accesses, or when the problem sizes are not divisible by the partitioning sizes required for achieving high performance. The third technique—auto-tuning (Section 4.3.3)—is used to determine best performing kernels, partitioning sizes, and other parameters for the various algorithms described.

An example demonstrating the need to address the cases mentioned is given in Figure 4.11. Shown is the performance of the matrix-matrix multiplication routine for a discrete set of matrix dimensions. Seen are performance deteriorations, e.g., more than 24 GFlops/s in double precision (around 30% of the peak performance) and even worse in single precision. The techniques in this section, used as a complement to the kernels presented in Section 4.2, aim to streamline the development of BLAS kernels for GPUs that are of uniformly high performance.

4.3.1   Pointer Redirecting

A few possibilities of dealing with matrix dimensions not divisible by the blocking factor can be explored. One approach is to have some "boundary" TBs doing selective computation. This will introduce several if-else statements in the kernel which will prevent the threads inside a TB to run in parallel. Figure 4.12 shows the GEMM performance following this approach.

Another approach is instead of preventing certain threads from computing (with if-else statements), to let them do similar work as the other threads in a TB, and discard saving their results at the end. This can lead to some illegal memory references as illustrated in Figure 4.13.

The *pointer redirecting* techniques are based on the last approach and include redirecting the illegal memory references to valid ones, within the matrix of interest, in a way that would allow the memory accesses to be coalescent. The specifics of the redirecting depend on the kernel, but in general, if a thread is to access invalid rows/columns of a matrix (beyond row/column M/N), the access is redirected towards the last row/column.

Figure 4.14(a) shows the pointer redirecting for matrix A in GEMM with $Op(A) = A$ and $Op(B) = B$. Threads t1, t2, t3, and t4 access valid memory location, and threads beyond that, e.g., t5, t6 access the memory accessed by t4. As a result no separate memory read operation will be issued and no latency will be experienced for this extra load. Figure 4.14(b) shows the data access pattern for matrix B – $nb \times N_{TBY}$ data of matrix B is loaded into shared memory by $N_{TX} \times N_{TY}$ threads in a coalesced manner. The left $nb \times N_{TBY}$ block is needed, but the right $nb \times N_{TBY}$ is only partially needed. As discussed before, the illegal memory accesses will be redirected to the last column of B. The redirection as done presents a simple solution that has little overhead and does not break the pattern of coalesced memory access.

Figures 4.15 and 4.16 show the performance results for GEMM using the pointer redirecting technique. In double precision the performance is improved by up to 24 GFlops/s and in single precision by up to 170 GFlops/s.

The technique is applied similarly to the other kernels. The case where the starting address of any of the operands is not a multiple of $16 * sizeof(element)$ (but the leading dimension is a multiple of 16) is also handled similarly—threads that must access rows "before" the first are redirected to access the first.

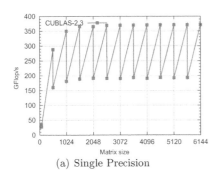

(a) Single Precision

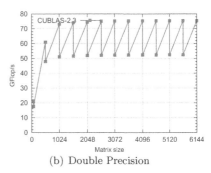

(b) Double Precision

FIGURE 4.11: Performance of GEMM on square matrices on a GTX 280.

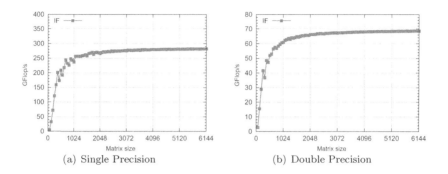

(a) Single Precision   (b) Double Precision

FIGURE 4.12: Performance of GEMM with conditional statements.

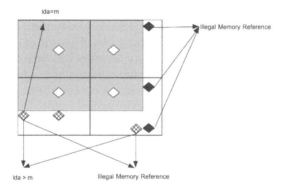

FIGURE 4.13: Possible illegal memory references in GEMM.

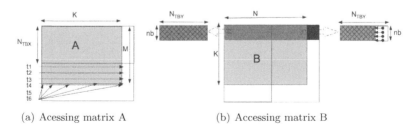

(a) Acessing matrix A   (b) Accessing matrix B

FIGURE 4.14: GPU GEMM ($C = \alpha AB + \beta C$) with pointer redirecting.

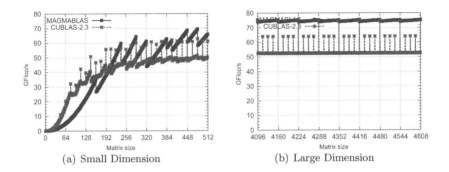

(a) Small Dimension   (b) Large Dimension

FIGURE 4.15: Performance of DGEMM on a GTX 280.

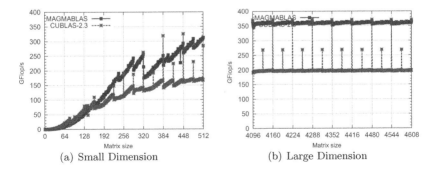

(a) Small Dimension   (b) Large Dimension

FIGURE 4.16: Performance of SGEMM on a GTX 280.

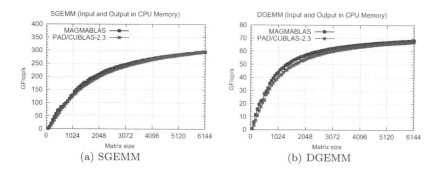

FIGURE 4.17: Performance of xGEMM with padding on a GTX 280.

4.3.2   Padding

If the input matrices are given on the CPU memory the performance can be enhanced by *padding*, given that enough memory is available for it on the GPU. Padding is the technique where a matrix of higher dimension (to make the new size divisible by N_{TB}) is allocated on the GPU memory, and the extra elements initialized by zero. Figure 4.17 shows the performance of xGEMM comparing the padding and pointer redirecting approaches when the data are in the CPU memory. The results show that for small matrix sizes the pointer redirecting gives better performance, and for larger matrices the two approaches are almost identical, as it is actually expected.

Padding is applied to certain CPU-interface MAGMA routines [3], e.g., the LU factorization. In general, users of the CPU interface are encouraged to apply "padding" to all routines, in the sense that users must provide at least working space matrices on the GPU with leading dimensions divisible by 16. Note that if the leading dimension is not divisible by 16 none of the techniques presented will have coalescent memory accesses, unless internally the data are copied into another padded matrix. This second form of padding does not require zeroing the extra space—just allocating it so that coalescent memory accesses are enabled.

4.3.3   Auto-Tuning

Automatic performance tuning (optimization), or auto-tuning in short, is a technique that has been used intensively on CPUs to automatically generate near-optimal numerical libraries. For example, ATLAS [13, 14] and PHiPAC [15] are used to generate highly optimized BLAS. In addition, FFTW [16] is successfully used to generate optimized libraries for FFT, which is one of the most important techniques for digital signal processing. With the success of auto-tuning techniques on generating highly optimized DLA kernels on CPUs, it is interesting to see how the idea can be used to generate near-optimal DLA kernels on modern high-performance GPUs.

Indeed, work in the area [17] has already shown that auto-tuning for GPUs is a very practical solution to easily port existing algorithmic solutions on quickly evolving GPU architectures and to substantially speed up even highly hand-tuned kernels.

There are two core components in a complete auto-tuning system:

Code generator The code generator produces code variants according to a set of pre-defined, parametrized templates/algorithms. The code generator also applies certain state-of-the-art optimization techniques.

Heuristic search engine The heuristic search engine runs the variants produced by the code generator and finds out the best one using a feedback loop, e.g., the performance results of previously evaluated variants are used as a guideline for the search on currently unevaluated variants.

Below is a review of certain techniques and choices of parameters that significantly impact the performance of the GEMM kernel. Therefore, these techniques and parameters must be (and have been) incorporated into the code generator of an auto-tuning GEMM system. The ultimate goal is to develop similar auto-tuning for all of the BLAS of interest.

Auto-tuning GEMM: Figure 4.7 depicts the algorithmic view of a GEMM code template. It was already mentioned that five parameters can critically impact performance (see Table 4.1 for a sample choice), and therefore are incorporated in a GEMM code generator. This choice though can be extended and enhanced with various optimization techniques:

Number of threads computing a row: Section 4.2.3.1 imposed the constraint $N_{TX} \times N_{TY} = N_{TBX}$ so that each thread in a TB is computing an entire row of the submatrix of C computed by the TB (denoted further as BC). This constraint can be lifted to introduce an additional template parameter. Depending upon the value of N_T each thread will compute either an entire row or part of a row. For example, suppose $N_{TBY} = 16$ and $N_{TBX} = 64$, and the TB has 16×4 threads, then each thread will compute exactly one row of BC. If the thread block has 16×8 threads, then each thread will compute half of a row.

A/B being in shared memory: As described in Section 4.2.3.1, whether A or B is put into shared memory plays a crucial factor in the kernel's performance. Different versions of GEMM ($Op(X)$ is X or X^T) require putting A and/or B into shared memory. This parameter of the auto-tuner is denoted by sh_{AB}. When only (part of) A is in shared memory each thread per TB computes an entire column or part of a column of BC. When both A and B are in shared memory the computation can be splitted in terms of rows or columns of the resulting submatrix of C.

Submatrix layout in shared memory: This parameter determines the layout of each $N_{TBX} \times nb$ submatrix of the matrix A (referred to as

BA from now on) or $N_{TBY} \times nb$ submatrix of the matrix B (referred to as BB from now on) in the shared memory, i.e., whether the copy of each block BA or BB in the shared memory is transposed or not. Since the shared memory is divided into banks and two or more simultaneous accesses to the same bank cause bank conflicts, transposing the layout in the shared memory may help reduce the possibility of bank conflicts, thus potentially improving the performance.

Amount of allocated shared memory: Two parameters, $offset_{BA}$ and $offset_{BB}$, relate to the actual allocation size of BA or BB in shared memory. When $N_{TBY} = 16$ and $nb = 16$, it requires 16×16 2D-array for BB in shared memory. Depending upon the computation sometimes it is better to allocate some extra memory so that the threads avoid bank conflict while accessing operands from shared memory data. It means allocating a 16×17 array instead of 16×16. So there is an offset of 1. It could be 0, 2, or 3 depending upon other parameters and the nature of computation. The auto-tuner handles this offset as a tunable parameter in internal optimization.

Prefetching into registers: As in CPU kernels, GPU kernels can benefit by prefetching into registers. For the access of matrices A and B, the auto-tuner inserts prefetch instruction for the data needed in the next iteration and checks the effect. Insertion of prefetch instruction leads to usage of registers which might limit the parallelism of the whole code. The auto-tuner investigates this with various combinations of prefetches: no prefetch, prefetch A only, prefetch B only, and prefetch both A and B, to finally pick the best combination.

Loop optimization techniques: Different state-of-the-art loop optimization techniques such as strip mining and loop unrolling are incorporated in order to extract parallelism and achieve performance. Another interesting loop optimization technique, namely *circular loop skewing*, was incorporated in the auto-tuner to deal with GPU global memory layout. Circular loop skewing is based upon a very simple idea of reordering the computation in the inner loop. In the context of GPUs, inner loops are considered the data parallel tasks that make up a kernel. These tasks are scheduled by CUDA (controlling the outer loop) on the available multiprocessors and the order of scheduling sometimes is crucial for the performance. Circular loop skewing techniques are incorporated to explore benefits of modified scheduling. Their most important use is in removing performance deteriorations related to *partition camping* (described above).

Precision: The code generator also takes precision as a parameter.

The code generator takes all these parameters as input and generates the

Kls	Prec	N_{tbx}	N_{tby}	nb	N_{tx}	N_{ty}	sh_{AB}	$Trns$	$op(A)$	$op(B)$	$skewing$
K1	S/DP	32	8	4	8	4	B	No	N	T	No
K2	S/DP	64	16	4	16	4	B	No	N	T	No
K3	S/DP	128	16	8	16	8	B	No	N	T	No
K4	S/DP	256	16	16	16	16	B	No	N	T	No
K5	DP	32	32	8	8	8	AB	No	T	N	No
K6	DP	64	16	16	16	4	B	Yes	N	N	No
K7	DP	128	16	8	16	8	B	Yes	N	N	No
K8	SP	64	16	4	16	4	B	No	N	T	All
K9	SP	64	16	4	16	4	B	No	N	T	Selective

TABLE 4.2: Different kernel configurations.

kernel, the timing utilities, the header file, and the Makefile to build the kernel. The code generator first checks the validity of the input parameters before actually generating the files. By validity it means (1) the input parameters conform to hardware constraints, e.g., the maximum number of threads per thread block $N_{TX} \times N_{TY} \leq 512$ in GTX 280, and (2) the input parameters are mutually compatible, e.g., $(N_{TBX} \times N_{TBY})\%(N_{TX} \times N_{TY}) = 0$, i.e., the load of BA's data into share memory can be evenly distributed among all the threads in a thread block, etc. By varying the input parameters, the auto-tuner can generate different versions of the kernel, and evaluate their performance, in order to identify the best one. Along the way the auto-tuner tries to optimize the code by using different optimization techniques such as prefetching, circular loop skewing and adjusting offset in shared memory allocation as described above. One way to implement auto-tuning is to generate a small number of variants for some matrices with typical sizes during installation time, and choose the best variant during run time, depending on the input matrix size and high-level DLA algorithm.

Performance results: Table 4.2 gives the parameters of different xGEMM kernels used in this section. The table also provides parameters for all the kernels used in Section 4.2.3.1. The *Trns* parameter denotes if the kernel was implemented by taking tranpose operation in both sides of the equation of the original operation, as:

$$C := \alpha \, A^T B^T + \beta C \quad \text{or} \quad C^T := \alpha \, BA + \beta C^T.$$

Figure 4.18 compares the performance of the xGEMM auto-tuner in double precision with the CUBLAS 2.3 for multiplying square matrices where $Op(A) = A^T$ and $Op(B) = B$. It can be seen that the performance of the auto-tuner is apparently 15% better than the CUBLAS 2.3 DGEMM. The fact that the two performances are so close is not surprising because the auto-tuned code and CUBLAS 2.3's code are based on the same kernel, and this kernel was designed and tuned for current GPUs (and in particular the GTX 280), targeting high performance for large matrices.

The global memory layout of current GPUs presents challenges as well as opportunities for auto-tuners. As shown in Figure 4.19(a), CUBLAS 2.3 SGEMM has performance deteriorations for certain problem sizes when

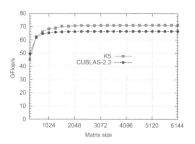

FIGURE 4.18: Performance of auto-tuned DGEMM kernel $(Op(A) = A^T,$ $Op(B) = B)$ on a GTX 280.

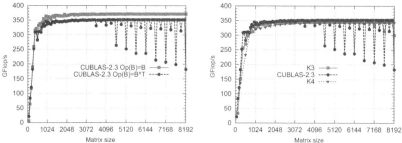

(a) Performance comparison of SGEMM kernel between $Op(B) = B$ and B^T with $Op(A) = A$.

(b) Auto-tuned kernel with tuned algorithmic parameter.

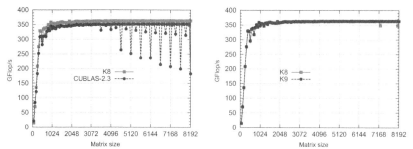

(c) Auto-tuned kernel with circular skewing in all dimensions.

(d) Auto-tuned kernel with selective circular skewing.

FIGURE 4.19: Performance of the auto-tuned SGEMM $(Op(A) = A,$ $Op(B) = B^T)$ kernel for square matrices on a GTX 280.

$Op(A) = A$ and $Op(B) = B^T$. Interestingly, when $Op(A) = A$ and $Op(B) = B$, the performance is very smooth. The reason for this is that GPU global memory is interleaved into a number of memory modules and the memory requests from all the concurrently running thread blocks may not be evenly distributed among the GPU memory modules. As a result the memory requests are sequentially processed and all the threads experience huge memory latency. This phenomenon is referred to as *partition camping* in NVIDIA terms. The auto-tuner found two kernels ($K3$, $K4$), as shown in Figure 4.19(b), that work significantly better in this situation. $K3$ and $K4$ work better because as partition size N_{TBX} is increased, the total number of accesses to global memory for matrix B's data is correspondingly 1/2 and 1/4 compared to that for kernel $K2$ (besides, TLP is increased). Kernels $K3$ and $K4$ perform fair compared to CUBLAS 2.3 in any dimension, and remarkably well for the problem sizes where CUBLAS 2.3 has performance deteriorations. Interestingly, the auto-tuner was successful in finding a better kernel by applying circular loop skew optimization in kernel $K2$. The performance is shown in Figure 4.19(c). Note that there are no performance deteriorations and performance is better than CUBLAS 2.3 for all matrix sizes. However, this technique does not work in all cases and may have to be applied selectively. The performance of such kernel (K9) is shown in Figure 4.19(d).

Finally, we point out that in the area of DLA, it is very important to have high-performance GEMMs on rectangular matrices, where one size is large and the other is fixed within a certain block size (BS), e.g., BS = 64, 128, up to about 256 on current architectures. For example, in an LU factorization (with look-ahead) it requires two types of GEMM, namely one for multiplying matrices of size N×BS and BS×N−BS, and another for multiplying N×BS and BS×BS matrices. This situation is illustrated on Figure 4.20, where we compare the performances of the CUBLAS 2.3 *vs* auto-tuned DGEMMs occurring in the block LU factorization of a matrix of size 6144×6144. The graphs show that the auto-tuned code significantly outperforms (up to 27%) the DGEMM from CUBLAS 2.0.

These results support experiences and observations by others on "how sensitive the performance of GPU is to the formulations of your kernel" [18] and that an enormous amount of well-thought experimentation and benchmarking [6,18] is needed in order to optimize performance.

4.4 Summary

Implementations of the BLAS interface are a major building block of dense linear algebra libraries, and therefore must be highly optimized. This is true for GPU computing as well, as evident from the MAGMA library, where the availability of fast GPU BLAS enabled a hybridization approach that stream-

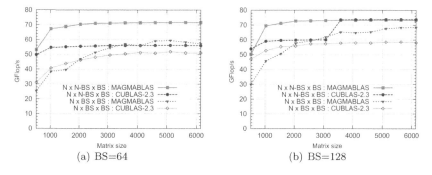

(a) BS=64   (b) BS=128

FIGURE 4.20: Performance comparison of the auto-tuned (solid line) *vs* CUBLAS 2.3 DGEMMs occurring in the block LU factorization (for block sizes BS = 64 on the left and 128 on the right) of a matrix of size 6144×6144. The two kernels shown are for multiplying $N \times BS$ and $BS \times N - BS$ matrices (denoted by $N \times N - BS \times BS$), and $N \times BS$ and $BS \times BS$ matrices (denoted by $N \times BS \times BS$). K6 was used when BS = 64 and K7 was used when BS = 128.

lined the development. This chapter provided guidance on how to develop these needed high-performance BLAS kernels for GPUs. Described were not only basic principles and important issues for achieving high performance, but also specifics on the development of each of the BLAS considered. In particular, the stress was on the two important issues—blocking and coalesced memory access—demonstrated in the kernels' design throughout the chapter. The main ideas were demonstrated on general and symmetric matrices, in both the transpose and non-transpose cases, for a selection of Level 1, 2, and 3 BLAS kernels that are crucial for the performance of higher-level DLA algorithms. Moreover, three optimization techniques that are GPU specific and crucial for developing high performance GPU kernels, were considered. The first two techniques, pointer redirecting and padding, are used in cases where the input data is not aligned to directly allow coalescent memory accesses, or when the problem sizes are not divisible by the partitioning sizes required for achieving high performance. The third technique, auto-tuning, is used to automate the process of generating and determining best performing kernels, partitioning sizes, and other parameters for the various algorithms of interest.

The implementations of variation of the kernels described are available as part of MAGMA BLAS library through the MAGMA site http://icl.eecs.utk.edu/magma/.

Bibliography

[1] K. Asanovic, R. Bodik, B. C. Catanzaro, J. J. Gebis, P. Husbands, K. Keutzer, D. A. Patterson, W. L. Plishker, J. Shalf, S. W. Williams, and K. A. Yelick. The Landscape of Parallel Computing Research: A View from Berkeley. Technical Report UCB/EECS-2006-183, Electrical Engineering and Computer Sciences Department, University of California at Berkeley, 2006.

[2] E. Anderson, Z. Bai, C. Bischof, L. S. Blackford, J. W. Demmel, J. J. Dongarra, J. Du Croz, A. Greenbaum, S. Hammarling, A. McKenney, and D. Sorensen. *LAPACK Users' Guide*. SIAM, Philadelphia, PA, 1992. http://www.netlib.org/lapack/lug/.

[3] S. Tomov, R. Nath, P. Du, and J. Dongarra. MAGMA version 0.2 Users' Guide. http://icl.cs.utk.edu/magma, November 2009.

[4] BLAS: Basic linear algebra subprograms. http://www.netlib.org/blas/.

[5] CUDA CUBLAS Library. http://developer.download.nvidia.com.

[6] V. Volkov and J. Demmel. Benchmarking GPUs to tune dense linear algebra. In *SC '08: Proceedings of the 2008 ACM/IEEE conference on Supercomputing*, pages 1–11, Piscataway, NJ, 2008. IEEE Press.

[7] Y. Li, J. Dongarra, and S. Tomov. A note on auto-tuning GEMM for GPUs. In *ICCS '09: Proceedings of the 9th International Conference on Computational Science*, pages 884–892, Berlin, Heidelberg, 2009. Springer-Verlag.

[8] NVIDIA CUDA Compute Unified Device Architecture—Programming Guide. http://developer.download.nvidia.com, 2007.

[9] S. Tomov, R. Nath, and J. Dongarra. Dense linear algebra solvers for multicore with GPU accelerators. UTK EECS Technical Report ut-cs-09-649, December 2009.

[10] S. Tomov and J. Dongarra. Accelerating the reduction to upper Hessenberg form through hybrid GPU-based computing. Technical Report 219, LAPACK Working Note 219, May 2009.

[11] James W. Demmel. Trading Off Parallelism and Numerical Stability, EECS Department, University of California, Berkeley, UCB/CSD-92-702, September 1992.

[12] Nicholas J. Higham. Stability of parallel triangular system solvers, *SIAM J. Sci. Comput.*, 16(2):400–413, 1995.

[13] R. Whaley, A. Petitet, and J. Dongarra. Automated Empirical Optimizations of Software and the ATLAS Project. *Parallel Computing*, 27(1–2):3–35, 2001.

[14] Jim Demmel, Jack Dongarra, Victor Eijkhout, Erika Fuentes, Antoine Petitet, Rich Vuduc, Clint Whaley, and Katherine Yelick. Self adapting linear algebra algorithms and software. Proceedings of the IEEE 93 (2005), no. 2, special issue on "Program Generation, Optimization, and Adaptation." Proceddings, vol. 93, 2, pp. 293–312.

[15] Jeff Bilmes, Krste Asanovic, Chee-Whye Chin, and James Demmel. Optimizing Matrix Multiply Using PHiPAC: A Portable, High-Performance, ANSI C Coding Methodology. International Conference on Supercomputing, 1997, pp. 340–347.

[16] Matteo Frigo and Steven G. Johnson. FFTW: An adaptive software architecture for the FFT. Proc. 1998 IEEE Intl. Conf. Acoustics Speech and Signal Processing, vol. 3, IEEE, 1998, pp. 1381–1384.

[17] Y. Li, J. Dongarra, and S. Tomov. A note on auto-tuning GEMMfor GPUs. In ICCS '09, pages 884–892, Berlin, Heidelberg, 2009. Springer-Verlag.

[18] Michael Wolfe. Compilers and more: Optimizing GPU kernels. HPC Wire, http://www.hpcwire.com/features/33607434.html, October 2008.

Part II

Sparse Linear Algebra

Chapter 5

Sparse Matrix-Vector Multiplication on Multicore and Accelerators

Samuel Williams

Lawrence Berkeley National Laboratory

Nathan Bell

NVIDIA Research

Jee Whan Choi

Georgia Institute of Technology

Michael Garland

NVIDIA Research

Leonid Oliker

Lawrence Berkeley National Laboratory

Richard Vuduc

Georgia Institute of Technology

5.1   Introduction

This chapter consolidates recent work on the development of high-performance multicore and accelerator-based implementations of *sparse matrix-vector multiplication* (SpMV). As an object of study, SpMV is an interesting computation for two key reasons. First, it appears widely in applications in scientific and engineering computing, financial and economic modeling, and information retrieval, among others, and is therefore of great practical interest. Secondly, it is both simple to describe but challenging to implement well, since its performance is limited by a variety of factors, including low computational intensity, potentially highly irregular memory access behavior, and a strong input dependence that be known only at run time. Thus, we believe SpMV is both practically important and provides important insights for understanding the algorithmic and implementation principles necessary to making effective use of state-of-the-art systems.

The key findings and results of this chapter are primarily the direct result of three recent publications [5, 7, 15]. This chapter focuses on synthesizing the main findings from across the three studies, emphasizing high-level design and implementation principles. Some of the data in this chapter are new, as they include recent hardware platforms not available in prior work (*e.g.*, Intel Nehalem, STI PowerXCell 8i, and NVIDIA GeForce GTX 285). However, we also must necessarily omit discussion of some platform-specific implementation details, as well as a more in-depth discussion of some of the research issues explored in the original studies (*e.g.*, auto-tuning). For such details, we recommend that the interested reader consult the original studies.

5.2   Sparse Matrix-Vector Multiplication:
Overview and Intuition

Sparse matrix-vector multiplication (SpMV) operations are of particular importance in computational science. They represent the dominant cost in many *iterative methods* for solving large-scale linear systems and eigenvalue problems which arise in a wide variety of scientific and engineering applications. In the course of solving a sparse linear system $Ax = b$, such methods generally require the computation of hundreds or perhaps thousands of SpMV operations with the matrix A. Sparse matrix-vector multiplication is the foundation of a broad class of solvers, including notable examples such as the conjugate gradients method (CG), the generalized minimum residual method (GMRES), and the biconjugate gradients stabilized method (BiCGstab), among many others [11]. The remaining components of these methods reduce to dense linear algebra operations that are readily handled by optimized BLAS implementations.

The specific SpMV operation we consider is $y \leftarrow y + Ax$, where A is an $M \times N$ sparse matrix, and x, y are dense vectors. We refer to x as the *source vector* and y as the *destination vector*. By "sparse," we mean that most of the entries of A are zero, and therefore compact representations of A can eliminate unnecessary storage and computation. However, the cost of a sparse representation is a more complex data structure since, unlike the dense case, it must explicitly track which non-zero entries are stored. As an example, Figure 5.1 illustrates the most common sparse matrix representation, called compressed sparse row (CSR) storage, and provides a base-line sequential SpMV implementation.

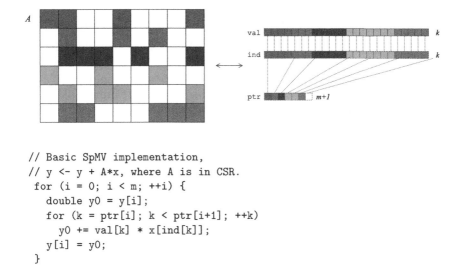

```
// Basic SpMV implementation,
// y <- y + A*x, where A is in CSR.
 for (i = 0; i < m; ++i) {
    double y0 = y[i];
    for (k = ptr[i]; k < ptr[i+1]; ++k)
      y0 += val[k] * x[ind[k]];
    y[i] = y0;
 }
```

FIGURE 5.1: Compressed sparse row (CSR) storage, and a basic CSR-based SpMV implementation.

To gain some intuition for SpMV performance, we make two observations. First, SpMV requires just two floating-point operations (flops) per non-zero entry of A—namely one multiplication and one addition. By comparison, the code given in Figure 5.1 will execute many more instructions on secondary tasks such as integer indexing, resulting in a relatively high overhead. Second, SpMV has relatively little reuse and is memory intensive. Temporal data locality is limited to the accesses of x and y; every element of A is used exactly once. Indeed, a first order estimate on SpMV performance is simply that it will be bounded from below by the time to read A, which in effect amounts to the time required to stream the matrix data structure from memory through the processor. On a modern processor, whose peak computational throughput is substantially higher than its peak memory bandwidth, we thus expect its computational efficiency to be low and ultimately limited by memory bandwidth. Consequently, much of the effort to optimize SpMV performance focuses on changing loop and data structures to maximize parallelism and reduce non-flop instruction overheads, to regularize memory access patterns, to minimize memory traffic, and to maximize locality. We discuss how this intuition, given current multicore and accelerator architectures described in Section 5.3, informs specific design principles and optimization techniques in Sections 5.4 and 5.5, respectively.

5.3   Architectures, Programming Models, and Matrices

Our SpMV optimizations are driven by the diversity in available hardware architectures, parallel programming models, and input matrices that arise in practice. This section summarizes the main characteristics of these platforms and inputs that influence the process of optimizing and tuning SpMV.

5.3.1   Hardware Architectures

We consider SpMV optimization in the context of three diverse multicore and accelerator platforms: a dual-socket quad-core server based on Intel's Xeon X5550 multicore processor ("Nehalem"); a dual-socket blade based on IBM's QS22 PowerXCell 8i processor; and NVIDIA's GeForce GTX 285 GPU. Table 5.1 summarizes the specifications of these systems.

These architectures differ along several dimensions, perhaps the most significant of which is the programmer's view of the memory system, as summarized by Table 5.2. In particular, we may broadly classify these architectures by the number of levels of memory (address spaces) and the nature or degree of software control. Differences in these design aspects affect how the programmer must (a) orchestrate data movement and (b) restructure the computation to achieve good spatial and temporal locality within each level of memory.

For example, our Intel Nehalem platform, typical of conventional cache-based mulitcore processors, uses an explicit two-level memory hierarchy: DRAM and registers. Programmers (often via compilers) will control the movement of data from DRAM to registers (through explicit loads and stores) and then from registers to functional units (other instructions). Cache hierarchies are often instantiated between DRAM and registers to accelerate performance, largely through automated placement of demand memory accesses within the caches. Automated caching simplifies programming, at the cost of a loss of transparency in when and how data move through the hierarchy as well as the cost of data movement. For the most part, programmers may modify

TABLE 5.1: Architectural summary of evaluated platforms. Note, all performance numbers are theoretical peak.

Core Architecture	Intel Nehalem	IBM Cell SPE	NVIDIA GT200 SM
Type	SMT out-of-order	dual-issue in-order	SIMT in-order
Clock (GHz)	2.66	3.20	1.47
DP Peak (GFlop/s)	10.66	12.80	2.96
Register File	$16 \times 128b$	$128 \times 128b$	$512 \times 1024b$
Local Store	—	256 KB	16 KB
L1 Data Cache	32 KB	—	—
L2 Cache	256 KB	—	—
Socket Architecture	**Xeon X5550 Nehalem**	**PowerXCell 8i Cell Blade**	**GTX 285 GeForce**
Cores per Socket	4	8 (+PPE)	30
Last Level Cache	8 MB L3	—	—
Primary memory parallelism paradigm	HW prefetch	DMA	Multithreading
Node Architecture	**Xeon X5550 Nehalem**	**PowerXCell 8i Cell Blade**	**GTX 285 GeForce**
Sockets per SMP	2	2	1 (+CPU)
DP Peak (GFlop/s)	85.33	76.80	88.84
DRAM Pin Bandwidth (GB/s)	51.20	51.20	159.00

their programs to elicit better cache behavior, though caches and hardware prefetchers try to render such modifications unnecessary.

To better support performance-oriented programming, accelerator architectures have tried to provide more explicit control of the memory hierarchy. Architectures like Cell's SPEs take a three-level approach with the addition of a software-controlled local store memory seated between DRAM and registers. For correct execution, the programmer must explicitly regiment transfers of data from DRAM to the local store (via DMA transfers), with the ability to rely on compilers to control data movement from the local store to registers.

GPU architectures have also adopted a three-level memory hierarchy. Programs access data from external DRAM (device memory) and store values in registers. They may also work with data in on-chip local store (shared memory), although unlike Cell's SPEs, this is not required. GPUs may also cache (device) DRAM to register transfers, although for this generation of GPU architecture these caches are for read-only data. For discrete GPU cards, such as

TABLE 5.2: Automation of the movement of data between address spaces. [1]Cached and prefetch accelerated. [2]Guided via language attributes. [3]Functions to interface with DMA engines.

Data Movement	Xeon X5550 Nehalem	PowerXCell 8i Cell Blade	GTX 285 T10P
LS↔regs	N/A	Compiler	Compiler[2]
DRAM↔LS	N/A	User[3]	Compiler[2]
DRAM↔regs	Compiler[1]	N/A	Compiler
Host↔DRAM	N/A	N/A	User[3]

those we benchmark here, the GPU DRAM is separate from the CPU DRAM (host memory) on the motherboard. Programs may transfer data between these memory systems via explicit DMA or memory mapping. For motherboard GPUs, both host and device memory are provided by the motherboard DRAM.

Intel Xeon X5550 (Nehalem): The recently released Nehalem includes minor enhancements to the Core microarchitecture, but dramatic changes to the cache and memory architecture. Each core supports two-way multithreading. When the per-thread instruction-level parallelism is low, running two threads per core can more efficiently utilize functional units. Cores implement two-way SIMD and separate add and multiply functional units, yielding a peak throughput of four double-precision flops per cycle per core.

Nehalem implements an inclusive (content of L2 cache includes that of the L1) cache hierarchy. In particular, each core has a private 32 KB L1 as well as a private 256 KB L2 cache. Moreover, all cores on a socket share an 8 MB L3 cache, in contrast to the previous processor generation's use of private 4 MB caches kept coherent via a snoopy (a cache coherency mechanism) frontside bus. Critically, Intel has integrated three DDR3 memory controllers on each chip and implemented a inter-chip network (QuickPath) that carries snoop and remote node access requests. Unfortunately, this non-uniform memory access (NUMA) architecture may suffer poor scalability on some challenging problems.

IBM QS22 PowerXCell 8i (Cell Blade): The IBM Cell processors used in this chapter represent the enhanced double-precision (eDP) variant of the Cell processor found in Sony's PlayStation3 game console. The Cell is based on a single-chip heterogeneous core design. In particular, each chip has one dual-threaded, dual-issue, conventional cache-based PowerPC core (the PPE) and eight, efficiency-optimized, local store-based SPEs. Each SPE executes all code from a small 256 KB, DMA-filled, local store and can execute one double-precision SIMD fused multiply add (FMA) per cycle. As such, the aggregate SPE performance greatly exceeds that of the PPE, meaning performance-critical and performance-intensive routines should be re-

TABLE 5.3: Programming models used by platforms. †Only in conjunction with libspe2.

Programming Model	Xeon X5550 Nehalem	PowerXCell 8i Cell Blade	GTX 285 GeForce
OpenMP	✓	—	—
PThreads	✓	✓†	—
CUDA	—	—	✓

implemented for the SPE architecture. Like Nehalem, all memory controllers are integrated on-chip, and the servers are dual-socket NUMA SMPs.

There are two principle differences between the QS22 blades and prior generations (*e.g.*, IBM's QS20). First, each SPE's double-precision performance has been dramatically improved to be half of single-precision performance. Secondly, the 512 MB of XDR DRAM per processor has been replaced with 16 GB of DDR2-800 DRAM per socket. In principle, both DRAM types should deliver the same bandwidth (25.6 GB/s per socket), but we have observed that this is rarely true. Rather, sustained bandwidth is often less than 20 GB/s per socket. Consequently, we expect better performance for large or floating-point intensive problems on QS22 than QS20, whereas we expect worse performance on small (less than 1 GB) memory-intensive problems.

NVIDIA GeForce GTX 285: The NVIDIA GPU considered in this chapter is based on the GT200 processor architecture. It differs markedly from the other systems in its direct support for massive fine-grained multithreading as the primary mechanism for hiding memory latency. This current-generation GPU consists of 30 streaming multiprocessors (SMs), each supporting up to 1024 co-resident threads, for a total of up to 30,720 threads per chip. Each SM has a very large 64 KB register file, providing 16,384 32-bit registers for its resident threads, and a 16 KB on-chip local store that can be shared amongst them. It schedules and executes its threads in SIMD groups of 32 called "warps." The on-board GDDR3 memory is connected to the GPU by a wide data path that delivers extremely high bandwidth, with a theoretical peak of 159 GB/s on the GTX 285.

5.3.2 Parallel Programming Models

As architectures continue to diversify, they have mandated specialized programming models. Thus, as shown in Table 5.3, our SpMV experiments employ three distinct shared memory parallel programming models, depending on the platform: OpenMP (Nehalem only), POSIX Threads (Nehalem and Cell), and CUDA (GTX 285 only).

OpenMP: OpenMP [3] is a pragma controlled, fork-join, shared memory parallel programming model. Its simplicity derives from the common use case of just identifying the key parallelizable loops and annotating them with a pragma. Although OpenMP provides some speedup for SpMV, we show that in practice a considerable amount of additional work can and must be performed in order to achieve high performance.

POSIX Threads: POSIX threads (or pthreads) [1] is a function-driven, fork-join, shared memory parallel programming model. For our purposes, when using pthreads, threads are created at the beginning of the application and used throughout in a bulk-synchronous SPMD model. Indeed, we created and used a number of fast, low-overhead, spin barriers designed to work in such a style.

Cell's SPEs are programmed using libspe. However, in many ways, it behaves like pthreads. Essentially, for every SPE thread, one creates a PPE thread. Often, that thread immediately spawns an SPE thread and promptly yields. As such in practice it is common for each application to have one pthread and 16 SPE threads running simultaneously.

CUDA: The CUDA platform [2] provides a simple and direct model for programming the GPU. A CUDA program consists of a one or more host threads, running on the CPU, that may launch parallel "kernels" on the GPU. Each kernel is a blocked SPMD computation: it executes a single sequential program across many parallel threads, which are additionally grouped into thread blocks. Threads within a block may synchronize freely at barriers, but separate blocks may not directly synchronize with each other. The decomposition of parallel work into kernels provides the only means for bulk synchronization between separate blocks.

Matching the hierarchical organization of threads is a hierarchy of disjoint memory spaces, including: (a) thread-private memory, which is typically stored in registers, (b) per-block shared memory, which is stored in fast on-chip local store, and (c) device memory visible to all threads, which is stored in external DRAM. GPU hardware provides a number of caching mechanisms for device memory. The read-only texture cache, utilized via language attributes, provides read-only caching of data optimized for access patterns typical in graphics applications. Future generations of GPU hardware will provide full read-write caching of device memory [9].

5.3.3   Matrices

SpMV performance depends strongly on the properties of the input matrix. The most influential factors include the matrix dimension, non-zero density (*e.g.*, non-zeros per row), variance in the number of non-zeros per row/column, and the specific non-zero pattern (*e.g.*, small dense subblocks, diagonal substructure, randomly distributed). In our experiments, we consider a variety of matrices that arise in real applications and that also vary with respect to these attributes, taken from various sources (see Vuduc [13, App. B]). Figure 5.2 summarizes this matrix benchmark set, and includes small "spyplots" of the non-zero pattern.

	Dense	Protein	Spheres	Cantilever	Wind Tunnel	Harbor	QCD	Ship	Economics	Epidemiology	Accelerator	Circuit	Webbase	LP
Spyplot														
Rows	2K	36K	83K	62K	218K	47K	49K	141K	207K	526K	121K	171K	1M	4K
Cols	2K	36K	83K	62K	218K	47K	49K	141K	207K	526K	121K	171K	1M	1M
NNZ	4.0M	4.3M	6.0M	4.0M	11.6M	2.4M	1.9M	4.0M	1.3M	2.1M	2.6M	0.9M	3.1M	11.3M
average NNZ/Row	2000	119	72	65	53	50	39	28	6	4	22	6	3	2825
Symmetric	-	✓	✓	✓	✓	-	-	✓	-	-	✓	-	-	-

FIGURE 5.2: Set of matrices used across all three platforms. Note: NNZ is the number of non-zeros. Although some matrices are symmetric, no implementation in this chapter exploits that property.

5.4 Implications of Architecture on SpMV

Given the diversity of architectural approaches and the sensitivity of SpMV performance on the input matrix (possibly known only at run time), we might reasonably conclude that there is not likely to be a single "best" SpMV implementation. In this section, we explore how some of the major architectural features will influence the design of an SpMV implementation.

5.4.1 Memory Subsystem

The reference CSR implementation of SpMV in Figure 5.1 is dominated by three memory access patterns: (a) a unit-stride read of the matrix non-zero values and column indices; (b) a matrix-dependent, and potentially random gather from the source vector; and (c) a unit-stride write of the destination vector. In principle, these are all memory demand requests, and although caches can filter many of them, in practice restructuring these accesses is essential to good performance.

Recall from Section 5.2 that in the best case of perfect temporal reuse of the source and destination vectors, a lower bound on the time to execute an SpMV is simply the time to read (stream) the matrix data structure. In double-precision CSR, this means roughly 8 bytes for the non-zero value plus 4 bytes for the integer column index, assuming a 32-bit `int`, or 12 bytes of traffic per non-zero. These requests will be compulsory misses on a cached memory hierarchy. At two flops per non-zero, performance in flops per second will be at most (DRAM bandwidth) divided by (12 bytes) times (2 flops), or $\frac{\text{bandwidth}}{6}$. The main solution is to further compress the matrix data structure through smaller indices, exploiting symmetry, or alternative matrix formats that can reduce indices (*e.g.*, exploiting dense block substructure via register blocking [8]).

If the source and destination vector working sets exceed the cache capacity, performance will be further diminished by capacity misses. Additionally, a highly irregular distribution of non-zeros will reduce the spatial locality of vector or cache line accesses, implying wasted bandwidth. In either case, careful orchestration of source and destination vector accesses is essential. Reordering rows and columns or use of explicit cache-level blocking of the matrix data structure (akin to cache blocking or tiling in the case of dense matrix operations) can help. Moreover, these techniques are applicable regardless of whether a particular architecture is actually cache based or instead uses a software-controlled local-store.

Analogous to the locality-enduced partitioning challenges in the distributed memory world, NUMA architectures warrant careful allocation and placement of data. Again, block-based organization of the matrix data structure, and allocation and placement of data accordingly are essential. Even for cacheable workloads, we may observe the same effects due to non-uniform cache architectures.

Finally, there are frequently additional memory (and even cache) latency tolerance mechanisms available, many of which can be controlled in a well-designed SpMV implementation. These mechanisms include hardware and software prefetchers, hardware multithreading, out-of-order execution, and DMA.

5.4.2   Processor Core

Though we intuitively expect the memory system to be the main bottleneck, it might not be for every combination of processor and matrix. In fact, modern efficiency-oriented cores may falter when dealing with the irregular computational structure associated with SpMV. In particular, consider that the reference SpMV implementation in Figure 5.1 has little or no instruction- or data-level parallelism (ILP and DLP, respectively); worse, its instruction mix is dominated by non-floating-point instructions. Attaining peak performance on conventional processor architectures requires just the opposite: high ILP, high DLP, and floating-point intensity. For example, on a Nehalem-class processor, one must express five-way ILP and two-way DLP collectively among the two threads per core. Clearly, one may trade ILP for DLP and thus express 10-way DLP via five instructions. Achieving peak performance on a GPU typically requires thousand-way thread-level parallelism (TLP) to fully utilize each SM and 32-way DLP within an SIMD warp to avoid divergence. On either machine, reorganization of loop structure (*e.g.*, segmented scan) as well as the data structure (*e.g.*, register blocking) may express more parallelism and ensure core performance is not an impediment to SpMV performance.

TABLE 5.4: Programming models used by platforms. [†]Only in conjunction with libspe2.

Optimization	Xeon X5550 Nehalem	PowerXCell 8i Cell Blade	GTX 285 GeForce
Partitioned Storage	✓	✓	✓
Register Blocking	✓	✓	✓
Index Compression	✓	16b only	—
Format Exploration	✓	BCOO only	✓
Cache Blocking	✓	✓	✓
TLB Blocking	✓	✓	—
SW Prefetch	✓	—	—
DMA	—	✓	—

Moreover, techniques like register blocking can asymptotically amortize the instruction or operation overhead to one load, multiply and add per non-zero.

5.5 Optimization Principles for SpMV

Given the architectures and intuition about SpMV outlined in previous sections, this section describes specific and effective performance optimization techniques. We organize these optimizations into three broad categories, based on their expected benefit: (a) efficient parallelization, (b) reducing memory traffic, or (c) orchestrating data movement. Table 5.4 summarizes these optimizations. There are many possible techniques within each category, of which we discuss only a subset in this chapter; refer to the original reference studies for additional details [5, 7, 15].

5.5.1 Reorganization for Efficient Parallelization

Multicore and accelerators are becoming massively parallel compute platforms. Although SpMV exhibits inherent loop-level parallelism, we must recast this parallelism as thread-, data-, or instruction-level parallelism and quite possibly, restructure the algorithm to express even more parallelism.

Naïve Parallelization: The simplest approach to parallelization is to simply allocate one or more rows to each thread. In such a scheme, there is typically far more thread-level parallelism than there is hardware support, and we might expect a thread scheduler to perform effective load balancing. This expection does indeed hold on GPU platforms, but for the Cell and Nehalem processors, thread creation and management is a relatively expensive operation. Consequently, on those platforms, we are driven to the other extreme where we match the expressed degree of software thread-level parallelism to available hardware thread-level parallelism. However, this thread-centric approach requires programmers perform explicit load balancing, which we do.

Segmented Scan: An alternative approach to parallelization is to treat each non-zero (or groups of consecutive non-zeros) as the unit of parallelism, rather than a row or rows. This so-called *segmented scan* implementation typically creates much more parallelism and obviates the load balancing problem [6]. Although this parallelism may be cast as thread-, data-, or instruction-level parallelism, they all require efficient conditional execution, which is a major challenge on the Cell and Nehalem platforms. As such, it was only implemented on the GPU platform.

SIMDization: All of the platforms provide some hardware mechanism to support short-vector oriented data parallelism. Although compilers can in principle extract and generate the code to exploit such SIMD units, we find that the state-of-practice in what the compiler provides flags what is possible. Therefore, we consider explicit SIMDization on the Cell and Nehalem platforms. Explicit SIMDization replaces pairs of memory or arithmetic operations on consecutive elements with a compiler intrinsic (e.g., _mm_mul_pd()). Clearly this manual process is extremely intrusive and should be used sparingly. On GPU platforms, in contrast, SIMDization is provided implicitly by the hardware, which packs consecutive threads of a block into 32-thread SIMD warps. Instead of using explicit vector operations, the programmer organizes the execution of threads to minimize execution and memory access divergence within warps.

Equivalent Representations: Rather than simply reorganizing loop structures to efficiently parallelize the kernel, we may also reorganize the matrix data structure itself. One approach is the ELLPACK/ITPACK [10] (ELL) storage format. This format organizes an $M \times N$ matrix having at most K non-zeros per row as two dense $M \times K$ arrays stored in column-major order, padding rows with fewer than K non-zeros with zeros. Parallelizing across rows is free of load imbalance, but at the cost of wasting work on the zero values inserted for padding. Clearly, matrices for which the maximum number of non-zeros per row is significantly larger than the average will perform poorly. For each architecture and matrix combination, one must analyze the potential benefit. Due to the massive parallelism demanded by the GPUs, ELLPACK (including our customized variants) was only considered for that platform.

Partitioned Storage: Nominally, in CSR one implements the sparse matrix on a shared memory architecture as three large arrays (one for values, one for column indices, one for row pointers). However, on NUMA and shared cache architectures such storage may be inefficient. Allocating the matrix all at once can lead to it being pinned to one set of memory controllers. Idle memory controllers can impair performance. Even applying the appropriate OpenMP pragma to the initialization of the matrix (for NUMA issues) can still result in suboptimal performance given bank and cache conflicts. An alternative solution implemented on the Cell and Nehalem platforms is to partition the matrix into submatrices and store each contiguously with the appropriate array padding to mitigate conflicts arising from thread contention in the cache and memory subsystem [16].

5.5.2 Orchestrating Data Movement

Across architectures, there are a variety of hardware mechanisms for orchestrating data movement, by which we mean placing or moving data in order to satisfy current or future demand accesses (loads and stores). The data reorganization techniques related to parallelism, such as partitioned storage (Section 5.5.1), can also influence these hardware-based data movement mechanisms. Beyond these, we can use additional hardware-specific operations or methods, namely prefetch and vector instructions, to explicity orchestrate data movement.

Hardware Stream Prefetching: Hardware stream prefetchers are an architectural component designed to hide memory latency. Typically, upon detection of a stream of cache misses (i.e., misses to consecutive cache lines), the prefetcher will engage and speculatively load the next lines. In doing so, the prefetcher can hide the true latency of a cache miss. Unfortunately, to minimize overhead, prefetchers use very simple heuristics and detect relatively simple unit-stride and strided patterns. Moreover, as they are outside of the core, they typically don't have access to a TLB and thus cannot cross a TLB page boundary. Eliciting good prefetcher behavior is essential to high performance on memory-intensive kernels. To that end, we often restructure access patterns into a limited (few per thread) number of unit-stride accesses. This is relatively easy for matrix and destination vector accesses when using CSR.

Software Prefetching: Occasional discontinuities in the address stream can halt a hardware stream prefetcher. Although such patterns are uncommon in purely streaming applications, SpMV contains a mix of streaming and random access behavior (*e.g.*, to the source vector) that can lead to cache misses or disruptions to the hardware prefetcher. To minimize these effects, we may insert a software prefetch (an instruction) for each cache line of the non-zero arrays. We may then tune to find the optimal prefetch distance — far enough ahead to hide memory latency but not so far as to evict useful data from the cache.

Direct Memory Access (DMA): Analogous to software prefetch, we may generate a DMA command to load a number of non-zeros or vector elements. Typically, we double-buffer these operations as to hide (rather than simply amortize) latency overhead. Just as software prefetch required tuning so as not to pollute the cache, on Cell we were forced to balance non-zero, source vector, and destination vector buffer sizes to maximize performance.

Vector: Vector instruction sets permit bulk loads and stores with a single instruction. Similarly, multiple loads to consecutive memory addresses may be coalesced into a single memory transaction. For SpMV, reorganizing memory accesses so that they may be coalesced will substantially improve performance. For example, utilizing a column-major layout for the 2D arrays in the ELL-PACK format ensures that consecutive threads, which process consecutive matrix rows, access memory in a coalesced manner.

5.5.3   Reducing Memory Traffic

When bandwidth constrained, one may improve SpMV performance by reducing memory traffic through additional data and loop structure reorganization. This section summarizes these techniques.

Cache Blocking: As noted in Section 5.4, the source or destination vectors could exhibit poor spatial and temporal locality. We can improve this behavior by translating cache blocking or tiling techniques commonly used for dense matrix computations to SpMV. However, where this technique amounts to loop restructuring in the dense case, the sparse case requires both loop restructuring and a change in data structure. In particular, we may partition the matrix into submatrices and store these submatrices individually. The most naïve solution would be to ensure that each submatrix spans a fixed number of columns — the cache block size. However, such a technique can be very inefficient and underutilize the cache (or local store) as not all source vector elements within that span may be used. Rather, we individually tune the size of each submatrix so that it touches a fixed number of source vector cache lines. That is, only non-empty columns count towards cache capacity. On a cache-based architecture, we may now perform SpMV on the submatrix and can ensure the working set size never exceeds cache capacity.

On Cell, a small change is performed. Rather than storing each column index in its entirety, we may separate out the cache block's column offset as well as high and low bits of the remainder. Within a cache block, there will be many duplicates of the high bits. As such, we may eliminate the duplicates from the volume of memory traffic and encode a list of unique cache lines that must be loaded. Prior to performing the submatrix SpMV operation, we use a DMA get list (`mfc_getl`) to gather these cache lines and pack them contiguously in the local store. Unlike the traditional DMA which operates only on contiguous data, a list DMA defines a list of disjoint addresses and stanza lengths that should be read from DRAM and packed contiguously in the local store. In essence they perform gather/scatter on arbitrary sized elements. As the offsets required to access the local store copy of the relevant source vector elements index the now packed elements, the offsets (column indices) are dramatically different, and guaranteed to be less than 256 KB. As such, when storing the matrix, not only may we encode the local store offsets instead of the DRAM offsets, but we may always use a 16-bit column index (the high 16 bits have been encapsulated into the DMA list). Thus, on Cell we maintain two copies of the matrix: the generic DRAM representation, and the local store optimized representation.

Cache blocking can also be applied on a GPU, either explicitly for each

SM's local store, or implicitly for the texture cache available in current generation NVIDIA GPUs. The texture cache mechanism permits tagging of arbitrary regions of device memory as read-only cacheable data. Accesses to texture locations by multiple threads may be satisfied by a single memory transaction, thus reducing external bandwidth demand, although it may not reduce memory fetch latency. Unlike local store, the texture cache can aggregate requests from threads even when they are not in the same thread block.

TLB Blocking: In many ways the performance impacts of cache misses translate to page cache (TLB) misses. As such, we may apply our cache blocking techniques to the TLB in which we mandate that each submatrix may touch not only a finite number of cache lines, but also a finite number of TLB pages. This technique often provides a performance boost on matrices for which cache blocking already benefited performance.

Index Compression: As cache blocking restricts the range of column indices for each submatrix, we only need to encode the offset from the first column for all indices and shift the pointer to the source vector. This results in column indices encoded with fewer bits and a reduction in memory traffic. This technique is implemented when possible on CPUs, but is always implemented on Cell because it is the local store offset that is being encoded.

Register Blocking: Register blocking exploits 2D similarity among column indices (geometric proximity in the sparsity pattern of non-zeros) to aggregate non-zeros into small dense matrices in which some elements are zero. Almost invariably, these matrices (blocks) are less than 16×16 and in this chapter they are always powers of two less than 8×8. The principal advantage of register blocking when memory bound is that the matrix structure requires only one index per block rather than one per non-zero. Asymptotically, this can reduce memory traffic by 33%, and thus boost performance by 50%.

Matrix Formats: In addition to the ubiquitous CSR and ELLPACK matrix storage formats, we also explored the simpler coordinate (COO) format. In coordinate, each non-zero is accompanied with both a column index and a row index. Nominally, non-zeros can appear in any order. As such, the resultant read/increment/write data dependency limits loop parallelization or software pipelining. However, we may optimize the format by sorting non-zeros by row. This allows for elimination of the data dependency in favor of a simple loop terminated when consecutive non-zeros have different row indices. The result resembles CSR, but is more amenable to segmented scan and conditional execution.

5.5.4   Putting It All Together: Implementations

In this chapter we present eight different SpMV kernels spanning five basic matrix formats (Table 5.5). We did not implement all 15 (five formats × three architectures) because some combinations were inappropriate for certain architectures. For each kernel, we implemented a different subset of the optimizations described in Sections 5.5.1–5.5.3. When we apply cache blocking-like optimizations (CPU, Cell, and the GPU's hybrid format), each submatrix may be individually optimized. In this section, we describe the architecture-specific peculiarities of each implementation grouped by format.

BCOO (CPU/Cell/GPU): Block coordinate (BCOO) was the most widely used matrix format. We implemented an SpMV kernel for each architecture. However, there were some differences. The CPU and Cell implementations were most similar. Both explored register blocks in powers of two from 1×1 to 8×8. The CPU implementation included exploration of prefetching as well as cache and TLB blocking. The Cell implementation uses DMA instead of prefetching, and always cache blocks for the maximum available local store capacity. When parallelized, one thread was created per hardware thread context or SPE. Cache and TLB blocking occur after parallelization based on non-zeros. Cell implements a degenerate form of segmented scan in which conditional stores are emulated in software via SIMD muxing.

The GPU implementation was somewhat different. It only implements 1×1 COO, has no need for prefetching, uses the texture cache for implicit exploitation of source vector temporal locality, and maximizes parallelism by individually assigning one CUDA thread to each non-zero. It then uses a segmented reduction reminiscent of Blelloch *et al.* [6] on the CM-2 and Cray C90 and the CUDA-based segmented scan implementation by Sengupta *et al.* [12].

BCSR (CPU/GPU): CSR flavors were implemented both on CPUs and GPUs. They were not implemented on Cell as the benefit would have been small and challenging to implement a fast 1×1 variant. The CPU implementation incorporates the same optimizations as its BCOO brethren, but the GPU implementations (there were two) are quite different. The first GPU CSR implementation, hereafter referred to as CSR(scalar), uses the standard loop and data structures. It parallelizes the computation by assigning one thread to process each row. Unfortunately, this simple approach is generally inefficient as threads within a warp do not access contiguous memory locations, thus preventing memory coalescing and degrading memory bandwidth efficiency. The second implementation, CSR(vector), assigns one 32-thread warp to each row, effectively strip mining the inner loop of the sequential SpMV computation. This method allows for memory coalescing at the cost of an intra-warp reduction at the end of each row. Independently, Baskaran and Bordawekar [4] implemented a similar approach, although they assign one half-warp to each row and pad each row to be a multiple of 16 in length. Their padding guarantees alignment, and hence slightly higher degrees of coalescing, albeit at the cost of potentially significant additional storage. This may incrementally improve performance in some cases, but shares the same fundamental performance characteristics.

GCSR (CPU): GCSR is a variant on BCSR and an alternative to BCOO. It is only beneficial when cache blocking produces submatrices with many empty rows distributed at random (i.e., hypersparse). GCSR augments the BCSR data structure with a row coordinate associated with each row

TABLE 5.5: Implementations as a function of machine. †Only 1×1 CSR.

Implementation	Xeon X5550 Nehalem	PowerXCell 8i Cell Blade	GTX 285 GeForce
BCOO	✓	✓	✓
BCSR	✓	—	✓†
GCSR	✓	—	—
Hybrid	—	—	✓
BELLPACK	—	—	✓

pointer. This coordinate is conceptually similar to column indices but as non-zeros are sorted by rows, far fewer are required.

Hybrid (GPU): Unlike the CPU/Cell implementations where hybrid implementations arise from local specialization based on individually optimized cache blocks, the "Hybrid GPU" implementation merges ELLPACK and COO to attain ELLPACK's high-performance potential with the performance invariability of COO. Given a parameter K, the matrix is partitioned into two portions. The first submatrix is stored in ELLPACK form with K non-zeros per row, padding rows with fewer than K non-zeros. The second submatrix is stored in COO form and holds the excess entries from rows with greater than K non-zeros. The splitting parameter K can be specified directly, or chosen automatically using an empirical heuristic which is currently $K = \max(4096, M/3)$.

BELLPACK (GPU): BELLPACK is a GPU-only implementation that applies one-dimensional row blocking, row permutation, and register blocking to ELLPACK. In CUDA terms, one-dimensional row blocking means that we assign one thread block per block row of the matrix, where the block row size is tunable. Combining 1 D row blocking and row permutation effectively reduces the padding required relative to conventional ELLPACK storage. Within a row block, warps are assigned to consecutive register-block rows. To achieve coalesced accesses, elements from different blocks within the same warp are interleaved. The register block size is tunable. There are no restrictions on the register block size, though a "poor" choice will lead to poor performance.

5.6 Results and Analysis

In this section, we present SpMV performance for our three platforms as a function of input (matrix), optimization (*e.g.*, register blocking), and approach to parallelization (*e.g.*, OpenMP vs. Pthreads). We then examine performance by architecture, to give both implementation and architectural insight.

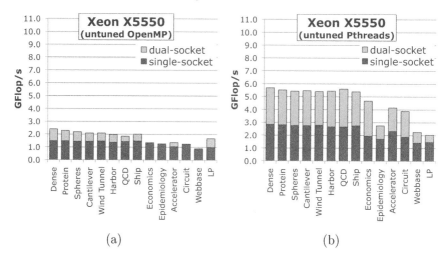

(a)   (b)

FIGURE 5.3: Untuned SpMV performance as a function of threading model, hardware concurrency, and matrix.

5.6.1   Xeon X5550 (Nehalem)

Figure 5.3(a) presents CSR SpMV performance as a function of threading model (OpenMP vs. Pthreads), matrix, and hardware (using one or both sockets of the dual-socket SMP). We observe that using one socket (via the OpenMP affinity environment variable) always delivers less than 1.5 GFlop/s. Intuitively, we expect the matrices with few non-zeros per row or large dimensions to deliver lower performance as a larger fraction of memory bandwidth is tasked with reading in row pointers and source vector elements. Nevertheless, the performance variability is quite low. Given this kernel should be bandwidth limited (ignore the 42.66 GFlop/s peak) with a per socket STREAM bandwidth less than 18 GB/s, one would naïvely expect performance less than 3 GFlop/s (two flops per 12-byte non-zero). The delivered performance is substantially less than this bound. Moreover, when the second socket is employed, performance improves by just 50% on some matrices and none on others.

This lack of socket scalability can well be explained by two facets of SpMV on this NUMA SMP. When the matrix was read from the disk, it was done so by one thread. The underlying first-touch policy placed data on the DIMMs with affinity to the core on which that thread was running. Unfortunately, this means that when SpMV is parallelized via OpenMP, only the memory controllers attached to those DIMMs were used. Thus half the SMP's bandwidth was thrown away. The lack of scalability for the simplest matrices is thus an artifact of limited bandwidth and high inter-socket latency. For the more challenging matrices (Epidemiology through Linear Programming), the fact that the vectors were allocated via the same first-touch policy similarly limits performance. Finally, some matrices (*e.g.*, Webbase) are particularly

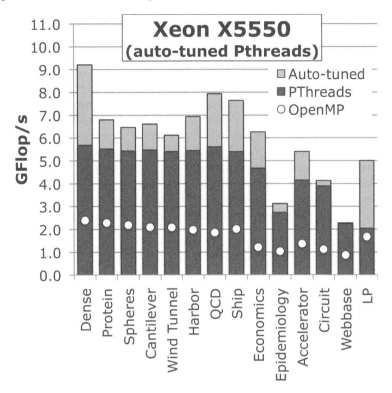

FIGURE 5.4: Nehalem pthread performance before and after auto-tuning. OpenMP included as reference.

poorly suited to iterative sparse methods because a naïve parallelization will induce all-to-all communication. That is, each socket updates its respective half of the vector, but on the next iteration each socket will need both halves. This forces an implicit data broadcast and exposes the limited inter-socket bandwidth.

In Figure 5.3(b) we observe that the pthreads implementation (with the NUMA-aware library matrix creation routines) delivers not only substantially better performance (better than 2.5×), but also better multi-socket scalability (typically 2×). The latter is well explained by proper NUMA allocation. However, even in the single-socket configuration, pthreads usually delivers twice the performance of OpenMP. This was quite surprising given we use the correct OpenMP affinity variables, load balancing shouldn't be an issue on at least some of the matrices, and OpenMP usually performs well for structured grid computations. Although the pthreads implementation did provide close to 3 GFlop/s per socket, we show that substantially better performance can be attained.

Figure 5.4 shows the performance benefits attained for the full SMP when auto-tuning is applied to the pthread implementation. We include OpenMP data as dots for comparison. As previously discussed, a vast number of optimizations were explored including register blocking, index compression, alternative matrix formats (BCOO, GCSR), prefetching, and cache/TLB blocking. We observe that auto-tuning often accelerated performance by 30% (typically from register blocking) and as much as 2.5× (the extreme case requiring TLB and cache blocking). Moreover, the conjunction of pthreads and auto-tuning consistently exceeded OpenMP performance by between 2.6× and 5.1×. Unfortunately, there are some matrices (*e.g.*, Epidemiology, Circuit, Webbase) for which our current tuning regimen seems illequiped. These are the problems that exhibit poor cache locality (both temporal and spatial) at any scale and demand substantial inter-core and inter-socket communication.

5.6.2   QS22 PowerXCell 8i

The Cell SpMV auto-tuner is built on the multicore auto-tuner we used for Nehalem. The principal differences are the threading model (superficial change from pthreads to pthreads+libspe) and the fact that due to the complexity of Cell programming, only a subset of the optimization space was implemented. Thus, on Cell the auto-tuner only implements BCOO (omitting BCSR/GCSR), always cache blocks for the available local store capacity, always compresses indices, and always uses DMA. However, the Cell auto-tuner does implement a variant on segmented scan in which each SPE runs a software pipelined, vector length of 1, BCOO segmented scan on its cache block. In essence, this transforms the doubly-nested BCSR or sequential BCOO implementations into a single very-fast pipelined loop. Note that the PPE performs no computation in the SpMV kernel, but rather is used for coordination.

Figure 5.5 presents baseline Cell performance (1×1 BCOO) using either one or both of the Cell chips on the QS22 SMP. It also shows auto-tuned performance using both chips. When examining single-socket performance, we see dramatically different behavior compared to Nehalem. Cell's performance is remarkably constant—a testament to the elimination of CSR's short loops (via segmented scan) coupled with a potentially compute-bound SPE. Although we expected the QS22's DDR2-based stream performance to be substantially lower than the QS20's XDR-based performance, the expected (bandwidth-only) performance bound of 3 GFlop/s per socket is somewhat higher than the observed performance (black bar). When the second socket is used, the matrices amenable to NUMA-parallelized attain a speedup of 2×. However, we observe a very similar behavior to that of Nehalem on the challenging matrices.

Principally, the Cell auto-tuner explores alternative register blockings while using only the BCOO format with index compression. Nevertheless, we observe that register blocking can dramatically improve performance for some matrices (over 2×). Nominally, register blocking should only improve perfor-

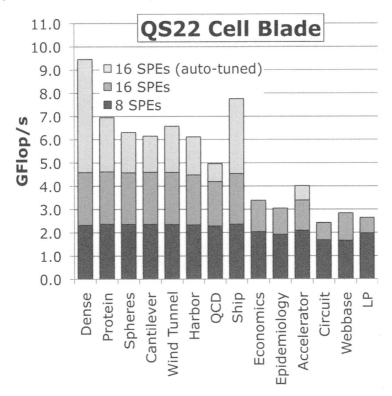

FIGURE 5.5: SpMV performance as a function of matrix, hardware concurrency, and optimization. Note, the untuned baseline is actually a DMA-enabled 1×1 COO implementation, and is thus by no means naïve.

mance by a factor of 1.5× (if memory-bound). We believe the discrepancy arises from a transition from compute bound to memory-bound. This may be confirmed as the attained performance (over 9 GFlop/s) aligns well with the bandwidth–arithmetic intensity product. Once again, we observe that our breadth of optimizations is insufficient for more than a third of the matrices. Clearly, there is still ample research material for SpMV.

5.6.3 GTX 285

Unlike the Nehalem and Cell implementations, there is not one single GPU auto-tuner. As such, in this section, we present the performance results of five different GPU SpMV implementations. Moreover, in all cases, we assume data (both matrices and vectors) remain resident in GPU *device* memory obviating the need for any PCIe transfers. This is a reasonable assumption for any local iterative sparse solver. Finally, all GPU computations are performed in

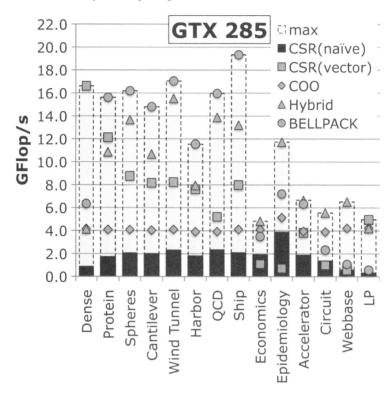

FIGURE 5.6: GPU performance as a function of matrix and implementation (dots). Dotted bars are used simply to visualize the best possible performance.

double precision and access to the source vector is accelerated via the on-chip, read-only texture cache.

Figure 5.6 presents each implementation's performance on each matrix. The black bars (naïve CSR) represent a straightforward loop parallelization of the standard 1×1 CSR implementation that assigns one thread per row. GPUs may require several thousand parallel threads in order to fully utilize the processor. This need for abundant parallelism is a challenge for a simple thread-per-row decompsition strategy when matrices have relatively few rows, as is the case with our Dense and LP examples which have 2,000 and 4,284 rows, respectively. Despite the appearance of substantial parallelism on the remainder, this simple implementation exhibits substantial memory divergence when accessing the CSR data structure, since as threads access different rows, they access disjoint rather than contiguous non-zeros. Thus it does not effectively utilize the available 159 GB/s of memory bandwidth since there is an order of magnitude difference between fully coalesced and uncoalesced memory access. All remaining implementations attempt to attain memory coalescing.

They are overlaid in Figure 5.6 via dots with light gray bars highlighting the maximum performance attained via any implementation.

The vectorized CSR implementation, which exhibits much less memory divergence, delivers substantially better performance than the scalar CSR implementation. In addition, it assigns one 32-thread warp per row and thus requires far fewer rows to develop sufficient find-grained parallelism. In the vectorized approach, the degree of memory divergence, and to some extent execution divergence, is determined by the distribution of non-zeros per row. When applied to the six finite-element matrices with an average 50 or more non-zeros per row the vectorized kernel achieves no less than 7.5 GFlop/s. On the other hand, matrices such as Webbase with approximately 3.1 non-zeros per row expose a weakness of the vectorized implementation—several threads of a warp will be idle when a row is much shorter than the warp width.

The segmented reduction-based COO implementation delivers very consistent, albeit lower, performance across the matrix suite. Although robust with respect to the number of rows and distribution of nonzeros per row, the COO implementation suffers from low arithmetic intensity (2 flops for 4+4+8 bytes) and the overhead of many inter-thread operations.

The Hybrid implementation, which combines the easy vectorization of ELLPACK with the robustness of COO, achieves good performance on most every matrix. For matrices with a naturally dense substructure, which is typical of matrices arising from finite element analysis, the register blocked BELLPACK implementation delivers better performance. For such matrices, it delivers 1.1–1.5× higher throughput, which is the range expected given the bandwidth it conserves by reducing the number of indices it must read.

The Nehalem and Cell results demonstrated that there is no one "best" implementation, but rather produce a family of implementations auto-tuned to fit specific matrices. These GPU performance results demonstrate the same trend: the best choice of SpMV kernel depends on the structure of the matrix. BELLPACK delivers superior performance for matrices with small dense blocks, the Hybrid format provides better results on matrices with more challenging sparsity patterns, and CSR(vector) worked best for the matrices with the most non-zeros per row.

5.7   Summary: Cross-Study Comparison

The implementations of SpMV described in this chapter represent the state-of-the-art in performance on the three target hardware platforms. Though SpMV is itself simple to state and to analyze, the diversity of architectural designs and input matrix characteristics means a complex combination of architecture- and matrix-specific techniques is essential to achieving this level of performance.

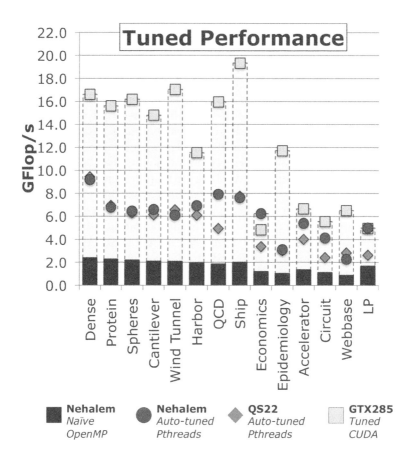

FIGURE 5.7: Performance comparison across architectures. Note: The naïve baseline (Nehalem/OpenMP) is shown as a black bar. The best implementation for each architecture—Nehalem (circle), Cell (diamond), GTX285 (square)—is shown as a color-coded dot with a light gray bar denoting the global best.

To summarize the main findings of this chapter, Figure 5.7 compares the best performance we attain on each platform across the suite of matrices. The common baseline performance is the naïve OpenMP parallelization running on Nehalem (the bottom black bars).

Clearly, all optimized implementations deliver substantially better performance. In addition, we observe that the GPU usually delivers better than twice Nehalem's performance, although this should come as no surprise given the GPU has more than triple the Nehalem SMP's bandwidth. As our intuition expected, bandwidth is the determining performance factor; however, note that achieving high levels of sustainable bandwidth requires a significant tuning effort.

Figure 5.7 also shows the gap between untuned and tuned performance, comparing the baseline OpenMP Nehalem implementation to its tuned Nehalem counterpart (circles). We see a 2–4× difference in performance. This observation suggests just how important selection of the appropriate programming model and performance tuning can be. Moreover, it implies the need for improved tools and techniques that simplify and automate the tuning process.

We observe several cases where Nehalem performance is comparable (Accelerator, Circuit) or exceeds (Economics, LP) GPU performance. Two of these matrices (Circuit and Economics) are the smallest in our test corpus, and may fit in Nehalem's aggregate SMP cache. However, the SMP nature impedes communication of vector elements between successive SpMVs, thus limiting the benefit. The other two have sparsity patterns that appear to pose problems for our GPU implementations, partially because the vector working set exceeds the texture cache's capacity.

One seeming change relative to the earlier literature shown in Figure 5.7 is Cell's relatively lackluster performance. However, we note that at the time of its initial release, the tuned Cell implementation delivered far better performance than any commodity CPU on the market at that time [14,15]. That is, due to the absence of any substantive Cell hardware development in the last 4 years, commodity multicore now delivers comparable performance.

Acknowledgments

The authors acknowledge Georgia Institute of Technology (Georgia Tech), its Sony-Toshiba-IBM Center of Competence, and the National Science Foundation (NSF) for the use of Cell Broadband Engine resources that have contributed to this research. The authors from Lawrence Berkeley National Laboratory (LBNL) are supported by the Director, Office of Science, of the U.S. Department of Energy (DOE) under contract number DE-AC02-05CH11231 and by NSF contract CNS-0325873, along with generous funding and equipment from Microsoft, Intel, and U.C. Discovery (under Awards #024263, #024894, and #DIG07-10227, respectively). The authors from Georgia Tech were supported in part by NSF award number 0833136, NSF CAREER award number 0953100, NSF TeraGrid allocation CCR-090024, joint NSF 0903447 and Semiconductor Research Corporation (SRC) Award 1981, a Raytheon Faculty Fellowship, and grants from the Defense Advanced Research Projects Agency (DARPA) and Intel Corporation. Any opinions, findings, and conclusions or recommendations expressed in this material are those of the authors and do not necessarily reflect those of DOE, NSF, SRC, DARPA, Microsoft, or Intel.

Bibliography

[1] The Open Group Base Specifications, Issue 6: POSIX Threads (pthread.h). IEEE Std 1003.1, 2004. http://www.opengroup.org/onlinepubs/009695399/basedefs/pthread.h.html.

[2] NVIDIA CUDA (Compute Unified Device Architecture): Programming Guide, Version 2.1. http://developer.download.nvidia.com/compute/cuda/2_1/toolkit/docs/NVIDIA_CUDA_Programming_Guide_2.1.pdf, December 2008.

[3] OpenMP: Application Program Interface, version 3.0, May 2008. http://www.openmp.org/mp-documents/spec30.pdf.

[4] Muthu Manikandan Baskaran and Rajesh Bordawekar. Optimizing sparse matrix-vector multiplication on GPUs using compile-time and run-time strategies. Technical Report RC24704 (W0812-047), IBM T.J. Watson Research Center, Yorktown Heights, NY, USA, December 2008.

[5] Nathan Bell and Michael Garland. Implementing a sparse matrix-vector multiplication on throughput-oriented processors. In *Proc. ACM/IEEE Conf. Supercomputing (SC)*, Portland, OR, USA, November 2009.

[6] Guy E. Blelloch, Michael A. Heroux, and Marco Zagha. Segmented operations for sparse matrix computations on vector multiprocessors. Technical Report, Carnegie Mellon University, Department of Computer Science, Pittsburgh, PA, USA, August 1993.

[7] Jee Whan Choi, Amik Singh, and Richard W. Vuduc. Model-driven autotuning of sparse matrix-vector multiply on GPUs. In *Proc. ACM SIGPLAN Symp. Principles and Practice of Parallel Programming (PPoPP)*, Bangalore, India, January 2010.

[8] Eun-Jin Im, Katherine Yelick, and Richard Vuduc. SPARSITY: Optimization framework for sparse matrix kernels. *Int'l J. of High Performance Computing Applications (IJHPCA)*, 18(1):135–158, February 2004.

[9] NVIDIA. NVIDIA's next generation CUDA compute architecture: Fermi^TM, v1.1. Whitepaper (electronic), September 2009. http://www.nvidia.com/content/PDF/fermi_white_papers/NVIDIA_Fermi_Compute_Architecture_Whitepaper.pdf.

[10] John R. Rice and Ronald F. Boisvert. *Solving Elliptic Problems Using ELLPACK*. Springer Verlag, 1984.

[11] Yousef Saad. *Iterative Methods for Sparse Linear Systems, Second Edition*. Society for Industrial and Applied Mathematics, April 2003.

[12] Shubhabrata Sengupta, Mark Harris, Yao Zhang, and John D. Owens. Scan primitives for GPU computing. In *Proc. ACM SIGGRAPH/EUROGRAPHICS Symp. Graphics Hardware*, San Diego, CA, USA, 2007.

[13] Richard W. Vuduc. *Automatic performance tuning of sparse matrix kernels*. PhD thesis, University of California, Berkeley, CA, USA, January 2004.

[14] Sam Williams, Leonid Oliker, Richard Vuduc, John Shalf, Katherine Yelick, and James Demmel. Optimization of sparse matrix-vector multiplication on emerging multicore platforms. In *Proc. ACM/IEEE Conf. Supercomputing (SC)*, 2007.

[15] Sam Williams, Richard Vuduc, Leonid Oliker, John Shalf, Katherine Yelick, and James Demmel. Optimizing sparse matrix-vector multiply on emerging multicore platforms. *Parallel Computing (ParCo)*, 35(3):178–194, March 2009. Extends conference version: `http://dx.doi.org/10.1145/1362622.1362674`.

[16] Samuel Webb Williams. *Auto-tuning performance on multicore computers*. UCB/EECS-2008-164, University of California, Berkeley, CA, USA, December 2008.

Part III

Multigrid Methods

Chapter 6

Hardware-Oriented Multigrid Finite Element Solvers on GPU-Accelerated Clusters

Stefan Turek, Dominik Göddeke, Sven H.M. Buijssen, and Hilmar Wobker

Institut für Angewandte Mathematik, TU Dortmund, Germany

6.1 Introduction and Motivation

The accurate simulation of real-world phenomena in computational science is often based on an underlying mathematical model comprising a system of partial differential equations (PDEs). Important research fields that we pursue in this setting are computational solid mechanics and computational fluid dynamics (CSM and CFD, see Section 6.3). Practical applications range from material failure tests, for instance crash tests in the automotive industry, to fluid and gas flow of any kind, for instance in chemical or medical engineering (e.g., simulation of blood flow in the human body to predict aneurysms) or flow around cars and aircrafts to minimize drag and lift forces. Moreover, the coupling of both models is essential for fluid structure interaction settings

113

(FSI) which represent problem fields of very high technological importance. Such configurations include polymer processing or microfluidic problems exhibiting very complex multiscale behavior due to nonlinear rheological or non-isothermal constitutive laws, and also due to self-induced oscillations of the structural parts in the flow field. In all these cases, the fluid part is mostly laminar, but highly viscous.

The corresponding flow models are based on the Navier-Stokes equations which seem to have a quite simple structure at first sight; nevertheless they constitute 'grand challenge' problems for mathematicians and physicists as well as engineers and computer scientists. They are (still today) subject to very intensive research activities, especially in the following fields:

- time dependent partial differential equations in complex domains
- strongly nonlinear systems of equations
- saddle-point problems due to the incompressibility constraint
- local changes of the problem character in space and time
- temporarily stiff systems of differential equations

These characteristics impose great challenges on almost all numerical algorithms and computational approaches. Among others, the following mathematical issues have to be taken into consideration if efficient simulation tools are to be designed:

- large, ill-conditioned nonlinear systems (*millions of unknowns*)
- locally varying time steps (*implicit schemes*)
- locally anisotropic spatial meshes (*complex geometries*)
- efficient parallel solvers (*decomposition-invariant scalability*)

From a mathematical and engineering point of view, finite element methods (FEM) are considered to be the most promising approaches for the numerical treatment of such PDE problems, due to their flexibility and accuracy, particularly for general settings and complex geometries including unstructured computational meshes. Moreover, they provide a complete framework which allows rigorous a posteriori error control and corresponding adaptive grid manipulations. While classical approaches provide a sharp quantitative estimation of the error only in certain specific configurations (and should therefore be better referred to as *error indicators*), sophisticated finite element techniques overcome these deficiencies and can be formulated in a very general framework such that adaptive error control strategies can also be applied to realistic flow and FSI configurations. In summary, the state-of-the-art finite element techniques form in combination with powerful and robust numerical solution schemes the underlying fabric of many modern simulation tools.

Looking at solutions rather than discretization techniques next, hierarchical (geometric) multigrid methods are more or less obligatory due to their asymptotic optimality, since many of the considered (flow) problems lead to huge, ill-conditioned problems where the condition number depends on the mesh width, resp., problem size: Even for a 'simple' Poisson problem, a multigrid solver with 'standard' choices of a smoother, data structure and numbering scheme for the unknowns executes faster than a (single-grid) Krylov subspace solver with very powerful elementary preconditioners for relevant problem sizes. Since the solution of Poisson-like subproblems is an essential building block of Navier-Stokes and also elasticity solvers, the use of multigrid techniques is mandatory for these problem classes.

As a conclusion, we can state that modern FEM software packages for general continuum mechanical PDE problems, especially in solid mechanics and fluid dynamics, are typically based on highly sophisticated discretization techniques, which have the potential to handle very general computational meshes. However, particularly in the case of huge realistic 3D problems, the realization of efficient parallel multigrid solvers, providing at least double precision accuracy due to typically bad condition numbers of the linear systems, is equally essential. Only the combination of all these mathematical components permits the design of flexible, robust and accurate simulation tools with high numerical efficiency.

Still, this is not enough to realize simulation software with correspondingly high computational efficiency. In numerics, hardware aspects used to be of minor importance since codes automatically ran faster with each new generation of processors. This trend has come to an end, as physical limitations (heat, leaking voltage, pin limits) have led to a paradigm change: Performance improvements are no longer driven by frequency scaling, but by parallelism and specialization. In fact, single-core performance already stagnates or even goes down. Quad-core CPUs are available off-the-shelf, and soon CPUs will have tens of parallel cores. Future manycore chip designs will likely be heterogeneous and contain general and specialized cores with nonuniform memory access characteristics (NUMA). Commodity multimedia processors such as the Cell BE or graphics processor units (GPUs) are considered as forerunners of this trend even though they can currently only be used as co-processors (accelerators) to the general-purpose CPU.

In order to achieve a significant percentage of the available peak performance, both on conventional and novel architectures, hardware characteristics must be taken into account in all stages of the implementation and code optimization. This includes the selection of appropriate data structures (e. g., to store matrices or to communicate data over the interconnects efficiently) and parallelization techniques, in particular when combining the coarse-grained parallelism on the cluster level and the medium- and fine-grained parallelism between the CPU cores, and within accelerator devices like GPUs. Furthermore, different approaches are necessary for different architectures. The same holds true for performance tuning techniques like spatial blocking to exploit cache hierarchies or to coalesce memory transfers into large, more efficient bulk transactions. On the other hand, the meticulous tuning of each application for each new hardware generation is prohibitively expensive, and techniques are required that encapsulate the hardware-awareness inside the underlying finite element discretization *and* solver toolkit, away from the applications.

We are convinced that in the field of high-performance finite element simulations, significant performance improvements can only be achieved by *hardware-oriented numerics* which is a quite young discipline in the field of computational science and engineering (CSE). The core paradigm of hardware-oriented numerics is that numerical and algorithmic foundation research must go hand in hand with (long-term) technology evolution: Prospective hardware trends enforce research into novel numerical techniques that are in turn better suited for the hardware. As an example, strong multigrid smoothers with optimal numerical properties often scale poorly in a parallel setting, due to their strong recursive coupling. Only correspondingly modified schemes that are potentially less numerically efficient on a single node (e. g., in terms of convergence rates) are able to achieve better overall performance in the user-relevant 'time-to-solution'metric. The ultimate goal of hardware-oriented numerics is thus to balance these metrics to achieve robust and ideally predictable close-to-peak performance. Only with the combination of the 'best' numerics and 'best' computational algorithms for a given hardware architecture is it possible to satisfy the aims of hardware-oriented numerics, namely to maximize the total efficiency, i.e., to realize the following 'vision' which is the underlying key idea for our finite element software project FEAST.

Hardware-oriented Numerics: Maximize Total Efficiency

High (guaranteed) accuracy for user-specific quantities with minimal number of degrees of freedom (#DOF) via fast and robust solvers—for a wide class of parameter variations—with optimal numerical complexity (O(#DOF)) while exploiting a significant percentage of the available sequential/parallel peak performance at the same time.

6.2 FEAST—Finite Element Analysis and Solution Tools

FEAST is our next-generation simulation toolkit, which prototypically implements a wide range of hardware-oriented numerics concepts. FEAST is designed for large-scale distributed memory simulations and uses MPI for communication. Here, we briefly present the key concepts, and refer to previous publications for details [2, 15, 18].

6.2.1 Separation of Structured and Unstructured Data

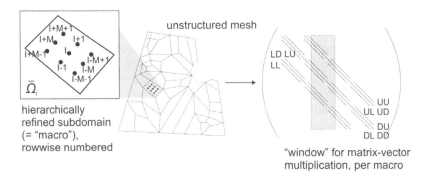

FIGURE 6.1: Locally structured, globally unstructured mesh in FEAST. Buijssen et al., 2009. *FEASTSolid and FEASTFlow: FEM Applications Exploiting FEAST's HPC Technologies, Springer.* 15 pp. With permission.

FEAST covers the computational domain with a collection of quadrilateral subdomains. The subdomains form an unstructured coarse mesh (cf. Figure 6.1 and also bottom of Figure 6.4), and each subdomain is refined in a generalized tensor product fashion. The resulting mesh is used to discretize the set of PDEs with finite elements. This approach caters to the contradictory needs of flexibility in the discretization and high performance: The unstructured coarse mesh retains flexibility in resolving geometric and simulation details, such as boundary layers or discontinuities. The tensor product property of the local meshes entails a linewise numbering of the unknowns which leads to a banded structure of the matrices and is exploited in optimized numerical linear algebra *and* multigrid smoothing and transfer components. Instead of keeping all data in one general, homogeneous data structure, FEAST stores only local FE matrices and vectors (corresponding to subdomains) and thus maintains a clear separation of structured and unstructured parts of the domain. Several subdomains can be grouped together and treated within one MPI process.

6.2.2 Parallel Multigrid Solvers

SCARC (*Scalable Recursive Clustering*), the solver concept at the core of FEAST, generalizes techniques from *multilevel domain decomposition* and *parallel multigrid*, combining their respective advantages into a very robust, and (numerically *and* computationally) efficient parallel solution scheme for (scalar) elliptic PDEs. Matrices and vectors are stored only locally as usual for distributed memory approaches, while at the same time a minimally overlapping decomposition ensures that on the one hand, the union of all local matrices always composes the 'virtual' global matrix, and on the other hand, only the minimally necessary amount of data needs to be shared via communication, i. e., only data associated with degrees of freedom lying on subdomain boundaries have to be exchanged, ensuring good scalability by design. Between the different mesh resolutions, the coupling is multiplicative, as in classical multigrid. Within one hierarchy level, the coupling is additive, i.e., the minimally overlapping subdomains are treated simultaneously and independently of each other. Global coarse grid problems are solved with UMFPACK [6] on a master node.

Instead of blockwise application of elementary local smoothers, the SCARC scheme employs full multigrid solvers acting locally on the individual subdomains to 'hide' local irregularities as much as possible from the outer solver; see the recursive data flow in Figure 6.2. The local solvers are typically configured to gain, e. g., one digit, and can fully exploit the underlying tensor product property by executing hardware-optimized code paths.

The resulting hierarchical solvers are very robust, exhibiting very good weak and strong scaling. In previously published work, we demonstrated—for the maximum available resources at that time—perfect weak scalability for the Poisson problem on up to 320 Xeon processors [10], and excellent strong scaling for applications from linearized elasticity and incompressible flow for an experiment that subsequently quadrupled the resources up to a maximum of 128 CPUs [19].

6.2.3   Scalar and Multivariate Problems

The guiding idea to treating multivariate problems with FEAST is to rely on the modular, highly optimized and extensively tested core routines for the scalar case in order to formulate robust schemes for a wide range of applications, rather than use the best suited numerical scheme for each application and repeatedly optimizing it for new architectures. Multivariate PDEs as they arise in CSM and CFD can be rearranged and discretized in such a way that the resulting equation systems consist of blocks that correspond to scalar subequations. We illustrate this exemplarily in Section 6.3.3 with help of the elasticity equation. Figure 6.2 summarizes the idea on a high level.

This special block-structure can be exploited in two ways: On the one hand, all standard linear algebra operations on the multivariate system (e. g., matrix-vector multiplications, defect computations, dot products) can be implemented as a series of operations for scalar systems, taking advantage of FEAST's highly tuned numerical linear algebra components. On the other hand, the process of solving multivariate linear equation systems can be brought down to the treatment of auxiliary scalar subsystems which can be efficiently solved by FEAST's optimized toolbox of parallel multigrid solvers. Once such a solution approach is set up, the hardware-awareness and all further improvements of FEAST's scalar library directly transfer to the solvers for multivariate systems. Such improvements can be the addition of better multigrid components (e. g., more robust local smoothers) or algorithmic adaptations to dedicated HPC architectures and hardware co-processors (see Section 6.2.4 and Figure 6.2). SCARC allows for an almost arbitrary (recursive) combination of multivariate and scalar as well as global and local solvers, thus providing great flexibility in tuning the linear solution schemes to the given problem.

6.2.4 Co-Processor Acceleration

Hardware accelerators such as GPUs are integrated into FEAST in a 'minimally invasive' way [10, 11, 13], encapsulating heterogeneities of the system on the compute node level, so that MPI sees a globally homogeneous system. Figure 6.2 illustrates how this concept fits into FEAST's general solution strategy.

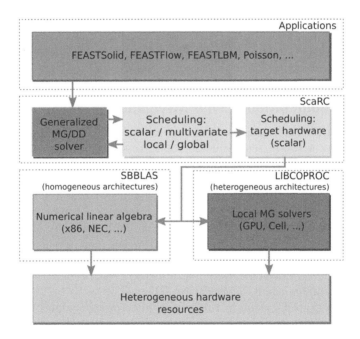

FIGURE 6.2: Illustration of the minimally invasive accelerator integration in FEAST.

The currently implemented prototype [4] offloads local scalar multigrid solvers, specifically tailored to the tensor product property, onto GPUs. This concentrates sufficient fine-grained parallelism in a separate task and thus minimizes the potential overhead of repeated co-processor configuration and data transfer in case of a system integration via relatively narrow busses such as PCIe. On accelerators that do not natively provide sufficient floating-point precision, a mixed precision iterative refinement technique is applied to ensure accuracy of the results [6]. The entire approach thus has one important benefit:

Application code does not need to be changed at all to benefit from co-processor acceleration.

6.3   Two FEAST Applications: FEASTSOLID and FEASTFLOW

In this section, we introduce two important classes of applications that have been built on top of FEAST: The solid mechanics code FEASTSOLID solves static and transient elasticity problems for small and finite deformations. The fluid dynamics code FEASTFLOW solves the transient incompressible Stokes and Navier-Stokes equations. Other applications, e. g., fluid structure interaction and Lattice Boltzmann methods, are actively being developed, but are not considered here.

6.3.1   Computational Solid Mechanics

In CSM the deformation of solid bodies under external loads is examined. We consider a two-dimensional (20) body covering a domain $\bar{\Omega} = \Omega \cup \partial\Omega$, where $\Omega \subset \mathbb{R}^d, d = 2, 3$, is a bounded, open set with boundary $\Gamma = \partial\Omega$. The boundary is split into two parts: the Dirichlet part Γ_D where displacements are prescribed and the Neumann part Γ_N where surface forces can be applied ($\Gamma_D \cap \Gamma_N = \emptyset$). Furthermore the body can be exposed to volumetric forces, e. g., gravity. FEASTSOLID is able to handle nonlinear finite elasticity problems and incompressible materials [20]. However, here we do not use these advanced problem-specific features and only treat the simple yet fundamental linearized 2D model problem of elastic, compressible material under static loading, assuming small deformations. This allows us to better quantify the solver aspects we focus on in this paper. In terms of complexity, this model problem of elasticity is situated between the standard Poisson problem and the Navier-Stokes equations.

Mathematically, we formulate the linearized elasticity equation in terms of the displacements $\boldsymbol{u}(\boldsymbol{x}) = \big(u_1(\boldsymbol{x}), u_2(\boldsymbol{x})\big)^{\mathsf{T}}$ of a material point $\boldsymbol{x} \in \bar{\Omega}$ as the only unknowns. The strains can be defined by the linearized strain tensor $\varepsilon_{ij} = \frac{1}{2}\Big(\frac{\partial u_i}{\partial x_j} + \frac{\partial u_j}{\partial x_i}\Big), i, j = 1, 2$, describing the linearized kinematic relation between displacements and strains. The material properties are reflected by the constitutive law, which determines a relation between the strains and the stresses. We use Hooke's law for isotropic elastic material, $\boldsymbol{\sigma} = 2\mu\boldsymbol{\varepsilon} + \lambda \operatorname{tr}(\boldsymbol{\varepsilon})\boldsymbol{I}$, where $\boldsymbol{\sigma}$ denotes the symmetric stress tensor and μ and λ are the so-called Lamé constants. The basic physical equations of elasticity are determined by equilibrium conditions. For a body in equilibrium, the inner forces (stresses) and the outer forces (external loads f) are balanced:

$$-\mathbf{div}\sigma = f, \qquad x \in \Omega.$$

Using Hooke's law to replace the stress tensor, the problem of linearized elasticity can be expressed in terms of the following elliptic boundary value problem, called the Lamé equation:

$$-2\mu \, \mathbf{div}\varepsilon(u) - \lambda \, \mathbf{grad} \, \mathrm{div} \, u = f, \qquad x \in \Omega, \tag{6.1a}$$

$$u = g, \qquad x \in \Gamma_D, \tag{6.1b}$$

$$\sigma(u) \cdot n = t, \qquad x \in \Gamma_N. \tag{6.1c}$$

Here, g is prescribed displacements on Γ_D, and t is given surface forces on Γ_N with outer normal n. To discretize the continuous problem, the domain $\bar{\Omega}$ is approximated by a collection of tensor product subdomains $\bar{\Omega}_i$ as described in Section 6.2.1. We consider the weak formulation of Equation (6.3) and apply a finite element discretization with conforming bilinear elements Q_1. For details on the finite element technique and on the elasticity problem, see for example the textbook by Braess [3].

6.3.2 Computational Fluid Dynamics

To tackle problems from CFD we discretize the Navier-Stokes equations. They describe the flow of Newtonian fluids like gases, water and many other liquids in a domain $\Omega \subset \mathbb{R}^d, d = 2, 3$ and are, under certain assumptions and simplifications, derived from the conservation laws for mass, momentum and energy. For the sake of simplicity we restrict ourselves to the 2D stationary case.

Confining the domain and imposing boundary conditions, i.e., in- and outflow conditions on the 'artificial' boundaries and slip or adhesion conditions at rigid walls, the following system of nonlinear equations is obtained under the assumption of constant kinematic viscosity $\nu > 0$ (independent of pressure and specific heat capacity) and constant temperature:

$$-\nu\Delta u + (u \cdot \nabla)u + \nabla p = f, \qquad x \in \Omega, \tag{6.2a}$$

$$\mathrm{div} \, u = \mathbf{0}, \qquad x \in \Omega, \tag{6.2b}$$

$$u = g, \qquad x \in \Gamma_D, \tag{6.2c}$$

$$\nu\partial_n u + p \cdot n = 0, \qquad x \in \Gamma_N. \tag{6.2d}$$

Here, u denotes the fluid velocity, p the pressure, n the outer normal vector and Γ_D and Γ_N the boundary parts with, respectively, Dirichlet and Neumann boundary conditions (i.e., inflow, outflow and adhesion conditions).

We discretize the equation with the Q_1/Q_1 bilinear element pair and use residual-based SUPG/PSPG stabilization (streamline upwind/Petrov-Galerkin and pressure-stabilization/Petrov-Galerkin) to account for the LBB deficiency of the Q_1/Q_1 pair and to stabilize convective terms [4,14]. To allow for anisotropic grid cells directional derivatives are incorporated in the stabilization terms [5]. For details on the CFD problem in general, see for example the textbook by Ferziger and Perić [7].

6.3.3   Solving CSM and CFD Problems with FEAST

In order to solve CSM and CFD problems using the FEAST intrinsics described in Section 6.2, the degrees of freedom of the discretized systems have to be ordered corresponding to the components of the underlying multivariate equations. We exemplarily illustrate this for the case of linearized elasticity, where the technique is sometimes called separate displacement ordering [1]. In the 2D case, the unknowns $\boldsymbol{u} = (u_1, u_2)^\mathsf{T}$ are the displacements in x- and y-direction. A corresponding operator-splitting of the left hand side of Equation (6.1a) yields

$$-\begin{pmatrix} (2\mu + \lambda)\partial_{xx} + \mu\partial_{yy} & (\mu + \lambda)\partial_{xy} \\ (\mu + \lambda)\partial_{yx} & \mu\partial_{xx} + (2\mu + \lambda)\partial_{yy} \end{pmatrix} \begin{pmatrix} u_1 \\ u_2 \end{pmatrix} = \begin{pmatrix} f_1 \\ f_2 \end{pmatrix}. \qquad (6.3)$$

The finite element discretization of the correspondingly arranged weak form leads to a linear equation system $\mathbf{Ku} = \mathbf{f}$ exhibiting a corresponding block structure,

$$\begin{pmatrix} \mathbf{K}_{11} & \mathbf{K}_{12} \\ \mathbf{K}_{21} & \mathbf{K}_{22} \end{pmatrix} \begin{pmatrix} \mathbf{u}_1 \\ \mathbf{u}_2 \end{pmatrix} = \begin{pmatrix} \mathbf{f}_1 \\ \mathbf{f}_2 \end{pmatrix}, \qquad (6.4)$$

where $\mathbf{f} = (\mathbf{f}_1, \mathbf{f}_2)^\mathsf{T}$ is the vector of external loads and $\mathbf{u} = (\mathbf{u}_1, \mathbf{u}_2)^\mathsf{T}$ the (unknown) coefficient vector of the finite element solution. Each global matrix/vector $\mathbf{K}_{ij}, \mathbf{u}_i, \mathbf{f}_i, (i,j = 1, 2)$, exists only 'virtually' through the corresponding local matrices/vectors defined over the local subdomains (cf. Section 6.2.1). The decisive advantage of this operator-splitting is that the matrices $\mathbf{K}_{11}$ and $\mathbf{K}_{22}$ correspond to scalar elliptic operators (cf. Equation (6.3)) which allows using FEAST's tuned scalar solvers (cf. Section 6.2.2).

We now illustrate how scalar subsystems can be utilized to solve the whole multivariate system by means of a basic preconditioned defect correction method:

$$\mathbf{u}^{k+1} = \mathbf{u}^k + \omega\tilde{\mathbf{K}}_{\mathrm{BGS}}^{-1}(\mathbf{f} - \mathbf{Ku}^k). \qquad (6.5)$$

This scheme acts on the global system (6.4) and thus couples the two sets of unknowns $\mathbf{u}_1$ and $\mathbf{u}_2$. $\tilde{\mathbf{K}}_{\mathrm{BGS}}$ is a block-Gauss-Seidel preconditioner that explicitly exploits the block structure of the matrix $\mathbf{K}$. One iteration of the global defect correction scheme consists of the following three steps:

1. Compute the global defect (cf. Section 6.2.3):

$$\begin{pmatrix} \mathbf{d}_1 \\ \mathbf{d}_2 \end{pmatrix} = \begin{pmatrix} \mathbf{f}_1 \\ \mathbf{f}_2 \end{pmatrix} - \begin{pmatrix} \mathbf{K}_{11} & \mathbf{K}_{12} \\ \mathbf{K}_{21} & \mathbf{K}_{22} \end{pmatrix} \begin{pmatrix} \mathbf{u}_1^k \\ \mathbf{u}_2^k \end{pmatrix} - \begin{pmatrix} \mathbf{f}_1 \\ \mathbf{f}_2 \end{pmatrix}.$$

2. Apply the block preconditioner

$$\tilde{\mathbf{K}}_{\text{BGS}} := \begin{pmatrix} \mathbf{K}_{11} & \mathbf{0} \\ \mathbf{K}_{21} & \mathbf{K}_{22} \end{pmatrix}$$

by approximately solving the system $\tilde{\mathbf{K}}_{\text{BGS}}\mathbf{c} = \mathbf{d}$. This is performed by two scalar solves and one (scalar) matrix-vector multiplication:

(a) Solve $\mathbf{K}_{11}\mathbf{c}_1 = \mathbf{d}_1$.

(b) Update RHS: $\mathbf{d}_2 = \mathbf{d}_2 - \mathbf{K}_{21}\mathbf{c}_1$.

(c) Solve $\mathbf{K}_{22}\mathbf{c}_2 = \mathbf{d}_2$.

3. Update the global solution with the (eventually damped) correction vector: $\mathbf{u}^{k+1} = \mathbf{u}^k + \omega\mathbf{c}$.

The solution of the two scalar subsystems (steps 2(a) and 2(b)) constitutes the largest amount of the total arithmetic work and fully exploits FEAST's tuned (and co-processor-accelerated) scalar solvers (cf. Section 6.4.3).

In the case of the Navier-Stokes equations, the whole solution process can be brought down to the solution of scalar systems as well. In a first step, the nonlinearities are resolved by a fixed point defect correction method such that repeatedly (in each nonlinear step) linear saddle-point systems of the form

$$\begin{pmatrix} \mathbf{A} + \mathbf{C}_1 & \mathbf{B} \\ \mathbf{B}^\mathsf{T} & \mathbf{C}_2 \end{pmatrix} \begin{pmatrix} \mathbf{u} \\ \mathbf{p} \end{pmatrix} = \begin{pmatrix} \mathbf{f} \\ \mathbf{g} \end{pmatrix} \tag{6.6}$$

are to be solved, where the matrices $\mathbf{C}_1, \mathbf{C}_2$ stem from the SUPG/PSPG stabilization terms. In 2D, the matrix $\mathbf{A} + \mathbf{C}_1$ has a 2×2 block structure similar to that of the stiffness matrix $\mathbf{K}$ in the elasticity case (see equation (6.4)). Our solution algorithm is a block Schur complement (BSC) approach as described by Murphy et al. [16]. It basically consists of a global BiCGStab solver acting on the whole saddle-point system which is block-preconditioned by

$$\begin{pmatrix} \tilde{\mathbf{A}} & \mathbf{0} \\ \mathbf{B}^\mathsf{T} & \tilde{\mathbf{S}} \end{pmatrix}. \tag{6.7}$$

Herein, $\tilde{\mathbf{A}}$ and $\tilde{\mathbf{S}}$ respectively denote preconditioners of $\mathbf{A} + \mathbf{C}_1$ and of the Schur complement matrix $\mathbf{S} = \mathbf{B}^\mathsf{T}(\mathbf{A} + \mathbf{C}_1)^{-1}\mathbf{B} - \mathbf{C}_2$. The former is realized by approximately solving subsystems of the kind $(\mathbf{A} + \mathbf{C}_1)\mathbf{c} = \mathbf{d}$ for which exactly the same solution strategy is applied as for the CSM system (6.4). The Schur complement preconditioner $\tilde{\mathbf{S}}$ is realized by a suitably weighted linear combination of pressure mass and Laplacian matrices (see Turek [17]). Applying the preconditioner means solving scalar subsystems which can again be done with FEAST's optimized SCARC schemes. Hence, the solution of the whole saddle-point system is mainly brought down to the solution of scalar systems again.

6.4   Performance Assessments

We demonstrate the performance and efficiency of our approach with a number of selected benchmark tests. These results have been gathered over the past two years on a wide range of different HPC installations and test clusters using CPUs and GPUs from different hardware generations.

6.4.1   GPU-Based Multigrid on a Single Subdomain

In the literature, reported speedups obtained on GPUs vary dramatically. Our first experiment assesses the (realistic) speedup we can obtain in our setting on a single, generalized tensor product subdomain. We use a GPU-based multigrid iteration to solve the fundamental scalar Poisson problem on one of the deformed subdomains near the inner boundary in the unstructured channel flow configuration shown in Figure 6.4 (bottom), prescribing an analytical right hand side so that we know the exact solution of the problem and can compute the error of the approximate solution. Single precision is insufficient to accurately solve this problem while we always achieve correct results with our mixed precision iterative refinement scheme [6].

TABLE 6.1: Multigrid solvers on a single subdomain (Time to solution in seconds and speedup).

Level	DOF	CPU double	GPU double	speedup	GPU mixed	speedup
5	1 089	0.0018	0.0156	0.1	0.0140	0.1
6	4 225	0.0059	0.0187	0.3	0.0206	0.3
7	16 641	0.0272	0.0260	1.1	0.0232	1.2
8	66 049	0.1460	0.0356	4.1	0.0284	5.2
9	263 169	0.7747	0.0656	11.8	0.0435	17.8
10	1 050 625	3.3609	0.1731	19.4	0.0944	35.6

The speedup comparisons in Table 6.1 have been obtained on a typical high-end GPU workstation as of June 2008 (Core2Duo E6750, fast DDR2-800 memory, NVIDIA GeForce GTX 280 GPU). Column 3 demonstrates that the CPU gets less and less efficient as soon as the problem does not fit entirely into cache anymore, while columns 4 and 6 show that the GPU needs reasonably large problem sizes to be fully saturated and hide all latencies of accesses to off-chip memory. The speedup we observe for these two configurations executing entirely in double precision is almost $20\times$ and includes the transfer of the right hand side to the device and the transfer of the solution back from the device. We do not include the transfer of the matrix data, because this is part of the matrix assembly and thus separated from our acceleration of linear solvers; see

Section 6.4.3. In addition, the mixed precision scheme on the device is almost twice as fast as native double precision (last two columns). It is noteworthy that we can solve a sparse linear system with 1 million unknowns in less than 0.1 seconds, using a fully assembled matrix, not merely a stencil.

6.4.2 Scalability

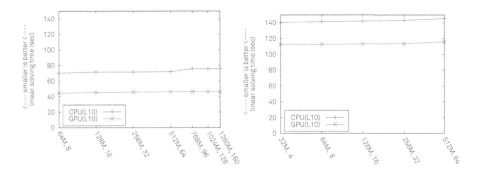

FIGURE 6.3: Weak scalability results for the Poisson (left) and the CSM (right) solvers (x-axis: #DOF and #nodes; y-axis: linear solving time in seconds). When performing the CSM tests, fewer nodes were available to us.

Our second experiment demonstrates the excellent weak scalability of our approach (we address strong scalability elsewhere [19]). The results shown in Figure 6.3 have been obtained on a cluster with two singlecore EM64T Xeon CPUs and a Quadro FX1400 GPU per node. We observe excellent scalability when simultaneously doubling the problem size and the number of nodes for both the Poisson problem and the FEASTSOLID application (using the same configuration as shown in Figure 6.4, top). As the GPU is one generation behind the Xeon processors and only has 128 MB of memory, it can barely hold the data associated with one subdomain, and we have to page data in and out of device memory for each local solve [10,13]. Therefore, the obtained speedups are not spectacular but still noteworthy. In particular, we are able to accurately solve a Poisson problem with more than 1.3 billion unknowns in slightly more than 40 seconds.

6.4.3 Application Speedup

In this section, we analyze how the (local) speedups of the solvers acting on a single subdomain (Section 6.4.1) translate to the application level. For the elasticity application FEASTSOLID, we solve a prototypical benchmark problem, a block subjected to an external load (see Figure 6.4, top). These results are obtained on a 16-node cluster (Opteron 2210 dualcore CPU, Quadro FX5600 GPU). Each of the 64 subdomains is refined up to level 10 for a total

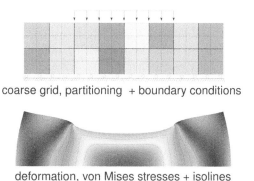

coarse grid, partitioning + boundary conditions

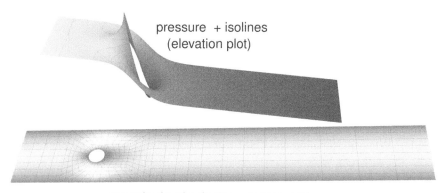

deformation, von Mises stresses + isolines

pressure + isolines
(elevation plot)

magnitude of velocity + coarse grid

FIGURE 6.4: Computed benchmark results for the elasticity (top) and fluid dynamics (bottom) solvers.

problem size of 128 million unknowns. Smoothing the local multigrid with simple Jacobi iterations is sufficient, the mesh is isotropic and the operator only mildly anisotropic. The GPU-accelerated solver achieves a speedup by a factor of 2.6 for refinement level 10, and 2.0 for level 9, respectively [13].

We benchmark the GPU-accelerated FEASTFLOW application on a small test cluster (four nodes, Opteron 2214 dualcore, GeForce 8800 GTX). These GPUs are slightly faster than the Quadros used in the tests of the elasticity application, but still two generations behind the GPU used in Section 6.4.1. For the full Navier-Stokes solver and a 'flow around a cylinder' benchmark (see Figure 6.4, bottom), a strong smoothing operator is required due to the nonlinearities. We resort to an alternating direction implicit variant of a tridiagonal smoother that couples each unknown with its left and right neighbor. The GPU implementation [4] of this smoother is based on cyclic reduction, while the CPU implementation can use the less expensive serial Thomas algorithm. For this configuration, only refinement level 9 fits into memory of the four nodes. The total speedup we observe in terms of 'time to solution' is 1.9× (results updated from [9]).

Discussion. We first emphasize that our reported speedups are very noteworthy, as they have been obtained without changing a single line of application code: Halving the execution time of an already carefully (numerically and implementationally) optimized code is usually impossible. Nonetheless, the question arises why the gap to the 'ideal' measurements (Section 6.4.1) is so large. The reason is inherent to our highly modular approach: We only accelerate the local scalar multigrid solvers, i.e., only a portion of the entire linear solver. The approach is thus prone to be limited by Amdahl's law. To quantify this effect and to assess how the achieved speedups of the local multigrid solver translate to the global application, we equip the code with timers measuring the local multigrid solves only. Looking at the elasticity solver first, we observe local speedups of 4.1 and 9.0 on level 9 and 10, respectively. The local work constitutes roughly 66 % of the entire linear solver, so the maximum achievable total speedup is 3× assuming infinite local acceleration. With a factor of 2.6× on level 10, we are reasonably close to this ideal speedup. For the Navier-Stokes solver, we measure a local speedup of 11.4×, resulting in an acceleration of the linear solver by a factor of 4.3. The accelerable percentage of the linear solver is approximately 84 %, limiting the speedup to a factor of 6.4. However, considering the entire nonlinear problem, all matrices have to be assembled in each nonlinear step (see Section 6.3.3), a task that is currently not accelerated by the GPU. Furthermore, the nonlinear defect correction loop is executed on the CPU. For this particular test problem, 40 % of the entire time to solution is spent in the assembly routine, and for the accelerated solver, this percentage increases to over 70 %. Consequently, the speedups are diminishing significantly, and measurements reveal that only 50 % of the entire solution process (baseline CPU version) can be accelerated, resulting in a theoretical upper bound of a factor of two. With a measured speedup of 1.9× for the entire application, we are sufficiently close to this optimum.

Conclusion and Outlook. Even when using a strong, robust multigrid smoother, our GPU-based local multigrid solver is more than one order of magnitude faster than a corresponding CPU implementation. Since we only accelerate the scalar local portions of the entire multivariate global linear solver, the total speedup on the application level is bound by the fraction of time spent in these local solves. The only possibility to improve speedups significantly while not suffering from the implications of Amdahl's law is to re-implement large portions of each application specifically for the GPU, which can be impractical or prohibitively expensive in practice. Nonetheless, we achieve speedups by a factor of two or more for simulations in CSM and CFD. In the spirit of 'hardware-oriented numerics' however, we are pursuing ideas to design solution schemes with a higher potential for acceleration, ideally more than 90 %. The linear solver in the Navier-Stokes benchmark is already close to this goal, exhibiting an acceleration potential of 84 %. For nonlinear problems however, the assembly of the linearized systems can constitute the dominant part of the computations, at least for the GPU-accelerated version. The second challenge in future work is thus to port the assembly process to GPUs in a similarly 'minimally invasive' way, so that we maintain the most important benefit of our approach: Application code does not have to be changed at all.

6.5   Summary

We have motivated the concept of *hardware-oriented numerics* and described our finite element and solver toolbox FEAST, which prototypically implements many of its aspects. For two important model applications from solid mechanics and fluid dynamics we have demonstrated how we break down the solution process into sequences of scalar solves. This approach has the important advantage that all performance improvements and in particular adaptions to specific hardware architectures are automatically transferred to the application level, without having to change application code at all. In the context of this chapter, we use GPUs as accelerators to the general-purpose CPU. We implemented scalar multigrid solvers on the GPU, which executes the scalar local subdomain solves within the multivariate global parallel solver. The resulting hybrid solvers scale very well, and we observe noteworthy speedups. However, as we only accelerate portions of the entire solution scheme, our speedups are limited by the remaining fraction, and local speedups of more than one order of magnitude do not (yet) translate fully to the application level.

Acknowledgments

This work has been funded in part by German Deutsche Forschungsgemeinschaft (projects TU 102/22-2, TU 102/27-1 and TU 102/11-3), and by German Bundesministerium für Bildung und Forschung in the SKALB project (grant 01IH08003D) of call 'HPC Software für skalierbare Parallelrechner'.

Bibliography

[1] Owe Axelsson. On iterative solvers in structural mechanics; separate displacement orderings and mixed variable methods. *Mathematics and Computers in Simulations*, 50(1–4):11–30, November 1999.

[2] Christian Becker. *Strategien und Methoden zur Ausnutzung der High-Performance-Computing-Ressourcen moderner Rechnerarchitekturen für Finite Element Simulationen und ihre Realisierung in FEAST (Finite Element Analysis & Solution Tools)*. PhD thesis, Universität Dortmund, May 2007. http://www.logos-verlag.de/cgi-bin/buch?isbn=1637.

[3] Dietrich Braess. *Finite Elements—Theory, Fast solvers and Applications in Solid Mechanics*. Cambridge University Press, 2nd edition, April 2001.

[4] Alexander N. Brooks and Thomas J. R. Hughes. Streamline upwind/Petrov-Galerkin formulations for convection dominated flows with particular emphasis on the incompressible Navier-Stokes equations. *Computer Methods in Applied Mechanics and Engineering*, 32(1–3):199–259, September 1982.

[5] Sven H.M. Buijssen. *Efficient Multilevel Solvers and High Performance Computing Techniques for the Finite Element Simulation of the Transient, Incompressible Navier–Stokes Equations*. PhD thesis, TU Dortmund, Fakultät für Mathematik, 2010.

[6] Timothy A. Davis. A column pre-ordering strategy for the unsymmetric-pattern multifrontal method. *ACM Transactions on Mathematical Software*, 30(2):165–195, June 2004.

[7] Joel H. Ferziger and Milovan Perić. *Computational Methods for Fluid Dynamics*. Springer, Berlin, 3rd edition, December 2001.

[8] Dominik Göddeke. *Fast and Accurate Finite-Element Multigrid Solvers for PDE Simulations on GPU Clusters*. PhD thesis, TU Dortmund, Fakultät für Mathematik, May 2010.

[9] Dominik Göddeke, Sven H.M. Buijssen, Hilmar Wobker, and Stefan Turek. GPU acceleration of an unmodified parallel finite element Navier-Stokes solver. In Waleed W. Smari and John P. McIntire, editors, *High Performance Computing & Simulation 2009*, pages 12–21, June 2009.

[10] Dominik Göddeke, Robert Strzodka, Jamaludin Mohd-Yusof, Patrick S. McCormick, Sven H.M. Buijssen, Matthias Grajewski, and Stefan Turek. Exploring weak scalability for FEM calculations on a GPU-enhanced cluster. *Parallel Computing*, 33(10–11):685–699, September 2007.

[11] Dominik Göddeke, Robert Strzodka, Jamaludin Mohd-Yusof, Patrick S. McCormick, Hilmar Wobker, Christian Becker, and Stefan Turek. Using GPUs to improve multigrid solver performance on a cluster. *International Journal of Computational Science and Engineering*, 4(1):36–55, November 2008.

[12] Dominik Göddeke, Robert Strzodka, and Stefan Turek. Performance and accuracy of hardware-oriented native-, emulated- and mixed-precision

solvers in FEM simulations. *International Journal of Parallel, Emergent and Distributed Systems*, 22(4):221–256, January 2007.

[13] Dominik Göddeke, Hilmar Wobker, Robert Strzodka, Jamaludin Mohd-Yusof, Patrick S. McCormick, and Stefan Turek. Co-processor acceleration of an unmodified parallel solid mechanics code with FEASTGPU. *International Journal of Computational Science and Engineering*, 4(4):254–269, October 2009.

[14] Thomas J.R. Hughes, Leopoldo P. Franca, and Marc Balestra. A new finite element formulation for computational fluid dynamics: V. Circumventing the Babuška-Brezzi condition. A stable Petrov-Galerkin formulation of the Stokes problem accomodating equal-order interpolations. *Computer Methods in Applied Mechanics and Engineering*, 59(1):85–99, November 1986.

[15] Susanne Kilian. *ScaRC: Ein verallgemeinertes Gebietszerlegungs-/Mehrgitterkonzept auf Parallelrechnern.* PhD thesis, Universität Dortmund, Fachbereich Mathematik, January 2001. http://www.logos-verlag.de/cgi-bin/buch?isbn=0092.

[16] Malcolm F. Murphy, Gene H. Golub, and Andrew J. Wathen. A note on preconditioning for indefinite linear systems. *SIAM Journal on Scientific Computing*, 21(6):1969–1972, May 2000.

[17] Stefan Turek. *Efficient Solvers for Incompressible Flow Problems: An Algorithmic and Computational Approach.* Springer, June 1999.

[18] Stefan Turek, Christian Becker, and Susanne Kilian. Hardware–oriented numerics and concepts for PDE software. *Future Generation Computer Systems*, 22(1–2):217–238, February 2004.

[19] Stefan Turek, Dominik Göddeke, Christian Becker, Sven H.M. Buijssen, and Hilmar Wobker. FEAST—Realisation of hardware-oriented numerics for HPC simulations with finite elements. *Concurrency and Computation: Practice and Experience*, February 2010. Special Issue Proceedings of ISC 2008, in press.

[20] Hilmar Wobker. *Efficient Multilevel Solvers and High Performance Computing Techniques for the Finite Element Simulation of Large-Scale Elasticity Problems.* PhD thesis, TU Dortmund, Fakultät für Mathematik, March 2010.

Chapter 7

Mixed-Precision GPU-Multigrid Solvers with Strong Smoothers

Dominik Göddeke

Institut für Angewandte Mathematik, TU Dortmund, Germany

Robert Strzodka

Max Planck Institut Informatik, Saarbrücken, Germany

7.1 Introduction

In this chapter, we present efficient fine-grained parallelization techniques for robust multigrid solvers, in particular for numerically strong, inherently sequential smoothing operators. We apply them to sparse ill-conditioned linear systems of equations that arise from grid-based discretization techniques like finite differences, volumes and elements. Our exemplary results demonstrate both the numerical and runtime performance of these techniques, as well as significant speedups over conventional CPUs.

We implement the parallelization techniques on graphics processors as representatives of throughput-oriented wide-SIMD many-core architectures:

GPUs offer a tremendous amount of fine-grained parallelism compared to commodity CPU designs, with up to 30 'cores' and more than 30,000 threads in flight simultaneously on current devices [3]. Our implementation uses NVIDIA CUDA, but the techniques we present are generally applicable to many-core architectures, e.g., using OpenCL [9], an open industry standard targeting diverse multi- and many-core architectures. We refer to the CUDA documentation [11] for an in-depth explanation of the terminology: 'memory coalescing' (block memory transfers), 'warps' and 'half-warps' (SIMD granularity for computation and memory access), shared memory (small on-chip scratchpad memory), thread blocks (groups of threads with on-chip data exchange and synchronization).

7.1.1 Numerical Solution of Partial Differential Equations

Many important real-world phenomena are modeled with partial differential equations (PDEs), and their fast and accurate solution is of very high relevance in practice. Examples include fluid dynamics (Navier-Stokes equations), elasticity (Lamé equations) and potential calculations in electrostatics (generalized Poisson equation), and many more. We restrict ourselves to the generalized Poisson problem as the fundamental model problem at the core of all these application domains.

For the solution on computers, the model is discretized. In case of finite elements, the computational domain is covered with a mesh of elements, and the solution is sought in a discrete function space over those elements. The adaptable parameters in this function space are called degrees of freedom. We generate finer and finer meshes by conforming (but potentially anisotropic) refinement, which yields a natural hierarchy of nested grids with more and more degrees of freedom. After discretization, assembly of the system matrix and eventual linearization, huge *sparse* linear systems of equations have to be solved. In the following, we restrict ourselves to bilinear conforming elements and a mesh of quadrilaterals; however, other finite element/volume/difference discretizations and computations in more than two space dimensions lead to similar sparse systems where our parallel techniques would also apply.

Geometric multigrid solvers are among the most efficient methods for the solution of such sparse systems, for two reasons: They are asymptotically optimal with linear complexity in the number of unknowns; and they converge independently of the mesh width, which is particularly important since the condition number of the problems under consideration increases proportionally to the level of refinement.

7.1.2 Hardware-Oriented Discretization of Large Domains

Throughout this chapter, we assume that the PDE is discretized on a generalized tensor product domain. Here we briefly sketch how structured subdomains can be used in the solution of large PDE problems.

Figure 7.1 (left) depicts a favorable discretization approach for large domains. We start with an unstructured coarse mesh which offers sufficient flexibility in the discretization of PDEs on complex domains. Each coarse element is independently refined in a tensor product fashion, which allows implicit neighbor localization, for example with a linewise numbering of the unknowns. This does not limit us to cartesian meshes; anisotropic refinement and r-adaptivity through mesh deformation (Figure 7.1 (right)) are possible. Each of these refined *local* meshes represents a subdomain in a large solver package for CPU/GPU-clusters called FEAST, which is described in Chapter 6.

The global mesh comprising the subdomains is block-structured. The zoomed view in Figure 7.1 illustrates a very important benefit of this 'locally structured globally unstructured' approach: The matrices corresponding to the subdomains consist of multiple bands. In case of bilinear conforming finite elements there are nine bands realizing the coupling of each unknown with itself and its eight surrounding neighbors. The bands are stored as separate arrays with an appropriate zero-padding. In our focus on fine-grained parallelization techniques, we describe in the following the solution of a generalized Poisson problem on a single subdomain, i. e., we discuss the most time consuming building block in this larger context. FEAST takes care of the data exchanges between the subdomains and the outer solver scheme.

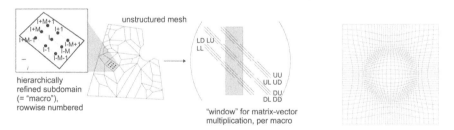

FIGURE 7.1: From left to right: Tensor product property of one subdomain, unstructured coarse mesh, system matrix structure and strongly deformed subdomain which still exhibits the tensor product property. Buijssen et al., 2009. *FEASTSolid and FEASTFlow: FEM Applications Exploiting FEAST's HPC Technologies, Springer.* 15 pp. With permission.

7.1.3 Mixed-Precision Iterative Refinement Multigrid

A typical subdomain contains one million unknowns. So even on a simple cartesian mesh, solving the linear systems in single-precision floating-point arithmetic does not give accurate results due to the high condition number [6]. We therefore apply a mixed-precision iterative refinement technique, which provably converges to double-precision accuracy but performs the majority of the work in fast-single precision. An outer double-precision iteration computes the residual in each step and uses it as the right hand side for an auxiliary problem. This system is approximately solved to one digit accuracy using single-precision multigrid, and the solution is added in double-precision to the global iterate. In general, mixed-precision schemes are very favorable in terms of performance, both for compute-intensive applications [10] (single precision is between two and eight times faster on current and upcoming CPUs and GPUs) and for memory-intensive calculations like the ones we consider in this chapter: The effective halving of bandwidth requirements results in an asymptotical speedup of a factor of two. Twice the number of values can be stored in the same (register, cache, main) memory, and twice the number of values can be transferred between memory levels in the same amount of time.

7.2 Fine-Grained Parallelization of Multigrid Solvers

A multigrid solver acting on a hierarchy of refined meshes comprises the following building blocks: Residual calculations, norm calculations (for convergence control), scaled vector-vector additions, prolongation and restriction (transfer of residuals and corrections between different mesh resolutions), the coarse grid solver and *smoothing operators*. Most of these operations parallelize in a straightforward way [4]: Defect calculations with banded matrices best use one thread per element in the residual vector, i.e., per matrix row; and grid transfer operations can be formulated as fixed coefficient interpolations. We use a Krylov subspace scheme to solve the coarse grid problems, which essentially needs the same type of operations as the multigrid solver. In the following, we can thus concentrate on parallelization techniques for multigrid smoothers that are efficient both in terms of runtime and convergence rates.

7.2.1 Smoothers on the CPU

In multigrid solvers, smoothing is typically realized by few steps of a damped, preconditioned defect correction iteration,

$$\mathbf{x}^{k+1} = \mathbf{x}^k + \omega \mathbf{C}^{-1}(\mathbf{b} - \mathbf{A}\mathbf{x}^k),$$

where ω is the damping parameter and $\mathbf{C}$ denotes the preconditioner. We can therefore further simplify the presentation in this section and only consider the application of the preconditioner, which is realized by a formal inversion, $\mathbf{c} = \mathbf{C}^{-1}\mathbf{d}$, where $\mathbf{d} = \mathbf{b} - \mathbf{A}\mathbf{x}^k$ denotes the defect. In the following, we use the terms *smoother* and *preconditioner* synonymously.

Figure 7.2 depicts a 5×5 mesh ($M = 5$, $N = 25$) and the sparsity structure of the matrix. The first subdiagonal of the matrix realizes the coupling of each entry with its left neighbor (its predecessor in the numbering), the first

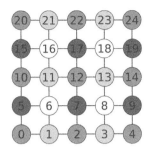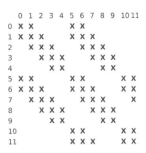

FIGURE 7.2: Numbering of the unknowns (left) and illustration of the sparsity structure of the system matrix (right, only the first half of the unknowns shown). See Section 7.2.3 for an explanation of the coloring scheme.

superdiagonal the coupling to the right, and analogously, the lower three bands and the upper three bands realize the coupling with the bottom and top values, respectively. It is important to note that all off-diagonals contain zeros at those entries corresponding to the beginning and end of each line in the mesh. This simple observation is crucial for the design of efficient implementations of the preconditioners.

Let us now define numerically more and more efficient smoothing operators that exploit the above structure, by starting with a formal decomposition of the system matrix into its nine bands (cf. Figure 7.1 for notation):

$$\mathbf{A} = (\mathbf{LL} + \mathbf{LC} + \mathbf{LU}) + (\mathbf{CL} + \mathbf{CC} + \mathbf{CU}) + (\mathbf{UL} + \mathbf{UC} + \mathbf{UU})$$

Adopting this notation, the *JACOBI* and *GSROW* preconditioners read:

$$\mathbf{C}^{\mathrm{JAC}} := \mathbf{CC} \qquad \mathbf{C}^{\mathrm{GSROW}} := (\mathbf{LL} + \mathbf{LC} + \mathbf{LU} + \mathbf{CL} + \mathbf{CC})$$

The suffix 'row' is used to indicate that the preconditioners correspond to a linewise numbering of the unknowns. The tridiagonal preconditioner requires the inverse of the center tridiagonal matrix:

$$\mathbf{C}^{\mathrm{TRIDIROW}} := (\mathbf{CL} + \mathbf{CC} + \mathbf{CU})$$

The *TRIGSROW* preconditioner combines the tridiagonal and Gauß-Seidel preconditioners:

$$\mathbf{C}^{\mathrm{TRIGSROW}} := (\mathbf{LL} + \mathbf{LC} + \mathbf{LU} + \mathbf{CL} + \mathbf{CC} + \mathbf{CU})$$

Due to the linewise numbering, the dependencies to the right within each line (**CU**) in *TRIDIROW* and *TRIGSROW* are implicit and the application of the preconditioner involves formally inverting a linear equation system. All other dependencies are explicit in case of sequential execution, but for a parallel execution we will need to decouple the dependencies further, e.g., for the parallel application of the *GSROW* preconditioner.

The same schemes can be formulated for a columnwise numbering, in which the dependencies along the columns are implicit, resulting in the elementary preconditioners *GSCOL*, *TRIDICOL* and *TRIGSCOL*. Alternate application of the row and column direction preconditioners leads to the class of *alternating direction implicit (ADI)* methods [12], and we denote the resulting preconditioners by *ADIGS*, *ADITRIDI* and *ADITRIGS*.

7.2.2   Exact Parallelization: Jacobi and Tridiagonal Solvers

The application of the *JACOBI* preconditioner corresponds to an elementwise scaling of the defect vector with the matrix diagonal. In this operation all vector entries can be updated independently of each other, so the parallelization is trivial.

The *TRIDIROW* preconditioner couples each degree of freedom with its immediate left and right neighbor, $\mathbf{C}^{\mathrm{TRIDIROW}}$ has a tridiagonal structure. The CPU implementation is a straightforward application of the *Thomas algorithm* [14], which in turn is Gaussian elimination in the tridiagonal matrix case. It comprises two phases, a forward sweep to simultaneously eliminate the lower subdiagonal and the main diagonal by distributing them to the superdiagonal and the right hand side, and a backward substitution to solve for the unknowns. Both sweeps are inherently sequential.

We make one simple yet important observation (cf. Figure 7.2): Each line in the mesh is completely independent of the other ones because there is no coupling between the lines in the preconditioner. In other words, *TRIDIROW* is a *linewise preconditioner*. This can also be seen in the matrix, which has zero entries at the corresponding positions in the first sub- and superdiagonal. So for an $M \times M$ domain we have M independent tridiagonal systems of M unknowns each. We exploit this independence in the parallelization by assigning each tridiagonal system to a thread block in CUDA. In a second parallelization stage we need to parallelize the execution inside each thread block which consists of parallel threads and shared on-chip memory for thread communication.

In the literature, three different parallel algorithms to solve such tridiagonal systems have been proposed since the 1960s: *cyclic reduction* [2,7], *parallel cyclic reduction* [8] and *recursive doubling* [13]. These algorithms have been targeted at vector architectures, and are thus good candidates for the implementation on GPUs, which can be interpreted as many-core processors where each processor has wide-SIMD characteristics [3]. Zhang et al. [15] analyze the GPU applicability of these approaches in detail. They conclude that cyclic reduction suffers from high-degree shared memory bank conflicts (see below) and poor thread utilization in lower stages of the solution process, while parallel cyclic reduction is not asymptotically optimal and recursive doubling is not optimal and additionally exhibits numerical stability issues. They propose a hybrid combination of cyclic reduction and parallel cyclic reduction to alleviate these deficiencies on modern GPUs. In parallel to their work, we have developed an alternative implementation of cyclic reduction which has much more favorable memory access patterns, and performs on par with their best hybrid solver despite poor thread utilization in lower stages of the reduction procedure.

In the following, we sketch the main ideas of our approach, and refer to our original publication for details [5]. Cyclic reduction proceeds by recursively halving the number of equations and the number of unknowns (*forward elimination*), until a system of two unknowns is reached. This system is then solved, and in the *backward substitution* phase, the remaining half of the unknowns (per level of the reduction tree) is determined using the previously solved values. We start by copying the three matrix bands and the right hand side into temporary vectors in shared memory, denoted by $\widehat{\cdot}$. In each step of the forward reduction, all odd-indexed equations are updated in parallel; equation i of the current system is a linear combination of the equations $i-1$, i and $i+1$:

$$k_1 = \frac{\widehat{CL}_i}{\widehat{CC}_{i-1}} \qquad k_2 = \frac{\widehat{CU}_i}{\widehat{CC}_{i+1}}$$
$$\widehat{CL}_i = -k_1\widehat{CL}_{i-1}$$
$$\widehat{CC}_i = \widehat{CC}_i - k_1\widehat{CU}_{i-1} - k_2\widehat{CL}_{i+1}$$
$$\widehat{CU}_i = -k_2\widehat{CU}_{i+1}$$
$$\widehat{d}_i = -\widehat{d}_i - k_1\widehat{d}_{i-1} - k_2\widehat{d}_{i+1}$$

After the solution on the coarsest level of this reduction, all even-indexed values are solved in parallel by substituting the already solved c_{i-1} and c_{i+1} into equation i:

$$c_i = \frac{\widehat{d}_i - \widehat{CL}_i c_{i-1} - \widehat{CU}_i c_{i+1}}{\widehat{CC}_i}$$

Cyclic reduction performs $23M$ arithmetic operations, almost three times as many as the Thomas algorithm. However, it only requires $2\log_2 M - 1$ steps in parallel. Cyclic reduction can be implemented entirely in-place, which is important due to the size constraints of fast on-chip shared memory. But this straightforward in-place algorithm suffers significantly from high-degree bank conflicts in shared memory in later stages of the elimination and early stages of the substitution sweeps [15]. The shared memory access stride is doubled in each step in the reduction, as can be seen in the equations above. Our solution to this problem is to group the values always in two contiguous arrays corresponding to odd and even indices.

Implementational Details: When the initial data are loaded into shared memory, the even- and odd-indexed values are already separated using modulo two indexing. With an appropriate padding between the arrays this read operation from global to shared memory is fully coalesced and free of bank conflicts. The output of each elimination step again writes into separate even and odd arrays. The new even-indexed values overwrite the location of the old odd-indexed ones in a contiguous fashion, whereas the new odd-indexed values are written into a new array. With appropriate padding there are absolutely no bank conflicts in the involved read and write operations. As a tradeoff to reduce the amount of memory needed, we can remove the padding and introduce two-way conflicts for the writes only. Unfortunately, any out-of-place cyclic reduction scheme makes the backward substitution more difficult, because the even-indexed values cannot be updated in-place. Instead, each substitution combines the updated even values and the already computed odd values with two stride-two writes to form the new odd values. Only on the first level no further combination is needed and all even-indexed values can be updated in-place, before the result is written back into global memory. Because of the special treatment of the first level the additional storage requirements of our efficient implementation are $M \cdot (1/4 + 1/8 + \ldots) = M/2$.

7.2.3 Multicolor Parallelization: Gauß-Seidel Solvers

$\mathbf{C}^{\text{GSROW}}$ is a lower triangular matrix (cf. Figure 7.2). Starting with the first row of the matrix which only affects one unknown, the CPU implementation performs a standard forward substitution sweep to compute $\mathbf{c} = (\mathbf{C}^{\text{GSROW}})^{-1}\mathbf{d}$. All values $\mathbf{c}_i$ depend on the previously updated values $\mathbf{c}_{i-1}, \mathbf{c}_{i-M+1}, \mathbf{c}_{i-M}$ and $\mathbf{c}_{i+M-1}$ corresponding to the coupling by the matrix bands $\mathbf{CL}, \mathbf{LU}, \mathbf{LC}$, and $\mathbf{LL}$ to the left and the mesh line below. The execution can be parallelized with so-called *wavefront* techniques [1], which result in limited parallellism and irregular data access patterns and thus do not lead to efficient GPU implementations.

A method to generate wide and regular parallelism is to use a *multicoloring* scheme [1]. A simple example is the *red-black* Gauß-Seidel scheme, which alternatingly updates all odd-indexed and then all even-indexed values in parallel in case of a five-point stencil in 2D or seven-point in 3D. In our setting we have a nine-point stencil in 2D, so we need four colors to decouple all dependencies. We split the nodes into four disjoint index sets by alternating colors between lines in the mesh, and additionally alternating colors between adjacent grid points per line, cf. Figure 7.2 (left). The application of the Gauß-Seidel preconditioner thus decomposes into four sweeps over the domain, each sweep is trivially parallel. We illustrate the procedure exemplarily for the nodes 12, 13, 17 and 18 (Figure 7.2). During the treatment of node 12 (red, first color), no neighbors have been treated yet, and it is updated with a standard Jacobi step. Node 13 (green, second color) can incorporate the previously updated nodes 12 and 14. Node 17 (blue, third color) takes into account the 'green' nodes 11, 13, 21, 22 and the 'red' nodes 12 and 22, and finally, node 18 (yellow) includes all eight neighbors.

In this algorithm, the first $N/4$ unknowns are coupled only with themselves

(**CC**), and the remaining three sets of $N/4$ unknowns each incorporate more and more updated values by shifting them to the right hand side at the beginning of each color update. Figure 7.3 summarizes the multicolor Gauß-Seidel (*MC-GSROW*) preconditioner. From a theoretical point of view the multicoloring corresponds to a renumbering of the unknowns, which, however, may change the convergence behavior [1].

---

1. For all red nodes in parallel:
$\mathbf{c}_i = \mathbf{d}_i/\mathbf{CC}_i$

2. For all green nodes in parallel:
$\mathbf{c}_i = (\mathbf{d}_i - \mathbf{CU}_i\mathbf{c}_{i+1} - \mathbf{CL}_i\mathbf{c}_{i-1})/\mathbf{CC}_i$

3. For all blue nodes in parallel:
$\mathbf{c}_i = (\mathbf{d}_i - \mathbf{LU}_i\mathbf{c}_{i-M+1} - \mathbf{LC}_i\mathbf{c}_{i-M} - \mathbf{LL}_i\mathbf{c}_{i-M-1}$
$- \mathbf{UU}_i\mathbf{c}_{i+M+1} - \mathbf{UC}_i\mathbf{c}_{i+M} - \mathbf{UL}_i\mathbf{c}_{i+M-1})/\mathbf{CC}_i$

4. For all yellow nodes in parallel:
$\mathbf{c}_i = (\mathbf{d}_i - \mathbf{CU}_i\mathbf{c}_{i+1} - \mathbf{CL}_i\mathbf{c}_{i-1}$
$- \mathbf{LU}_i\mathbf{c}_{i-M+1} - \mathbf{LC}_i\mathbf{c}_{i-M} - \mathbf{LL}_i\mathbf{c}_{i-M-1}$
$- \mathbf{UU}_i\mathbf{c}_{i+M+1} - \mathbf{UC}_i\mathbf{c}_{i+M} - \mathbf{UL}_i\mathbf{c}_{i+M-1})/\mathbf{CC}_i$

FIGURE 7.3: Multicolored *MC-GSROW* preconditioner.

Implementational Details: The *efficient* implementation of multicolored Gauß-Seidel preconditioners is surprisingly challenging, due to the inherent stride-two memory access pattern implied by the 'checkerboard' numbering. In our general framework in which different smoothers can be plugged into the same multigrid solver, we want to avoid smoother dependent changes to the global data structures, so our solution is to fuse the computations per mesh line into one CUDA kernel that treats both associated colors. We use blocks of 128 (or fewer) threads for each line in the mesh, and let each thread compute two values, the last thread additionally treats the last value as M is a power of two plus one. All loads from global memory are staged through shared memory, and with linear addressing, these loads are fully coalesced. Unfortunately, this mapping now introduces bank conflicts in shared memory when reading the values to compute the final result for each color. This can be avoided by separating the values during the loads, so that all even- and odd-indexed values are stored contiguously, similar to our implementation of the tridiagonal solver (cf. Section 7.2.2). To guarantee correctness, each block loads data with a halo of one to each side. The resulting kernels are thus completely free of bank conflicts and off-chip memory is accessed in a fully coalesced fashion. In other words, the available on- and off-chip memory bandwidth is used in the most efficient manner.

7.2.4 Combination of Tridiagonal and Gauß-Seidel Smoothers

$\mathbf{C}^{\mathrm{TRIGSROW}}$ couples each grid point with its immediate left, bottom and right neighbors, see Figure 7.2. To apply this preconditioner, we iterate over the lines of the mesh, shift the already known values from the lower line to the right hand side (Gauß-Seidel part) and solve the remaining tridiagonal system exactly. In contrast to the *TRIDIROW* preconditioner, the lines depend sequentially on each other, so to recover parallelism among the lines we again apply a multicoloring technique that decouples the dependencies.

In case of two colors, the *TRIDIROW* preconditioner is first applied to all even lines in parallel, and later when the odd lines are computed they can incorporate the new values from the even lines. To strengthen the coupling between lines, more colors can be used. With four colors, only every fourth line is treated with the *TRIDIROW* preconditioner, resulting in four parallel sweeps, that couple each block of four lines more closely because each color incorporates the newly computed lower line. In the limit of M colors, the sequential variant is recovered. Beside the coupling strength, the actual choice on the number of colors depends also on the underlying hardware, namely, how much independent parallel work is needed to saturate the compute device. In our experiments on current GPUs, we have found that the choice of four colors is the best compromise between runtime and numerical performance.

The implementation of this *MC-TRIGSROW* preconditioner is a combination of code from the previous sections: The *TRIDIROW* kernel from Section 7.2.2 is extended by a small code piece that incorporates the results from other mesh lines into the right hand side of the linear equation system, a task that has already been discussed in Section 7.2.3.

7.2.5 Alternating Direction Implicit Method

To apply the ADI variant of a preconditioner $\mathbf{C}$ to the vector $\mathbf{d}$, three steps are performed: First $\mathbf{C}^{\mathrm{ROW}}$ is applied, then the new defect is computed and finally $\mathbf{C}^{\mathrm{COL}}$ is applied. Consequently, in terms of work performed, one ADI step corresponds to two steps of an elementary preconditioner.

In a preprocessing stage, the bands of the preconditioner $\mathbf{C}^{\mathrm{COL}}$ corresponding to a columnwise numbering of the degrees of freedom are generated from the initial matrix that corresponds to a linewise numbering of the degrees of freedom. This is justified because the same matrix is needed for all smoothing steps. The values in the rearranged bands do not have to be re-assembled. For instance, the band $\mathbf{LC}^{\mathrm{ROW}}$ is read with stride M and copied, with an appropriate starting offset, contiguously into the new band $\mathbf{CL}^{\mathrm{COL}}$. The preconditioners $\mathbf{C}^{\mathrm{ROW}}$ and $\mathbf{C}^{\mathrm{COL}}$ thus have exactly the same structure, and we can use the implementations for the rowwise variants developed throughout this chapter without any modifications for the columnwise and ADI cases, we just have to provide the corresponding set of matrix bands as kernel arguments.

The conversion of a vector from its row- to columnwise representation must be performed on the device, before and after the application of $\mathbf{C}^{\mathrm{COL}}$. The CUDA kernel for this operation is surprisingly challenging, because a naïve code has either coalesced reads and stride-M writes or vice versa. We instead use a tile-based approach where each block of threads rearranges one tile: Each thread reads four sequential values of a 32×32 tile, using two-dimensional addressing of a block with 256 threads. All reads are fully coalesced, and the values are stored in shared memory, directly at their permuted locations in the tile. Because of the sequential addressing, these writes are also free of bank conflicts. In the same way, each thread stores a portion of the rearranged tile back to off-chip memory; these writes are also fully coalesced.

7.3   Numerical Evaluation and Performance Results

7.3.1   Test Procedure

We evaluate our multigrid solvers with the different smoothers on a generalized Poisson problem:

$$-\text{div} \left(\mathbf{G} \text{ grad } \mathbf{u} \right) = \mathbf{f}$$

Setting $\mathbf{G}$ to the identity yields the standard Poisson problem. This example allows us to evaluate both grid and operator anisotropies as they often occur in practice. To avoid special convergence effects that might occur on axis-aligned domains only, our test domain is a general quadrilateral with the four corner coordinates $(0,0)$, $(1.5,0)$, $(0.75,1)$ and $(0.25,0.75)$. We employ mesh anisotropies (e.g., typical for boundary layer resolution in fluid dynamics) by applying anisotropic mesh refinement: For each refined element in the bottommost line and leftmost column, the new midpoint x_c is calculated by recursively applying the formula $x_c = x_l + \nu \cdot \frac{x_r - x_l}{2}$ with a given *anisotropy factor* ν ($\nu = 1.0$ yields uniform refinement) and x_l and x_r denoting the coordinates of the left and right edge of an element before subdivision (analogously for the y-component). All other elements are refined uniformly. This leads to matrices with locally condensed anisotropies and an according increase in the condition number. We set $\nu = 0.8$ in the first refinement step, $\nu = 0.6$ in the last refinement, and $\nu = 0.7$ otherwise. Such anisotropies are captured well by ADI linewise preconditioners.

Furthermore, we introduce anisotropies into the operator via the 2×2 matrix $\mathbf{G}$. With the function $H(x,y) := (\sin \pi x \cdot \sin 7\pi y)$ we create a vector field $\mathbf{v} = (v_1, v_2) := (-\frac{\partial H}{\partial y}, \frac{\partial H}{\partial x})$. Then we define the corresponding *anisotropic diffusion* matrix $\mathbf{G} := \mathbf{R} \cdot \mathbf{S} \cdot \mathbf{R}^T$:

$$\mathbf{R}(x,y) := \frac{1}{\|\mathbf{v}(x,y)\|_2} \left(\begin{array}{cc} v_1(x,y) & v_2(x,y) \\ -v_2(x,y) & v_1(x,y) \end{array} \right), \quad \mathbf{S}(x,y) := \left(\begin{array}{cc} \|\mathbf{v}(x,y)\|_2 & 0 \\ 0 & 1 \end{array} \right)$$

Anisotropies of this kind are generally not captured well by linewise preconditioners.

7.3.2   Solver Configuration and Hardware Details

We use a multigrid solver with various smoothing operators and configure it to reduce the initial residuals by one digit. We always perform four elementary pre- and postsmoothing steps (i. e., two ADI steps) in a *V* cycle, and a damping parameter ω that has been empirically determined as 'optimal' in the CPU runs. This solver executes in single precision, and is used in an outer iterative refinement loop that computes in double precision, see Section 7.1.3. The outer iteration is configured to gain eight digits.

These solvers are either executed entirely on an NVIDIA GeForce GTX 280 GPU (30 multiprocessors, 16 kB shared memory per multiprocessor, 141.7 GB/s bandwidth to GDDR3 offchip memory), or entirely on an Intel Core2Duo E6750 CPU (Conroe, 4 MB level-2 cache, 2.66 GHz, 10.6 GB/s overall bandwidth to DDR2-800 memory). GPU timings include the initial transfer of the right hand side, and a final transfer of the solution to the CPU, but do not include the transfer of matrix data. The CPU code is single threaded. Both these decisions reflect the typical usage scenario of these solvers in the context of FEAST, see Section 7.1.2. We have previously demonstrated that the mixed-precision scheme yields exactly the same results as a double precision solver, even for very ill-conditioned problems [5,6]. Due to the limited amount of shared memory, the smoothers involving our tridiagonal solver can only be executed up to $M = 513$.

7.3.3   Numerical Evaluation

We measure the pure numerical performance of the schemes in terms of the convergence rate ρ, which quantifies the reduction of the residuals after convergence (eight digits gained) in k iterations:

$$\rho := \left(\frac{\|\mathbf{A}\mathbf{x}^k - \mathbf{b}\|_2}{\|\mathbf{A}\mathbf{x}^0 - \mathbf{b}\|_2} \right)^{1/k}$$

Figure 7.4 depicts the convergence rates of the outer solvers on the CPU and the GPU, for increasing level of refinement. A multigrid solver equipped with the *JACOBI* smoother fails to converge unless aggressive damping by $\omega \le 0.3$ is applied, which makes the schemes incompetetive (results therefore not shown). The sequential *GSROW* on the CPU and the parallel *MC-GSROW* on the GPU converge in exactly the same way. The *ADITRIDI* preconditioner yields qualitatively the same behavior on the CPU and the GPU, and is marginally better on the former. The *ADITRIGS* smoother gives overall best results, and for large problem sizes the CPU variant is slightly better than the GPU implementation that uses four colors.

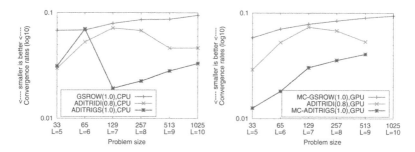

FIGURE 7.4: Convergence rates of the outer solver, on the CPU (left) and the GPU (right), logarithmic scale.

7.3.4 Runtime Efficiency

These convergence rates do not tell the entire story, because we use a relative stopping criterion for the inner solver, and so they suffer from granularity effects (cf. *ADITRIGS* CPU, $L = 5, 6, 7$ in Figure 7.4). To enable a more honest comparison that is furthermore independent of the problem size and the target architecture, we compute the time (in μs) per unknown per digit gained,

$$t_{\text{rel}} := -\frac{t_{\text{total}} \cdot 10^6}{N \cdot k \cdot \log_{10} \rho},$$

where t_{total} denotes the total time in seconds of the outer solver to converge in k iterations. Smaller t_{rel} values are better. As efficiency is usually measured in a metric where larger values mean better efficiency, we define the *total runtime efficiency* of a given solver as the reciprocal of t_{rel}. Figure 7.5 depicts the resulting efficiency of the three solver configurations on the CPU and the GPU.

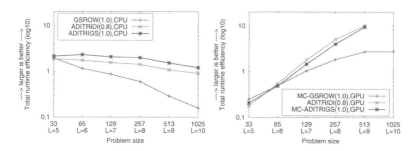

FIGURE 7.5: Efficiency of the solvers, on the CPU (left) and the GPU (right), logarithmic scale.

In this metric, the superiority of the strong smoothing operators becomes obvious: On the CPU, *ADITRIGS* is 30–40 % more efficient than *ADITRIDI*, and for the two largest problem sizes, 5.2 and 7.7 times faster than *GSROW*. This is an immediate consequence of the relative inner stopping criterion; a multigrid solver equipped with *GSROW* needs many more cycles to gain one digit than the two more powerful smoothers. The exact iteration numbers (sum of all inner multigrid cycles) are 130, 18 and 12, respectively. On the GPU, the results are different: *ADITRIDI* is the most efficient configuration, 5–20 % faster than the (numerically more powerful) *MC-ADITRIGS* smoother, and 2.8 and 3.7 times faster than *MC-GSROW* (exact iteration numbers: 141, 18 and 12). The difference between the two ADI preconditioners is a consequence of the different amount of independent parallel work and thus of the runtime rather than the numerical performance. In other words, the difference is caused by the degree of device saturation, which determines the ability to hide memory stalls. The *ADITRIDI* smoother launches M kernels per application, the *MC-ADITRIGS* smoother four times $M/4$, and consequently only for $M = 129$ there are actually enough independent thread blocks to keep all multiprocessors busy, and only for $M = 513$ the device can hide latencies of off-chip memory accesses adequately. The beneficial numerical properties of the stronger smoother are thus offset by its less efficient execution. Therefore, we expect that on future devices with more on-chip memory, *MC-ADITRIGS* will be the most efficient smoother for larger problem sizes.

The importance of device saturation can also be seen in Figure 7.6, which illustrates the speedup achieved by the GPU solvers over their CPU counterparts. The CPU becomes less and less efficient the more the computations move out of the cache, while the GPU on the other hand favors large problem sizes. The break-even point between CPU and GPU is $M = 129$ for the Gauß-Seidel and tridiagonal preconditioners, and $M = 257$ for *ADITRIGS*. For the largest admissible problem sizes, we achieve speedup factors of 17.7, 9.4 and 6.3, respectively. The speedups of all schemes rise steadily with the problem size which lets us expect even larger speedups on future devices with more on-chip shared memory, the currently limiting factor for the maximum domain size solvable by *ADITRIDI* and *MC-ADITRIGS*.

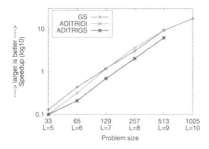

FIGURE 7.6: Speedup of the GPU over the CPU, logarithmic scale.

7.3.5 Smoother Selection

The previous sections mainly compare the different smoothers with respect to total runtime efficiency, which is independent of architecture particularities and motivated by the tradeoff between the numerical and runtime performance. We have presented results for only one problem configuration but can report that many more configurations also yield the observed smooth tradeoff, where a gradual increase of coupling, i.e., a reduction of independent parallel work, means a gradual increase of numerical performance and a gradual decrease of runtime performance. However, sometimes the gradual tradeoff becomes a disproportionate tradeoff, e.g., a strong smoother becomes indispendable because of its capability to solve ill-conditioned problems on which weaker smoothers fail or run disproportionately slow, as already observed to some extent for the Gauß-Seidel results on large domains.

7.4 Summary and Conclusions

Numerically strong smoothers and preconditioners are characterized by sequential data dependencies, and do not parallelize in a straightforward manner. For linewise preconditioners, exact parallel algorithms exist, and we have presented an efficient implementation of a cyclic reduction tridiagonal solver. For other preconditioners, traditional *wavefront techniques* can be applied, but their irregular and limited parallelism makes them a bad match for GPUs. Therefore, we discussed *multicoloring* techniques to recover parallelism in these preconditioners, by decoupling some of the dependencies at the expense of reduced numerical performance. However, by carefully balancing the coupling strength (more colors) with the parallelization benefits, the multicolored *MC-GSROW* and *MC-ADITRIGS* retain almost all of the sequential numerical performance. Due to their advantageous numerical properties, solvers equipped with strong preconditioners are between four and eight times more efficient than with simple Gauß-Seidel preconditioners, and we achieve speedups factors between 6 and 18 with the GPU implementations over carefully tuned CPU variants.

Acknowledgments

This work has been funded in part by German Deutsche Forschungsgemeinschaft (project TU102/22-2), and by German Bundesministerium für Bildung und Forschung in the SKALB project (grant 01IH08003D) of call 'HPC Software für skalierbare Parallelrechner.'

Bibliography

[1] Richard Barrett, Michael Berry, Tony F. Chan, James W. Demmel, June Donato, Jack J. Dongarra, Victor Eijkhout, Roldan Pozo, Charles Romine, and Henk A. van der Vorst. *Templates for the Solution of Linear Systems: Building Blocks for Iterative Methods*. SIAM, 2nd edition, November 1994.

[2] Bill L. Buzbee, Gene H. Golub, and Clair W. Nielson. On direct methods for solving Poisson's equations. *SIAM Journal on Numerical Analysis*, 7(4):627–656, December 1970.

[3] Kayvon Fatahalian and Mike Houston. A closer look at GPUs. *Communications of the ACM*, 51(10):50–57, October 2008.

[4] Dominik Göddeke. *Fast and Accurate Finite-Element Multigrid Solvers for PDE Simulations on GPU Clusters*. PhD thesis, TU Dortmund, Fakultät für Mathematik, May 2010.

[5] Dominik Göddeke and Robert Strzodka. Cyclic reduction tridiagonal solvers on GPUs applied to mixed precision multigrid. *IEEE Transactions on Parallel and Distributed Systems, Special Issue: High Performance Computing with Accelerators*, March 2010.

[6] Dominik Göddeke, Robert Strzodka, and Stefan Turek. Performance and accuracy of hardware-oriented native-, emulated- and mixed-precision solvers in FEM simulations. *International Journal of Parallel, Emergent and Distributed Systems*, 22(4):221–256, January 2007.

[7] Roger W. Hockney. A fast direct solution of Poisson's equation using Fourier analysis. *Journal of the ACM*, 12(1):95–113, January 1965.

[8] Roger W. Hockney and Chris R. Jesshope. *Parallel Computers*. Adam Hilger, November 1981.

[9] Khronos OpenCL Working Group. The OpenCL Specification, version 1.0. http://www.khronos.org/opencl, December 2008.

[10] Julie Langou, Julien Langou, Piotr Luszczek, Jakub Kurzak, Alfredo Buttari, and Jack J. Dongarra. Exploiting the performance of 32 bit floating point arithmetic in obtaining 64 bit accuracy (revisiting iterative refinement for linear systems). In *SC '06: Proceedings of the 2006 ACM/IEEE Conference on Supercomputing*, November 2006. Article No. 113.

[11] NVIDIA Corporation. NVIDIA CUDA programming guide version 2.3. http://www.nvidia.com/cuda, July 2009.

[12] Donald W. Peaceman and Henry H. Rachford Jr. The numerical solution of parabolic and elliptic differential equations. *Journal of the Society for Industrial and Applied Mathematics*, 3(1):28–41, March 1955.

[13] Harold S. Stone. An efficient parallel algorithm for the solution of a tridiagonal linear system of equations. *Journal of the ACM*, 20(1):27–38, January 1973.

[14] Llewellyn Hilleth Thomas. Elliptic problems in linear difference equations over a network, 1949. Watson Scientific Computing Laboratory Report, Columbia University, New York.

[15] Yao Zhang, Jonathan Cohen, and John D. Owens. Fast tridiagonal solvers on the GPU. In *Proceedings of the 15th ACM SIGPLAN Symposium on Principles and Practice of Parallel Programming (PPoPP 2010)*, pages 127–136, January 2010.

Part IV

Fast Fourier Transforms

Chapter 8

Designing Fast Fourier Transform for the IBM Cell Broadband Engine

Virat Agarwal

IBM T.J. Watson Research Center, Yorktown Heights

David A. Bader

Georgia Institute of Technology, Atlanta

8.1   Introduction

The Cell Broadband Engine (or the Cell/B.E.) [6, 16, 17, 34] is a novel high-performance architecture designed by Sony, Toshiba, and IBM (STI), primarily targeting multimedia and gaming applications. The Cell/B.E. consists of a traditional microprocessor (called the PPE) that controls eight SIMD co-processing units called synergistic processor elements (SPEs), a high-speed memory controller, and a high-bandwidth bus interface (termed the element interconnect bus, or EIB), all integrated on a single chip. The Cell is used in Sony's PlayStation 3 gaming console, Mercury Computer System's dual Cell-based blade servers, IBM's QS20 Cell Blades, and the Roadrunner supercomputer.

In this chapter we present the design of an efficient parallel implementation of Fast Fourier Transform (FFT) on the Cell/B.E. FFT is of primary importance and a fundamental kernel in many computationally intensive scientific applications such as computer tomography, data filtering, and fluid dynamics. Another important application area of FFTs is in spectral analysis of speech, sonar, radar, seismic, and vibration detection. FFTs are also used in digital filtering, signal decomposition, and in solution of partial differential equations. The performance of these applications relies heavily on the availability of a fast routine for Fourier transforms.

In our design of Fast Fourier Transform on the Cell (FFTC) we use an iterative out-of-place approach to solve 1D FFTs with 1K to 16K complex input samples. We describe our methodology to partition the work among the SPEs to efficiently parallelize a *single FFT computation* where the source and output of the FFT are both stored in main memory. This differentiates our work from the prior literature and better represents the performance that one realistically sees in practice. The algorithm requires a synchronization among the SPEs after each stage of FFT computation. Our synchronization barrier is designed to use inter SPE communication without any intervention from the PPE. The synchronization barrier requires only $2 \log p$ stages (p: number of SPEs) of inter SPE communication by using a tree-based approach. This significantly improves the performance, as PPE intervention not only results in a high communication latency but also in sequentialization of the synchronization step. We achieve a performance improvement of over 4 as we vary the number of SPEs from 1 to 8. We attain a performance of 18.6 GFLOP/s for a single-precision FFT with 8K complex input samples and also show significant speedup in comparison with other architectures. Our implementation is generic for this range of complex inputs. The source code is freely available from our CellBuzz project in SourceForge (`http://sourceforge.net/projects/cellbuzz/`).

This chapter is organized as follows. We first present the related research work in Section 8.2. We describe FFT and the algorithm we choose to parallelize in Section 8.3. The novel architectural features of the Cell processor are reviewed in Section 8.4. We then present our design to parallelize FFT on the Cell and optimize for the SPEs in Section 8.5.

8.2   Related Work

The formal description of Fourier theory and the FFT can be found in a variety of articles and textbooks [14, 22, 29]. Over the past few decades many people have studied the Discrete Fourier Transform (DFT), and have presented algorithms for the fast computation of the DFT, such as Cooley-Tukey FFT algorithm [15], prime-factor FFT algorithm [8], Brunn's FFT algorithm [6], Rader's FFT algorithm [32]. Several articles describe the vectorization of the FFT algorithm [1, 2, 27, 35]. There are other articles that present variations to the FFT algorithm [7, 26, 30, 33, 37]. Hardware-based FFTs have also been widely studied [28].

The literature also contains several publications related to FFTs on the Cell/B.E. processor and other multicore architectures. Table 8.1 provides an

TABLE 8.1: Related work

Reference	Underlying algorithm	Size	Performance	Base architecture	Advantages/limitations
Williams et al. [20]	Naive radix-2 FFT algorithm	$4K$, $16K$, $64K$ complex 1D, $1K^2$, $2K^2$ complex 2D	30-42 GFlop/s (1D), 36-41 Gbps (2D), single precision	IBM Cell/B.E.	Performance estimates.
Cico, Cooper and Greene [13]	Single precision FFT algorithm	$4K$, $16K$, $64K$ complex 1D, $1K^2$, $2K^2$ complex 2D	22.1 GFlop/s for a single FFT, 176.8 GFlop/s for 8 FFTs in parallel ($8K$), 90.8 GFlop/s for $64K$	IBM Cell/B.E.	Code tuned for specific sizes, for $1K$ and $8K$ data reside in SPE for performance measurements.
Chow, Fossum and Brokenshire [12]	Cell-optimized vectorized radix-2 FFT algorithm	16M complex	46.8 GFlop/s	IBM Cell/B.E.	Works only for input with 16M complex samples.
FFTW [19]	Variety of FFT algorithms are used and are machine optimized	1D, 2D, and 3D FFTs of various sizes	size specific	Intel/AMD/IBM/Sun commodity/multicore processors as well as IBM Cell/B.E.	User friendly library, but higher performance can be obtained than FFTW by tuning for specific FFT sizes.
Chellappa, Franchetti and Püeschel [9]	Spiral optimized Cooley-Tukey FFT algorithm	For DFT on multiple SPEs, 16-32K complex	5-36 GFlop/s single precision (for the input sizes in previous column)	IBM Cell/B.E.	Automatic optimized code generation using spiral, large FFTs currently not supported.
Chen et al. [10]	Radix-2 Cooley-Tukey FFT algorithm	64K complex 1D, 256x256 complex 2D	20.72 GFlop/s (1D), 20 GFlop/s (2D)	IBM Cyclops-64	Performance estimates, specific to the given sizes.
Govindaraju and Manocha [20]	Stockham autosort Cooley-Tukey-based FFT algorithm	0-8M complex values (1D)	16 GFlop/s for 4M single precision 1D	NVIDIA 8800 GPU	Limited by GPU/CPU interconnect.

admittedly cursory overview of recent algorithms/implementations in this literature. We categorize these solutions with respect to the underlying algorithm used, size of FFT, performance as reported in the publication, architecture on which the algorithm has been tested for, and limitations.

8.3   Fast Fourier Transform

FFT is an efficient algorithm that is used for computing the DFT. Some of the important application areas of FFTs have been mentioned in the previous section. There are several algorithmic variants of the FFTs that have been well studied for parallel processors and vector architectures [2–5].

In our design we utilize the naive Cooley-Tukey radix-2 Decimate in Frequency (DIF) algorithm. The pseudo-code for an out-of-place approach of this algorithm is given in Alg. 1. The algorithm runs in $\log N$ stages and each stage requires $\mathrm{O}(N)$ computation, where N is the input size.

The array w contains the *twiddle factors* required for FFT computation. At each stage the computed complex samples are stored at their respective locations thus saving a bit-reversal stage for output data. This is an iterative algorithm which runs until the parameter *problemSize* reduces to 1. Fig. 8.1 shows the butterfly stages of this algorithm for an input of 16 sample points (4 stages).

Apart from the theoretical complexity, another common performance metric used for the FFT algorithm is the floating-point operation (FLOP) count. On analyzing the sequential algorithm, we see that during each iteration of the innermost *for* loop there is one complex addition for the computation of first output sample, which accounts for 2 FLOPs. The second output sample requires one complex subtraction and multiplication which accounts for 8 FLOPs. Thus, for the computation of two output samples during each innermost iteration we require 10 FLOPs, which suggests that we require 5 FLOPs for the computation of a complex sample at each stage. The total computations in all stages are $N \log N$ which makes the total FLOP count for the algorithm as $5N \log N$.

Algorithm 1: Sequential FFT algorithm

Input: array $A[0]$ of size N

1   $NP \longleftarrow 1$;
2   $problemSize \longleftarrow N$;
3   $dist \longleftarrow 1$;
4   $i1 \longleftarrow 0$;
5   $i2 \longleftarrow 1$;
6   **while** $problemSize > 1$ **do**
7   *Begin Stage*;
8     $a \longleftarrow A[i1]$;
9     $b \longleftarrow A[i2]$;
10    $k = 0, jtwiddle = 0$;
11   **for** $j \leftarrow 0$ **to** $N - 1$ **step** $2 * NP$ **do**
12      $W \longleftarrow w[jtwiddle]$;
13   **for** $jfirst \leftarrow 0$ **to** NP **do**
14        $b[j + jfirst] \leftarrow a[k + jfirst] + a[k + jfirst + N/2]$;
15        $b[j+jfirst+Dist] \leftarrow (a[k+jfirst]-a[k+jfirst+N/2])*W$;
16      $k \leftarrow k + NP$;
17      $jtwiddle \leftarrow jtwiddle + NP$;
18   swap$(i1, i2)$;
19    $NP \leftarrow NP * 2$;
20    $problemSize \leftarrow problemSize/2$;
21    $dist \leftarrow dist * 2$;
22   *End Stage*;

Output: array $A[i1]$ of size N

8.4   Cell Broadband Engine Architecture

The Cell/B.E. processor is a heterogeneous multicore chip that is significantly different from conventional multiprocessor or multicore architectures. It consists of a traditional microprocessor (the PPE) that controls eight SIMD co-processing units called SPEs, a high-speed memory controller, and a high-bandwidth bus interface (termed the EIB), all integrated on a single chip. Fig. 8.2 gives an architectural overview of the Cell/B.E. processor. We refer the reader to [6,18,21,25,31] for additional details.

The PPE runs the operating system and coordinates the SPEs. It is a 64-bit PowerPC core with a vector multimedia extension (VMX) unit, 32 KB L1 instruction and data caches, and a 512 KB L2 cache. The PPE is a dual-issue, in-order execution design, with two-way simultaneous multithreading. Ideally,

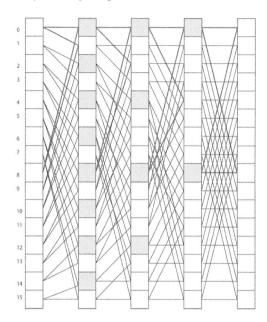

FIGURE 8.1: Butterflies of the ordered DIF FFT algorithm. From D.A. Bader and V. Agarwal, FFTC: Fastest Fourier Transform for the IBM Cell Broadband Engine, *14th IEEE International Conference on High Performance Computing* (HiPC), Springer-Verlag LNCS 4873, 172–184, Goa, India, December 2007; and D.A. Bader, V. Agarwal, and S. Kang, Computing Discrete Transforms on the Cell Broadband Engine, *Parallel Computing*, Elsevier 35(3):119–137, 2009. With permission.

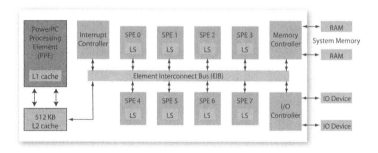

FIGURE 8.2: Cell Broadband Engine Architecture. From D.A. Bader and V. Agarwal, FFTC: Fastest Fourier Transform for the IBM Cell Broadband Engine, *14th IEEE International Conference on High Performance Computing* (HiPC), Springer-Verlag LNCS 4873, 172–184, Goa, India, December 2007; and D.A. Bader, V. Agarwal, and S. Kang, Computing Discrete Transforms on the Cell Broadband Engine, *Parallel Computing*, Elsevier 35(3):119–137, 2009. With permission.

all the computation should be partitioned among the SPEs, and the PPE only handles the control flow.

Each SPE consists of a synergistic processor unit (SPU) and a memory flow controller (MFC). The MFC includes a DMA controller, a memory management unit (MMU), a bus interface unit, and an atomic unit for synchronization with other SPUs and the PPE. The SPU is a micro-architecture designed for high performance data streaming and data-intensive computation. It includes a 256 KB *local store* (LS) memory to hold the SPU program's instructions and data. The SPU cannot access main memory directly, but it can issue DMA commands to the MFC to bring data into the LS or write computation results back to the main memory. DMA is non-blocking so that the SPU can continue program execution while DMA transactions are performed.

The SPU is an in-order dual-issue statically scheduled architecture. Two SIMD [12] instructions can be issued per cycle: one compute instruction and one memory operation. The SPU branch architecture does not include dynamic branch prediction, but instead relies on compiler-generated branch hints using *prepare-to-branch* instructions to redirect instruction prefetch to branch targets. Thus branches should be minimized on the SPE as far as possible.

The MFC supports naturally aligned transfers of 1, 2, 4, or 8 bytes, or a multiple of 16 bytes to a maximum of 16 KB. DMA list commands can request a list of up to 2,048 DMA transfers using a single MFC DMA command. Peak performance is achievable when both the effective address and the local storage address are 128 bytes aligned and the transfer is an even multiple of 128 bytes. In the Cell/B.E., each SPE can have up to 16 outstanding DMAs, for a total of 128 across the chip, allowing unprecedented levels of parallelism in on-chip communication. Kistler et al. [25] analyze the communication network of the Cell/B.E. and state that applications that rely heavily on random scatter and/or gather accesses to main memory can take advantage of the high-communication bandwidth and low latency.

With a clock speed of 3.2 GHz, the Cell processor has a theoretical peak performance of 204.8 GFlop/s (single precision). The EIB supports a peak bandwidth of 204.8 GB/s for intra-chip transfers among the PPE, the SPEs, and the memory and I/O interface controllers. The memory interface controller (MIC) provides a peak bandwidth of 25.6 GB/s to main memory. The I/O controller provides peak bandwidths of 25 GB/s inbound and 35 GB/s outbound.

8.5   FFTC: Our FFT Algorithm for the Cell/B.E. Processor

There are several architectural features that make it difficult to optimize and parallelize the Cooley-Tukey FFT algorithm on the Cell/B.E. The algorithm is branchy due to presence of a doubly nested *for* loop within the outer *while* loop. This results in a compromise on the performance due to the absence of a branch predictor on the Cell. The algorithm requires an array that consists of the $N/2$ complex twiddle factors. Since each SPE has a limited local store of 256 KB, this array cannot be stored entirely on the SPEs for a large input size. The limit in the size of the LS memory also restricts the maximum input data that can be transferred to the SPEs. Parallelization of a single FFT computation involves synchronization between the SPEs after every stage of the algorithm, as the input data of a stage are the output data of the previous stage. To achieve high performance it is necessary to divide the work equally among the SPEs so that no SPE waits at the synchronization barrier. Also, the algorithm requires $\log N$ synchronization stages which impacts the performance. It is difficult to vectorize every stage of the FFT computation. For

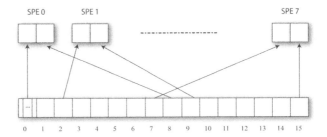

FIGURE 8.3: Partition of the input array among the SPEs (e.g. 8 SPEs in this illustration).

vectorization of the first two stages of the FFT computation it is necessary to shuffle the output data vector, which is not an efficient operation in the SPE instruction set architecture. Also, the computationally intensive loops in the algorithm need to be unrolled for best pipeline utilization. This becomes a challenge given a limited LS on the SPEs.

8.5.1   Parallelizing FFTC for the Cell

As mentioned in the previous section for best performance it is important to partition work among the SPEs to achieve load balancing. We parallelize by dividing the input array held in main memory into $2p$ chunks, each of size $\frac{N}{2p}$, where p is the number of SPEs.

During every stage, SPE i is allocated chunk i and $i + p$ from the input array. The basis for choosing these chunks for an SPE lies in the fact that these chunks are placed at an offset of $N/2$ input elements. For the computation of an output complex sample we need to perform complex arithmetic operations between input elements that are separated by this offset. Fig. 8.3 gives an illustration of this approach for work partitioning among 8 SPEs.

The PPE does not intervene in the FFT computation after this initial work allocation. After spawning the SPE threads it waits for the SPEs to finish execution.

8.5.2 Optimizing FFTC for the SPEs

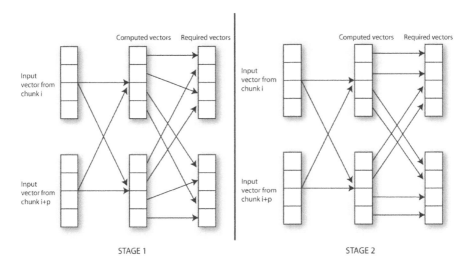

FIGURE 8.4: Vectorization of the first two stages of the FFT algorithm. These stages require a shuffle operation over the output vector to generate the desired output. From D.A. Bader, V. Agarwal, and K. Madduri, "On the Design and Analysis of Irregular Algorithms on the Cell Processor: A Case Study on List Ranking," *21th IEEE International Parallel and Distributed Processing Symposium* (IPDPS), Long Beach, CA, March 2007; D.A. Bader, V. Agarwal, K. Madduri, and S. Kang, "High Performance Combinatorial Algorithm Design on the Cell Broadband Engine Processor," *Parallel Computing*, 33(10–11):720–740, 2007. With permission.

After dividing the input array among the SPEs, each SPE is allocated two chunks each of size $\frac{N}{2p}$. Each SPE fetches this chunk from main memory using DMA transfers and uses double-buffering to overlap memory transfers with computation. Within each SPE, after computation of each buffer, the computed buffer is written back into main memory at offset using DMA transfers.

The detailed pseudo-code is given in Alg. 2. The first two stages of the FFT

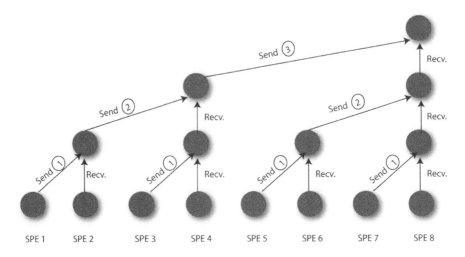

FIGURE 8.5: Stages of the synchronization barrier using inter-SPE commu-nication. The synchronization involves sending inter-SPE mailbox messages up to the root of the tree and then sending back acknowledgment messages down to the leaves in the same topology. From D.A. Bader and V. Agarwal, FFTC: Fastest Fourier Transform for the IBM Cell Broadband Engine, *14th IEEE International Conference on High Performance Computing* (HiPC), Springer-Verlag LNCS 4873, 172–184, Goa, India, December 2007; and D.A. Bader, V. Agarwal, and S. Kang, Computing Discrete Transforms on the Cell Broadband Engine, *Parallel Computing*, Elsevier 35(3):119–137, 2009. With permission.

algorithm are duplicated that correspond to the first two iterations of the outer *while* loop in sequential algorithm. This is necessary as the vectorization of these stages requires a shuffle operation (*spu_shuffle()*) over the output to re-arrange the output elements to their correct locations. Please refer to Fig. 8.4 for an illustration of this technique for stages 1 and 2 of the FFT computation.

The innermost *for* loop (in the sequential algorithm) can be easily vector-ized for $NP > 4$, that corresponds to the stages 3 through $\log N$. However, it is important to duplicate the outer *while* loop to handle stages where $NP < buffersize$, and otherwise. The global parameter *buffersize* is the size of a single DMA get buffer. This duplication is required as we need to stall for a DMA transfer to complete, at different places within the loop for these two cases. We also unroll the loops to achieve better pipeline utilization. This significantly increases the size of the code thus limiting the unrolling factor.

SPEs are synchronized after each stage, using *inter-SPE communication*. This is achieved by constructing a binary synchronization tree, so that syn-chronization is achieved in $2 \log p$ stages. The synchronization involves the use of inter-SPE mailbox communication without any intervention from the PPE. Please refer to Fig. 8.5 for an illustration of the technique.

This technique performs significantly better than other synchronization techniques that either use chain-like inter-SPE communication or require the PPE to synchronize between the SPEs. The chain-like technique requires $2p$ stages of inter-SPE communication whereas with the intervention of the PPE latency of communication reduces the performance of this barrier.

Algorithm 2: Parallel FFTC algorithm: View within SPE

Input: array in PPE of size N
Output: array in PPE of size N

1   $NP \longleftarrow 1$;
2   $problemSize \longleftarrow N$;
3   $dist \longleftarrow 1$;
4   $fetchAddr \longleftarrow$ PPE *input array*;
5   $putAddr \longleftarrow$ PPE *output array*;
6   $chunkSize \longleftarrow \frac{N}{2*p}$;
7   Stage 0 (SIMDization achieved with shuffling of output vector);
8   Stage 1 ;
9   **while** $NP < buffersize$ & $problemSize > 1$ **do**
10 | *Begin Stage*;
11 | Initiate all DMA transfers to get data;
12 | Initialize variables;
13 | **for** $j \leftarrow 0$ **to** $2 * chunkSize$ **do**
14 | | Stall for DMA buffer;
15 | | **for** $i \leftarrow 0$ **to** $buffersize/NP$ **do**
16 | | | **for** $jfirst \leftarrow 0$ **to** NP **do**
17 | | | | SIMDize computation as $NP > 4$;
18 | | Update $j, k, jtwiddle$;
19 | Initiate DMA put for the computed results
20 | swap($fetchAddr, putAddr$);
21 | $NP \leftarrow NP * 2$;
22 | $problemSize \leftarrow problemSize/2$;
23 | $dist \leftarrow dist * 2$;
24 | *End Stage*;
25 | Synchronize using Inter-SPE communication;
26   **while** $problemSize > 1$ **do**
27 | *Begin Stage*;
28 | Initiate all DMA transfers to get data;
29 | Initialize variables;
30 | **for** $k \leftarrow 0$ **to** $chunkSize$ **do**
31 | | **for** $jfirst \leftarrow 0$ **to** $\min(NP, chunkSize - k)$ **step** $buffersize$ **do**
32 | | | Stall for DMA buffer;
33 | | | **for** $i \leftarrow 0$ **to** $buffersize$ **do**
34 | | | | SIMDize computation as $buffersize > 4$;
35 | | Initiate DMA put for the computed results;
36 | Update $j, k, jtwiddle$;
37 | swap($fetchAddr, putAddr$);
38 | $NP \leftarrow NP * 2$;
39 | $problemSize \leftarrow problemSize/2$;
40 | $dist \leftarrow dist * 2$;
41 | *End Stage*;
42 | Synchronize using Inter-SPE communication;

8.6   Performance Analysis of FFTC

For compiling, instruction-level profiling, and performance analysis we use the IBM Cell Broadband Engine SDK 3.0.0-1.0, gcc 4.1.1 with level 3 optimization. From *'/proc/cpuinfo'* we determine the clock frequency as 3.2 GHz with revision 5.0. We use *gettimeofday()* on the PPE before computation on the SPE starts, and after it finishes. For profiling measurements we iterate the computation 10,000 times to eliminate the noise of the timer.

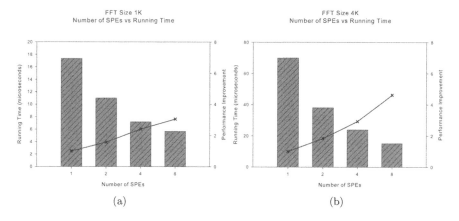

FIGURE 8.6: Running time of our FFTC code on 1K and 4K inputs as we increase the number of SPEs. From D.A. Bader and V. Agarwal, FFTC: Fastest Fourier Transform for the IBM Cell Broadband Engine, *14th IEEE International Conference on High Performance Computing* (HiPC), Springer-Verlag LNCS 4873, 172–184, Goa, India, December 2007; and D.A. Bader, V. Agarwal, and S. Kang, Computing Discrete Transforms on the Cell Broadband Engine, *Parallel Computing*, Elsevier 35(3):119–137, 2009. With permission.

For parallelizing a single 1D FFT on the Cell, it is important to divide the work among the SPEs. Fig. 8.6 shows the performance of our algorithm with varying the number of SPEs for 1K and 4K complex input samples. Our algorithm obtains a scalability of about 4x with 8 SPEs.

Our design requires a barrier synchronization among the SPEs after each stage of the FFT computation. We focus on FFTs that have from 1K to 16K complex input samples. For relatively small inputs and as the number of SPEs increases, the synchronization cost becomes a significant issue since the time per stage decreases but the cost per synchronization increases. With instruction level profiling we determine that the time required per synchronization stage using our tree-synchronization barrier is about 1 microsecond (3,200 clock cycles). We achieve a high-performance barrier using inter-SPE mailbox communication which significantly reduces the time to send a message, and

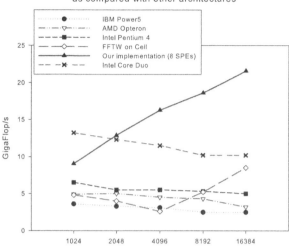

FIGURE 8.7: Performance comparison of FFTC with other architectures for various input sizes of FFT. The performance numbers are from benchFFT from the FFTW website. From D.A. Bader and V. Agarwal, FFTC: Fastest Fourier Transform for the IBM Cell Broadband Engine, *14th IEEE International Conference on High Performance Computing* (HiPC), Springer-Verlag LNCS 4873, 172–184, Goa, India, December 2007; and D.A. Bader, V. Agarwal, and S. Kang, Computing Discrete Transforms on the Cell Broadband Engine, *Parallel Computing*, Elsevier 35(3):119–137, 2009. With permission.

by using the tree-based technique we reduced the number of communication stages required for the barrier ($2 \log p$ steps).

Fig. 8.7 shows the single-precision performance for complex inputs of FFTC, our optimized FFT, as compared with the following architectures:

- **IBM Power 5:** IBM OpenPower 720, Two dual-core 1.65 GHz POWER5 processors.

- **AMD Opteron:** 2.2 GHz Dual Core AMD Opteron Processor 275.

- **Intel Core Duo:** 3.0 GHz Intel Xeon Core Duo (Woodcrest), 4MB L2 cache.

- **Intel Pentium 4:** Four-processor 3.06 GHz Intel Pentium 4, 512 KB L2.

We use the performance numbers from benchFFT [19] for the comparison with the above architectures. We consider the FFT implementation that gives best performance on these architectures for comparison.

main: | **compute_twiddle_factors:**

clks	Lab..	Even Pipe...	Odd Pipe...
1...	139:il		140:cwx ...
1...	141:il		142:shuf...
1...	143:nop ...		144:hbrp
1...	145:rot...		146:shuf...
1...	147:a		148:cwx
1...	149:a		150:cwx
1...	151:a		152:stqx...
1...	153:a		154:stqx...
1...	155:a		156:lqd ...
1...	157:rot...		158:lqd ...
1...	159:rot...		160:cwx ...
1...	161:il		162:cwx ...
1...	163:il		164:lnop...
1...	165:ila		166:cwx ...
1...	167:a		168:shuf...
1...	169:rot...		170:shuf...
1...	171:a		172:cwx ...
1...	173:ila		174:cwx ...
1...	175:il		176:stqx...

main: | **compute_twiddle_factors:**

clks	Lab...	Even Pipe...	Odd Pipe...
6...			540:stqx...
6...			541:stqx...
6...			542:lqx ...
6...			543:lqx ...
6...			544:lqx ...
6...			545:lqx ...
6...			546:lqx ...
6...			547:lqx ...
6...	548:fs		
6...			
6...			
6...	549:fs		
6...	550:fm		
6...			
6...			
6...			
6...			
6...			551:lnop...
6...	552:fm		
6	LC		554:brp

FIGURE 8.8: Analysis of the pipeline utilization using the *IBM Assembly Visualizer for Cell Broadband Engine*. The top figure shows full pipeline utilization for certain parts of the code and the bottom figure shows areas where the pipeline stalls due to data dependency.

The Cell/B.E. has two instruction pipelines, and for achieving high performance it is important to optimize the code so that the processor can issue two instructions per clock cycle. This level of optimization requires inspecting the assembly dump of the SPE code. For achieving pipeline utilization it is required that the gap between dependent instructions needs to be increased. We use the *IBM Assembly Visualizer for Cell/B.E.* tool to analyze this optimization. The tool highlights the stalls in the instruction pipelines and helps the user to reorganize the code execution while maintaining correctness. Fig. 8.8 shows the analysis of pipeline utilization. The left and right columns contain the instruction distribution among the two pipelines. For each instruction the 'x' in the column shows which cycle the execution starts for that instruction and the 'x's below show the number of cycles the instruction takes to execute in the pipeline. Some portions utilize these pipelines effectively (top figure) whereas there are a few stalls in other parts of the code which still need to be optimized (bottom figure).

8.7   Conclusions

In summary, we present FFTC, our high-performance design to parallelize the 1D FFT on the Cell/B.E. processor. FFTC uses an iterative out-of-place approach and we focus on FFTs with 1K to 16K complex input samples. We describe our methodology to partition the work among the SPEs to efficiently parallelize a single FFT computation. The computation on the SPEs is fully vectorized with other optimization techniques such as loop unrolling and double buffering. The algorithm requires a synchronization among the SPEs after each stage of FFT computation. Our synchronization barrier is designed to use inter-SPE communication only without any intervention from the PPE. The synchronization barrier requires only $2 \log p$ stages (p: number of SPEs) of inter-SPE communication by using a tree-based approach. This significantly improves the performance, as PPE intervention not only results in a high communication latency but also results in sequentializing the synchronization step. We achieve a performance improvement of over 4 as we vary the number of SPEs from 1 to 8. We also demonstrate FFTC's performance of 18.6 GFlop/s for an FFT with 8K complex input samples and show significant speedup in comparison with other architectures. Our implementation outperforms Intel Core Duo (Woodcrest) for input sizes greater than 2K and to our knowledge we have the fastest FFT for these ranges of complex input samples.

Acknowledgments

This work was supported in part by an IBM Shared University Research (SUR) award and NSF Grants CNS-0614915, CAREER CCF-0611589, and DBI-0420513. We would also like to thank Sidney Manning (IBM Corporation) for providing valuable input during the course of our research. We acknowledge our Sony-Toshiba-IBM Center of Competence for the use of Cell Broadband Engine resources that have contributed to this research.

Bibliography

[1] R.C. Agarwal and J.W. Cooley. Fourier transform and convolution subroutines for the ibm 3090 vector facility. *IBM J. Res. Dev.*, 30(2):145–162, 1986.

[2] R.C. Agarwal and J.W. Cooley. Vectorized mixed radix discrete Fourier transform algorithms. *Proc. of the IEEE*, 75(9):1283–1292, 1987.

[3] M. Ashworth and A.G. Lyne. A segmented FFT algorithm for vector computers. *Parallel Computing*, 6(2):217–224, 1988.

[4] A. Averbuch, E. Gabber, B. Gordissky, and Y. Medan. A parallel FFT on an MIMD machine. *Parallel Computing*, 15:61–74, 1990.

[5] D.H. Bailey. A high-performance FFT algorithm for vector supercomputers. *Intl. Journal of Supercomputer Applications*, 2(1):82–87, 1988.

[6] G. Bruun. z-transform DFT filters and FFT's. *Acoustics, Speech and Signal Processing, IEEE Transactions on*, 26(1):56–63, Feb 1978.

[7] C. Burrus and P. Eschenbacher. An in-place, in-order prime factor FFT algorithm. *IEEE Transactions on Acoustics, Speech and Signal Processing*, 29(4):806–817, Aug 1981.

[8] S.C. Chan and K.L. Ho. On indexing the prime factor fast Fourier transform algorithm. *Circuits and Systems, IEEE Transactions on*, 38(8):951–953, Aug 1991.

[9] S. Chellappa, F. Franchetti, and M. Püeschel. Computer generation of fast Fourier transforms for the Cell Broadband Engine. In *ICS '09: Proceedings of the 23rd International Conference on Supercomputing*, pages 26–35, New York, NY, USA, 2009. ACM.

[10] L. Chen, Z. Hu, J. Lin, and G.R. Gao. Optimizing the fast Fourier transform on a multi-core architecture. In *Parallel and Distributed Processing Symposium, 2007. IPDPS 2007. IEEE International*, pages 1–8, March 2007.

[11] T. Chen, R. Raghavan, J. Dale, and E. Iwata. Cell Broadband Engine Architecture and its first implementation. Technical Report, November 2005.

[12] A.C. Chow, G.C. Fossum, and D.A. Brokenshire. A Programming Example: Large FFT on the Cell Broadband Engine. *Tech. Conf. Proc. of the Global Signal Processing Expo (GSPx)*, 2005.

[13] L. Cico, R. Cooper, and J. Greene. Performance and Programmability of the IBM/Sony/Toshiba Cell Broadband Engine Processor. White paper, 2006.

[14] J. Cooley, P. Lewis, and P. Welch. Application of the fast Fourier transform to computation of fourier integrals, Fourier series, and convolution integrals. *IEEE Transactions on Audio and Electroacoustics*, 15(2):79–84, Jun 1967.

[15] J.W. Cooley and J.W. Tukey. An algorithm for the machine calculation of complex Fourier series. *Mathematics of Computation*, 19(90):297–301, 1965.

[16] IBM Corporation. Cell Broadband Engine technology. `http://www.alphaworks.ibm.com/topics/cell`.

[17] IBM Corporation. The Cell project at IBM Research. `http://www.research.ibm.com/cell/home.html`.

[18] B. Flachs and *et al.* A streaming processor unit for a Cell processor. In *International Solid State Circuits Conference*, volume 1, pages 134–135, San Fransisco, CA, USA, February 2005.

[19] M. Frigo and S.G. Johnson. FFTW on the Cell Processor. `http://www.fftw.org/cell/index.html`, 2007.

[20] N.K. Govindaraju and D. Manocha. Cache-efficient numerical algorithms using graphics hardware. *Parallel Comput.*, 33(10-11):663–684, 2007.

[21] H.P. Hofstee. Real-time supercomputing and technology for games and entertainment. In *Proc. SC*, Tampa, FL, November 2006 (Keynote Talk).

[22] L.B. Jackson. *Signals, systems, and transforms*. Addison-Wesley Longman Publishing Co., Inc., Boston, MA, USA, 1991.

[23] C. Jacobi, H.-J. Oh, K.D. Tran, S.R. Cottier, B.W. Michael, H. Nishikawa, Y. Totsuka, T. Namatame, and N. Yano. The vector floating-point unit in a synergistic processor element of a Cell processor. In *Proc. 17th IEEE Symposium on Computer Arithmetic*, pages 59–67, Washington, DC, USA, 2005. IEEE (ARITH '05) Computer Society.

[24] J.A. Kahle, M.N. Day, H.P. Hofstee, C.R. Johns, T.R. Maeurer, and D. Shippy. Introduction to the Cell multiprocessor. *IBM J. Res. Dev.*, 49(4/5):589–604, 2005.

[25] M. Kistler, M. Perrone, and F. Petrini. Cell multiprocessor communication network: Built for speed. *IEEE Micro*, 26(3):10–23, 2006.

[26] D. Kolba and T. Parks. A prime factor FFT algorithm using high-speed convolution. *IEEE Transactions on Acoustics, Speech and Signal Processing*, 25(4):281–294, Aug 1977.

[27] D.G. Kornand and J.J. Lambiotte. Computing the Fast Fourier Transform on a Vector Computer. *Mathematics of Computation*, 33:977–992, 1979.

[28] Y.-T. Lin, P.-Y. Tsai, and T.-D. Chiueh. Low-power variable-length fast Fourier transform processor. *Computers and Digital Techniques, IEE Proceedings -*, 152(4):499–506, July 2005.

[29] S.J. Orfanidis. *Introduction to signal processing*. Prentice-Hall, Inc., Upper Saddle River, NJ, USA, 1995.

[30] M.C. Pease. An adaptation of the fast Fourier transform for parallel processing. *J. ACM*, 15(2):252–264, 1968.

[31] D. Pham and *et al.* The design and implementation of a first-generation Cell processor. In *International Solid State Circuits Conference*, volume 1, pages 184–185, San Fransisco, CA, USA, February 2005.

[32] C.M. Rader. Discrete Fourier transforms when the number of data samples is prime. *Proceedings of the IEEE*, 56(6):1107–1108, June 1968.

[33] R. Singleton. An algorithm for computing the mixed radix fast Fourier transform. *IEEE Transactions on Audio and Electroacoustics*, 17(2):93–103, Jun 1969.

[34] Sony Corporation. Sony release: Cell architecture. `http://www.scei.co.jp/`.

[35] P. N. Swarztrauber. Vectorizing the FFTs. *Parallel Computations*, 1982.

[36] S. Williams, J. Shalf, L. Oliker, S. Kamil, P. Husbands, and K. Yelick. The potential of the Cell processor for scientific computing. In *Proc. 3rd Conference on Computing Frontiers (CF '06)*, pages 9–20, New York, NY, USA, 2006. ACM Press.

[37] C.-X. Zhong, G.-Q. Han, and M.-H Huang. Some new parallel fast Fourier transform algorithms. In *Parallel and Distributed Computing, Applications and Technologies, 2005. PDCAT 2005. Sixth International Conference on*, pages 624–628, December 2005.

This chapter is based on the authors' previously published work:

- D.A. Bader and V. Agarwal, "FFTC: Fastest Fourier Transform for the IBM Cell Broadband Engine," *14th IEEE International Conference on High Performance Computing* (HiPC), Springer-Verlag LNCS 4873, 172–184, Goa, India, December 2007.

- D.A. Bader, V. Agarwal, and S. Kang, "Computing Discrete Transforms on the Cell Broadband Engine," *Parallel Computing*, 35(3):119–137, 2009.

Chapter 9

Implementing FFTs on Multicore Architectures

Alex Chunghen Chow

Dell Inc.

Gordon C. Fossum

IBM Corporation

Daniel A. Brokenshire

IBM Corporation

9.1   Introduction

This chapter demonstrates how a traditional algorithm, Fast Fourier Transform (FFT), can exploit typical parallel resources on multicore architecture platforms to achieve near-optimal performance. FFT algorithms are of great importance to a wide variety of applications ranging from large-scale data analysis and solving partial differential equations, to the multiplication of large integers. The fundamental form of FFT can be traced back to Gauss (1777-1855) and his work in the early 19th century [17]. After computers emerged in the mid-20th century, this computationally efficient algorithm was rediscovered [8] and subsequently mapped to different modern computer architectures.

Modern processors and their Single Instruction Multiple Data (SIMD) architectures allow a programmer to efficiently execute massively data-parallel algorithms. The FFT algorithm is data-parallel in nature. But its data access transactions require a programmer to arrange input data and intermediate results in specific patterns to avoid unnecessary data movement. The associated computations also need to be properly scheduled to avoid unnecessary memory load and store operations [2, 15, 25].

Finding an optimal FFT algorithm for a given architecture is non-trivial. The scheduling of computation and data movement in order to reduce data movement overhead (e.g. register spill) is an NP-hard problem [23]. With the advent of multicore architectures, the additional levels of memory hierarchy make it even harder to find an optimal FFT implementation.

This chapter first reviews the fundamentals of the FFT algorithm including its computation, data movement and data preparation. Computations are the numeric operations required in an FFT algorithm. Data movement refers to the operations that move the data between the different data storage locations in the system. Data preparation refers to the steps required to organize the data for efficient utilization by the computation units. These three components provide the framework for high-performance design of FFTs.

Each of these components are then considered with respect to each level of a typical multicore memory hierarchy including the register file memory, private core memory, shared core memory and system memory. Each memory level has different considerations for computation, data movement and data preparation. Particular FFT optimization techniques for each aspect are explained. These techniques have been proven to provide significant performance improvement in many FFT implementations. The same paradigm can be similarly applied to the development of other types of multicore applications.

The principles of automatic FFT generation leverage the strategies for efficient design of an FFT over the available computational resources. And finally, a case study of a large 3-D FFT across a large cluster of multicore systems is presented.

Because of the many multicore hardware features available in the Cell Broadband Engine[TM](Cell/B.E.) architecture [18], the Cell/B.E. processor is frequently referenced to demonstrate typical parallel hardware resources and constraints existing in modern multicore architectures.

9.2 Computational Aspects of FFT Algorithms

An FFT is an efficient algorithm for computing discrete Fourier tranforms (DFTs). The size-N DFT is defined by the formula below where x_0, x_1, ... x_{N-1} are complex numbers.

$$X_k = \sum_{n=0}^{N-1} x_n e^{-\frac{2\pi i}{N}nk} \quad k = 0, ..., N-1 \tag{9.1}$$

In the case of radix-2, decimation-in-time, FFT, the DFT formula is re-expressed as two size-N/2 DFTs from the even-number-indexed elements (E_j) and the odd-number-indexed elements (O_j).

$$
\begin{aligned}
X_k &= \sum_{n=0}^{N-1} x_n e^{-\frac{2\pi i}{N}nk} \\
&= \sum_{m=0}^{\frac{N}{2}-1} x_{2m} e^{-\frac{2\pi i}{N}(2m)k} + \sum_{m=0}^{\frac{N}{2}-1} x_{2m+1} e^{-\frac{2\pi i}{N}(2m+1)k} \\
&= \sum_{m=0}^{M-1} x_{2m} e^{-\frac{2\pi i}{M}mk} + e^{-\frac{2\pi i}{N}k} \sum_{m=0}^{M-1} x_{2m+1} e^{-\frac{2\pi i}{M}mk} \\
&= \begin{cases} E_k + e^{-\frac{2\pi i}{N}k}O_k & for\ k < M \\ E_{k-M} - e^{-\frac{2\pi i}{N}(k-M)}O_{k-M} & for\ k \geq M \end{cases}
\end{aligned}
\tag{9.2}
$$

Because of the periodicity of the root of unity, it holds true that $E_{k-M} = E_k$ and $O_{k-M} = O_k$. This fact provides a recursive way to divide and conquer the original $O(N^2)$ DFT computational complexity and yield an $O(Nlog_2N)$ complexity.

The two size-$N/2$ DFTs come from the even and odd elements, respectively, along the time series. Such FFTs are called a decimation-in-time (DIT). Another form of FFT is decimation-in-frequency (DIF) [20] which has exactly the same computational complexity.

Regardless of the choice of FFT, DIT or DIF, the computation consists of repeatedly applying a simple computational kernel over an array of elements as defined by equation (9.2). This computation kernel is customarily represented as a *butterfly* dataflow graph as shown in Figure 9.1. By combining these butterflies with their complex roots of unity W (traditionally referred to as twiddle factors [15]), a complete FFT is achieved.

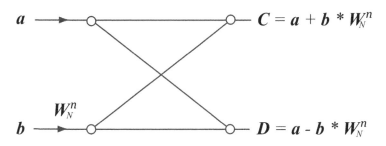

FIGURE 9.1: Radix-2 butterfly dataflow graph.

9.2.1 An Upper Bound on FFT Performance

Based on the computation counts of all butterflies, an upper bound on the performance for a particular target system can be established. Even though there are several choices for near-optimal FFT algorithms with different numbers of operations [9–11, 19], no proof is yet known for the mathematical question of what is the minimal number of floating-point operations (FLOPS) required to compute a DFT of a given size N.

For modern multicore processors which support fused multiply-add (FMA) instructions, there is a good choice. The simplest radix-2 FFT algorithm not only offers the highest usage of FMA instructions, its simplicity in implementation also serves as a good vehicle to demonstrate a practical optimization approach for multicore platforms [21].

The basic DIT radix-2 butterfly is used as the building block for most implementations, where (a, b) is the input, (C, D) is the output and W_i is the twiddle factor. Without FMA instructions, a butterfly requires 10 floating-point real number add or multiply operations. With FMA instructions, the butterfly is reduced to 6 floating-point multiply-add operations.

$$D_r = ((a_r - b_r * W_{Nr}^n) - b_i * W_{Ni}^n) \tag{9.3}$$
$$D_i = ((a_i - b_i * W_{Nr}^n) - b_r * W_{Ni}^n) \tag{9.4}$$
$$C_r = 2.0 * a_r - D_r \tag{9.5}$$
$$C_i = 2.0 * a_i - D_i \tag{9.6}$$

Subscripts r and i denote real and imaginary parts of a complex number, respectively.

Additional floating-point multiplications can be eliminated for certain simple twiddle factors where $n = 0$, $\frac{N}{2}$, $\frac{N}{4}$, etc. When applying this optimization on butterflies with $0, \frac{N}{2}, \frac{3N}{4}$ twiddle factors, the number of operations is reduced from 6 add/subtractions per butterfly to 4; a savings of 2 operations per butterfly. The operation count for an FFT of size $N \geq 8$ thus becomes:

$$3 * N * \log_2(N) - 4 * N + 8 \tag{9.7}$$

All the butterfly computations at the same recursive level are independent from each other and can be executed in parallel. SIMD instructions can be applied here to achieve multiples of performance gain. Using the standard FLOP formula, $5 * N * \log_2(N)/time$ (taken from benchFFT) a typical core that supports 4-way, single precision, SIMD, FMA operations is capable of achieving a FLOP rating that paradoxically exceeds the true processor capability. Figure 9.2 graphically depicts the potential performance upper bound.

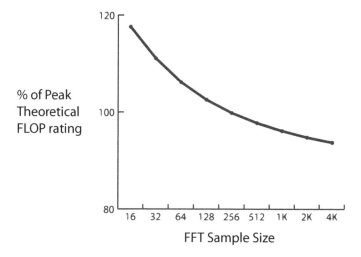

FIGURE 9.2: Upper bound percent theoretical peak FLOPS rating as a function of the FFT size.

9.3   Data Movement and Preparation of FFT Algorithms

Another aspect that affects the FFT performance is the overhead of moving and preparing data for the butterflies. Such overhead is not directly expressed in the formulas in equation (9.2) and has often been ignored by many FFT studies.

Because the size of the input array of an FFT is usually larger than that of the processor's register file, the data have to spill over more than one level of memory hierarchy. Moving data among the levels of the memory hierarchy is inevitable. More levels of memory hierarchy exist in multicore platforms than those of traditional single-core platforms. This data movement overhead can affect final performance more than the computational portion on many of today's computer systems.

A good algorithm not only needs to minimize the frequency of unnecessary data movement, it also needs to make data movement as efficient as possible. Usually, the algorithm must match the memory granularity defined by the hardware, placing additional restrictions on the data movement design.

Even when data are not transferred from one memory level to another, their location in the storage may not match what is required by the computation unit. For example, a typical single-precision SIMD operation requires four adjacent data elements to be operated by the same SIMD instruction. The sequential layout is not always the case for FFT arrays.

Figure 9.3 shows the dataflow graph of a small size-8 DIT FFT where $W_8^k = e^{-\frac{2\pi i}{8}k}$ are the twiddle factors. The graph illustrates one of the many ways to store the data array between stages. In this particular example, the index of the final output array is bit-reversed. This results from the ordering of the sub-FFT computations and also the recursive decimation of the even and odd parts of the sub-arrays. SIMD instructions likely cannot operate over such a decimated array without preparation of the data into a SIMD-consumable format.

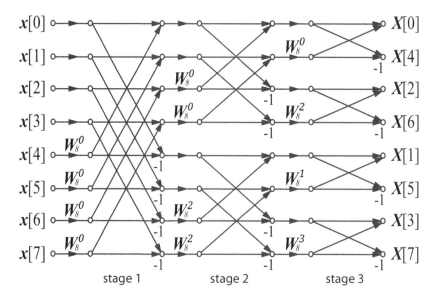

FIGURE 9.3: A size-8 DIT FFT dataflow graph.

Finding the most effective data movement and proper data layout for FFT is similar to the problem a compiler uses to map a directed acyclic graph (DAG) onto a finite number of registers while avoiding register spills. It has been demonstrated that, if the DAG is not ordered, this is an NP-hard problem [23]. With the many butterfly computations, their input and output data and dependencies and a set of hardware constraints, it is not surprising that FFT designers find it difficult to fully optimize their implementations over multicore platforms.

Proper scheduling of the data movement along the memory hierarchy and hardware-friendly data preparation are, in fact, two of the most rewarding steps in the algorithm design. This chapter discusses techniques and algorithms in these two areas. The same performance design principles can scale from instruction level, up to single core level and up to multicore and multinode systems.

9.4 Multicore FFT Performance Optimization

Because data movement is a significant factor affecting performance, it is important to optimize the FFT stages according to the system's memory hierarchy. At each level in the hierarchy, one should exploit all available hardware parallelism and at the same time minimize the effect of data access to the next level of memory hierarchy. By using this heuristic approach, one can achieve a performance close to the upper bound previously established. In a sense, the design process schedules the DAG and reduces the complexity of the design problem from NP-hard to linear [3] while ensuring that the final performance is nearly optimal.

One hardware parallelism technique proven to be effective is exploiting concurrency between the computational operations and data movement operations. Such a feature is common in modern CPU design. By allowing the core to move data while computing on another block of data, the overhead of data movement can be mostly hidden behind the time the computational instructions are being executed. Effectively, this results in virtually no overhead for data movement and near-optimal performance is achieved.

Such concurrency exists in all levels of the memory hierarchy of Cell/B.E. architecture [18]. At the register level, the Synergistic Processing Element (SPE) load and store instructions are executed on an instruction pipeline which is different from the one used by the computational instructions. These two activities can occur at the same time. At the private core memory level (e.g. local store), the direct memory access engine can transfer blocks of data to and from system memory at the same time the SPEs are executing their programs. While all these activities are occurring, the PowerPC core can further transfer data between the network and system memory without affecting computation and data movements at the other levels.

For large FFTs that cannot completely fit into one level of memory, the principle of general factorizations of the Cooley-Tukey FFT algorithm can be applied to allow the larger FFT to be divided into a set of smaller sub-FFTs for execution on the specific memory hierarchy level. Because the computational complexity of FFT is $O(Nlog_2N)$, the larger the N that can execute on a particular level, the less data, $O(1/log_2N)$, that need to transferred per computation unit. Therefore, it is best to execute as large an FFT as possible on each memory level.

9.5   Memory Hierarchy

Many factors contribute to the construction of a multi-level memory hierarchy. As the speed of the computation unit increases due to improved architecture and semiconductor process technology, the relative distance from the data storage to the computation units also increases. The proximity of the computation unit thus becomes limited. CPU designers can only fit a relatively small amount of data storage around the computation units and must therefore utilize more distant memory to hold the entire data set.

Multicore architectures generally have more levels of memory hierarchy than traditional processor designs. Not only do the designers provide memory close to the execution units (e.g. register file and private core memory), they also provide shared memory or cache (e.g. level 1 cache) among the closely located cores that allows the cores to do high-speed common data accesses. These near memories have drastically different access speeds than those located farther away, such as level 2 and level 3 cache, or even system memory.

Moving data across the memory hierarchy levels can sometimes be expensive. When a programmer properly avoids unnecessary data movement across the memory levels, dramatic performance improvements often result. Therfore, the programmer must understand the dataflow pattern of a particular function before successfully mapping the dataflow to the memory hierarchy.

The generic FFT algorithm has a large amount of parallelism. However, it does not specify the sequence of the butterfly computations or data movement within a given stage. For example in the size-8 DIT FFT (see Figure 9.3) a programmer can choose to perform the first four butterflies in the first stage and then the next four in the second stage and so on.

Assuming for the moment that this algorithm is to be run on a system with a private core memory capacity of 4 points, the algorithm will have to move the 8-point array between the private core memory and its next higher memory level a total of three times. But if we choose to compute positions 0, 2, 4 and 6 of stage 1 and stage 2 first, and then do the same for positions 1,

3, 5 and 7, the number of times the data must be moved is reduced to two. If the cost of the data movement between memory hierarchy levels is high, the performance can be significantly improved by simply rearranging the order of butterfly computations.

This level of knowledge cannot be derived automatically by a compiler from the implementation of a generic FFT. A programmer usually has to explicitly sequence the computation and its data movement. Many traditional parallel FFT implementations are designed just to avoid such data movement overhead [25].

Decades of FFT research have provided many types of algorithms and implementations for parallel machines [14]. These implementations adopted many innovative steps to make use of the memory hierarchy. Because of the fewer levels of the memory hierarchy in previous architectures, the legacy FFT implementation also cannot be easily ported to take advantage of these more numerous levels of memory hierarchy. Many of the parallel algorithms need to be rewritten or combined with other FFT techniques in order to scale over these levels without performing redundant data movement.

The required data movement across levels derives from the fact that the input arrays and the arrays that hold the intermediate results are larger than the size of a particular memory level. In order to host the larger array, the data must be moved in from the larger memory store to the smaller but faster memory, and later moved out again after computation. The principle of general factorizations within the Cooley-Tukey FFT algorithm can be applied here to allow the larger FFT to be divided into a set of smaller sub-FFTs for execution on the specific memory level.

It is also important to reuse the resident data as much as possible at the specific memory level before they are moved out to the next level. As previously demonstrated, combining the computation of the decimation stages, can reduce the data movement. If one is able to factor a 2^m-sized FFT, instead of a size-2 butterfly on the same level, this effectively reduces the data movement requirement m-fold. It is quite straightforward to conclude that one should use maximally factored FFTs for each particular memory hierarchy in order to reduce data movement.

There is however an upper bound in the benefit of a larger size. In many multicore architectures, computation and data movement can be done in parallel. By overlapping the computation and data movement, one can hide the cost or latency of data movement. The load/store of the data used by the butterfly can benefit from this parallelism greatly. One can load/store the data of the next/previous butterflies respectively while computing the current butterfly. If the computation of a butterfly takes longer than the load and store of the data, the FFT algorithm is *compute-bound*, otherwise, it is *memory bound*.

The computational complexity of FFT is $O(N \log_2 N)$ whereas the data movement overhead is $O(N)$. As N increases, the time needed to compute the butterflies increases faster than the time needed to move the data. At certain N, when the computation time is greater than the data movement time, increasing N at that memory level will no longer benefit the final performance because all the data movement latencies are hidden behind the computation, even though the amount of the data transfer is still reduced.

9.5.1 Registers and Load and Store Operations

The nearest memory in the memory hierarchy is the register file. With some architectures having up to 128 named registers [18], the register file can be exploited to achieve significant data reuse.

9.5.1.1 Applying SIMD Operations

SIMD is a very cost-effective hardware construct to increase the performance for data parallel applications. The SIMD instructions however require that the data elements of the same kind be arranged sequentially together so that the same SIMD operations can be applied. The register file memory hierarchy level doesn't experience the expense of data movement; however, preparing data for SIMD operations can introduce overhead that affects performance significantly.

FFTs have a unique characteristic that the neighboring data go through the same butterfly computations. The algorithm appears to be a perfect candidate for SIMD operations. However, observing the connections among the butterflies (see Fgure 9.3), the connection is not quite uniform across all stages. Assuming that it is a single-precision floating-point FFT, the first stage can be easily accomplished by one SIMD butterfly. However, the second and third stages do not have sufficient neighboring elements to achieve efficient SIMD computation. Reordering the data is required to arrange the elements for a SIMD butterfly. This is expensive, compared to the simple butterfly computation.

Also mentioned previously, the decimation process of the FFT algorithm results in a bit-reversed index of the final output array elements. The pattern suggests that SIMD operations are not applicable to the final stage butterflies.

The general factorization of the Cooley-Tukey algorithm allows one to break down a larger FFT into smaller FFTs. A technique for rearranging the array in an efficient manner that allows SIMD operations throughtout the entire FFT is possible [24]. For an FFT of size $N = N_1 * N_2$, the computation steps are:

(1) Perform N_1 FFTs of size N_2

(2) Multiply by complex roots of unity twiddle factors and transpose

(3) Perform N_2 FFTs of size N_1

For cases where N_1 and N_2 are both greater than and also multiples of the size of SIMD vector, both steps 1 and 3 can be performed in an SIMD manner because there are adequate SIMD multiples of independent FFTs in both steps. However, the input data of the smaller FFTs must still need to be reordered.

The arrangement of the data sequence is illustrated using index manipulations. Since most multicore SIMD processors operate on four single-precision elements at the same time, the smallest size FFT which can be implemented in an SIMD operation is 16. A 16-point FFT can be factored into two stages of four 4-point FFTs. A smaller size will result in too few independent FFTs for efficient SIMD utilization.

Figure 9.4 shows the array address bit manipulations performed for each of the three steps.

	msb			lsb
Initial data array	bit 3	bit 2	bit 1	bit 0
After step 1	bit 2	bit 3	bit 1	bit 0
After step 2	bit 1	bit 0	bit 2	bit 3
After step 3	bit 0	bit 1	bit 2	bit 3

FIGURE 9.4: Array bit manipulations.

In step 1, one single 4-point SIMD FFT is performed. Because bit 1 and bit 0 are for the adjacent four elements, they remain adjacent at the end of first step. The 4-point SIMD FFT operates on vector elements indexed by bits 3 and 2. Due to the bit-reversed indexing at the output of the FFT, bit 2 and bit 3 must be reversed at the output of step 1.

In step 2, the array is multiplied by proper twiddle factors and a 4×4 matrix transpose is applied. This transpose can readily be accomplised using a permute or shuffle instruction. Thus, indices 1 and 0 are now the indices of the higher 2 bits.

Step 3 consists of another four 4-point SIMD FFTs. At the same time, the lower two bits (bits 2 and 3) remain adjacent and bit 1 and bit 0 are reversed. The resulting reversed index bits match the desired position of the output array. By proper manipulation of index bits through matrix transposition, full use of SIMD operations can be achieved in all three steps. The same factorization and transpose principle can be applied to any 2-based, single-precision FFT size larger than 16.

9.5.1.2   Instruction Pipeline

An instruction requires certain duration to complete its execution. In order to maximize the use of hardware circuits, instruction execution is often divided into stages that allow more than one instruction executed concurrently in different stages of an instruction pipeline. Typically, an independent operation can be issued every cycle with the result available several cycles later.

This situation becomes more complicated when dependencies exist among the instructions. For example, an instruction may need the computation result from the previous one. The hardware pipeline at this level stalls the depending instruction until the required condition is met.

The computational instruction of a butterfly has two independent parts of computation, the real part and the imaginary part. These two groups of independent instructions can enter the pipeline and improve the efficiency of the core. However, one single SIMD butterfly is not sufficient to consume all the pipeline stages of the CPU.

The pipeline efficiency can be improved by increasing the number of independent butterflies being simultaneously computed. This can easily be achieved for larger sized FFTs which can be factored into larger numbers of sub-FFTs.

9.5.1.3   Multi-Issue Instructions

Most CPUs can issue more than one instruction on independent pipelines every cycle. For example, the Cell/B.E. SPU has two execution pipelines. Computational instructions execute on the even pipeline whereas the load and store instructions execute on the odd pipeline. In other words, the SPU can compute based on the register file values while load and store instructions simultaneously transfer data between the register file and private core memory. This enables a very important technique that allows a programmer to hide the cost of data movement behind the execution of computation even at this closest memory hierarchy level.

In order for data movement latency hiding to be effective, the computation time must be larger than the data movement time.

As previously explained, the complexity of FFT computation gives a good hint on how to balance the computation and the required data movement. Because the computational complexity of FFT is $O(N \log_2 N)$, a larger N to be executed at a particular level, the less data, $O(1/\log_2 N)$, needed to transfer per computation unit. The size of N will also depend on the memory characteristics at that level.

A butterfly using FMA operations requires 6 floating-point instructions, but 10 load/store instructions including the twiddle factors. The computation is not long enough to cover the data movement. Such implementation will be bounded by the data movement. However, if a radix-4 butterfly is used, then there are 24 compute instructions and 24 load/store instructions. This gave us a critical match between computation and data movement. A higher order radix butterfly (e.g. radix-8) may still be required to further reduce pressure on the memory subsystem.

Without the penalty of the data preparation or movement, the FLOPS achieved at register level is often very close to the upper bound shown in

Figure 9.2. The larger the register file, the more room the programmer has to increase the size of the sub-FFT and reduce the load/store pressure from/to private core memory.

9.5.2 Private and Shared Core Memory, and Their Data Movement

For large FFTs, the data may not fit within the private core memory. Most memory systems provide mechanisms to transfer blocks of data between the private core memory and system memory.

The modern design of memory controllers also allows the execution unit to operate independent of the data transfer. Some architectures, like the Cell/B.E., provide application control of a direct memory access (DMA) engine that can be used to programmatically move data between system memory and private core memory. Other architectures provide cache prefetch instructions or prefetch engines to facilitate application-scheduled data movement. This parallelism again allows programmers to hide the overhead of the data movement behind computations.

9.5.2.1 Factorization

The same optimization techniques at the register level can again be applied at this level. The Cooley-Tukey general factorization is used to divide the large FFTs into smaller ones that fit in the private core memory. The smaller ones however need to be as large as possible so that the data movement to the next level of the memory hierarchy is minimized. If the computation requirement is not balanced against the data movement overhead, it is likely that the implementation will be memory bound at this level.

9.5.2.2 Parallel Computation on Shared Core Memory

One very important multicore architecture design is the availability of an on-chip shared core memory accessible by all cores. Such memory benefits performance greatly. It not only provides a fast storage for commonly used data among the cores, the same hardware construct also serves as the foundation of fast synchronization protocol among the executing cores.

In the case of an FFT, shared core memory is the level where more than one core contributes to the computation of one FFT. The proper scheduling of sub-FFTs onto the cores is a very important task in performance planning. Because of the parallel natures among all the sub-FFTs at the same stage, job distribution is actually of a lesser issue than the preparation of the data for the consumption of the cores.

Because of the added computation capacity from the multiple cores, FFT performance above this level is likely to be bounded by data movement. Cores may constantly wait for data movement to complete. Application developers often combine the butterfly computations with other computations such as image filters to make use of these unused cycles to improve the final application performance.

9.5.3   System Memory

The distance and latency to access the system memory is orders of magnitude longer than accessing the registers or even private core memory. The memory system also introduces other kinds of constraints such as memory bank and pages. Reducing the data preparation and movement overhead becomes an even more important design consideration [2].

Index bit manipulation technique [16], such as the stride-by-1 algorithm [1, 7], allows the FFT array to be partitioned naturally and at the same time scheduling the data access pattern to meet hardware constraints.

For most multicore memory systems, the data transfer between private core memory, shared core memory and system memory has an optimal sized hardware granularity. Accessing data blocks sizes that are not multiples of the hardware granularities reduces the data transfer efficiency significantly. Thus, for an FFT implementation to achieve an optimal performance, the number of contiguous sample points must be arranged to be the multiples of the access granularity.

9.5.3.1   Index-Bit Reversal of Data Block Addresses

The bit reverse addressing beyond the memory granularity is still needed, though, to account for the decimation process. Different addressing techniques can be applied here depending on the available memory addressing mode of the target instruction system architecture. The most straightforward approach is to use straight line load and store sequences with absolute addressing instructions. In many cases, the performance gain and design simplicity out weigh the relatively small amount of code size increase.

The data block load and store operations can usually be performed independent of the computations; techniques such as double buffering can be used to effectively hide the data movement overhead. Double buffering uses two buffers so that the load and store operations move data on one buffer while the computations operate on the other buffer. This technique can be applied throughout the memory hierarchy where computation and data movement activity can operate in parallel.

9.5.3.2   Transposition of the Elements

Using the same technique as in the SIMD FFT algorithms, a matrix transpose is needed at some point to move the lower index bits to the upper portion and preserve the availability of contiguous samples. Instead of 2 bits for 4 contiguous SIMD samples, one may now use 16, 32 or even more contiguous samples.

The transposition is potentially a very expensive operation because it should be applied at the element level which is below the hardware granularity or SIMD vector level. Shuffle, permute or similar instructions are extremely useful in performing this transposition.

9.5.3.3 Load Balancing

Load balancing and synchronization are usually important performance considerations for parallel applications. The factorized FFTs can be naturally divided into a relatively small number of factors which can be equally distributed to the available cores. Therefore, load balancing is not a significant issue nor is the synchronization. Each factored group of sub-FFTs can do a minimal synchronization before entering into the computation of another factored group. The synchronization overhead among the cores is very small compared to the overall computations.

Overall, proper factorization against the availability of the private core memory and shared core memory, balancing the computation against the data movement, the choice of the index-bit manipulation and transposition operations are the common performance challenges to any large FFT implementation on a multicore platform.

9.6 Generic FFT Generators and Tooling

With the many levels of memory hierarchy, good planning of data movement arguably is the most important consideration for an FFT implementation on multicore platforms. For the many newer multicore platforms, it is often necessary for FFT developers to redesign FFT data movement and data preparation in order to accommodate different memory hierarchies. When platform-specific resources are exploited, architecture-aware approaches usually achieve superior results [4, 22].

The previous sections outlined a series of bottom-up techniques to achieve good data movement design. The outlined design approach makes it possible to design FFT performance tools and code generators that can be quickly retargeted against new multicore architectures and thus minimize software development cost.

9.6.1 A Platform-Independent Expression of Performance Planning

The outlined approach essentially provides a performance planning algorithm based on the implicit regularity of FFTs and the principle of data movement reduction. The performance planning heuristically reduces the large mapping space of an FFT DAG into a much smaller set [13] of near optimal solutions.

Traditional programming environments however do not provide a platform-independent way to capture the performance planning knowledge for the data movement on different levels of the memory hierarchy. Programmers are frequently forced to hardcode such planning knowledge in a platform-specific manner which may be expensive to port to other multicore platforms [12]. The platform-specific details include number of usable memory hierarchy levels, memory size at each level, physical characteristics and data movement primitives (e.g. load/store, caching, DMA, etc.). All these are different from platform to platform.

Thus, to enable a multicore FFT generator, one must first devise an expression system to capture the reusable performance planning knowledge. For example, System Entity Structure (SES) has been proposed as a platform-neutral vehicle to describe FFT performance planning knowledge [6]. SES is an effective method of Knowledge Representation and data engineering in support of Ontology development languages and environments based on SES Axioms [26]. When coupled with discrete event simulation formalism, DEVS [5], the whole framework enables an FFT performance evaluation system for different target platforms.

The FFT performance planning algorithm is described bottom-up hierarchically as an SES tree. The butterflies are the leaf nodes whereas each tree level maps to a potential level in a memory hierarchy.

The SES tree is general enough to allow the abstraction of all known physical computation (e.g. add, multiply and FMA), data movement (e.g. load, store and DMA) and memory constructs (e.g. register file and system memory) in a platform-independent manner.

9.6.2   Reducing the Mapping Space

Even though the resulting FFT SES description still represents a large number of possible implementations, the knowledge captured by the SES tree provides a partial scheduling of the DAG. The scheduling reduces the complexity of finding an optimal mapping to a linearly solvable task.

For a new platform, the platform-specific implementation of the FFT computation and data movement can then be associated with the respective primitive nodes. Because there are only a handful of primitive nodes of computation, data movements and data preparation, the cost of implementing the new primitives is manageable.

There may be more than one acceptable implementation with different computation kernel and data movement costs. The performance evaluation framework, based on DEVS formalism, is applied here to facilitate the selection of an optimized combination of computation kernel and data movement.

9.6.3 Code Generation

Once a solution is determined, the target code can then be generated by traversing the nodes of the tree. The SES tree allows the use of a multi-entity node to represent a set of nodes of the same type. Such representation maps well to the looping construct when needed.

Wherever sufficient space is available at a level of the memory hierarchy, it becomes possible to generate straight line code without causing the code size to exceed the system allowance. Straight line code conveniently eliminates the cost of the looping construct, which often has significant performance benefit.

The straight line code can also use direct-addressing mode to avoid address computation at runtime, with obvious performance implications for compute-bound code. The absolute addresses can be easily generated by the generator.

The generated code can either be in C or assembly language. Note that the data movement, data preparation and register allocation are already planned by the SES tree. The instruction scheduler here does not need to deal with the difficult problem of reducing register spills. Instead, the scheduler only needs to schedule the instructions to maximize the use of multi-issue pipelines, which is a much simpler task.

9.7 Case Study: Large, Multi-Dimensional FFT on a Network Clustered System

This section considers the computation of a 3-D, 16K × 16K × 16K, double-precision, complex-to-complex FFT, using 4096 networked nodes, each with a minimum of 17 or 18 GB of memory. The goal is to minimize the cost of moving data between the nodes. Each element of this large 3-D array is a double-precision complex number, consisting of an 8-byte real component followed by an 8-byte imaginary component.

There is a basic problem in all multi-dimensional FFT computations which has to be addressed, and it derives from two facts:

(1) A node must have all of the data along a particular direction, before it can do an FFT on that 1-D array. This means that any node performing an FFT in any direction needs all 16,384 elements along that line present in its memory.

(2) Moving data between compute devices can be expensive if the data are not organized in large contiguous chunks. *Large* is typically 512K bytes and up for network communication between nodes, and perhaps a cache line (e.g. 128 bytes) for communication between cores in a given node.

One good solution to this problem is to lay out the data in a tiled format rather than in long stripes. To deal with a large cluster, each of whose nodes has lots of cores, a doubly-tiled format is proposed consisting of small (2 × 2 × 2) tiles to facilitate movement between cores within a node, and large (256 × 256 × 4) tiles to facilitate movement between nodes in the network.

As is well known, a 3-D FFT is accomplished by performing 1-D FFTs in one direction, then performing 1-D FFTs in a second direction, and finally along the third direction. These three sets of FFTs will be referred to as "phases," and they will be accomplished in the standard order, first along the X-direction, then Y, and finally Z.

The number of FFTs to be done by each node in each of the three phases is 65,536 (16K * 16K / 4K). There are 4096 nodes and each one is initially loaded with 4096 of these 256 × 256 × 4 tiles, so that each node has a thin plate (four elements in z) spanning the entire 16K × 16K domain in x and y (see Figure 9.5(a)).

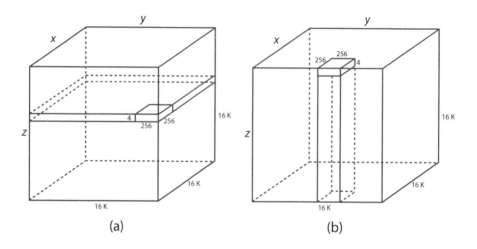

FIGURE 9.5: Graphical depiction of the FFT phases. (a) X-direction and Y-direction phase. (b) Z-direction phase. Figures are not drawn to scale.

Here's the particular layout of memory for tile t $(0 \leq t < 4096)$ in node n $(0 \leq n < 4096)$. Tile t in node n contains the following (x, y, z) elements of the array:

$$256 * (t \bmod 64) \leq x < 256 * ((t \bmod 64) + 1) \tag{9.8}$$

$$256 * (t \gg 6) \leq y < 256 * ((t \gg 6) + 1) \tag{9.9}$$

$$4 * n \leq z < 4 * (n + 1). \tag{9.10}$$

This layout is admittedly arcane, but it ensures that *no data movement needs to occur between nodes* as we transition from phase 1 (X-direction) to phase 2 (Y-direction). The idea behind the transition from phase 2 to phase 3 is this: as each node completes a tile column in the Y-direction, it can move the tiles in that column to the correct Z-direction (phase 3) nodes, such that the array element (x, y, z) is moved to node $(x \gg 8) + 64 * (y \gg 8)$. Thus, while the data were laid out in thin sheets along x and y for phases 1 and 2, they will be organized into fat columns along z for the third and last phase (see Figure 9.5(b)). As there are 64 such tile columns to be computed in each node, nearly complete overlap of computation and data movement can be accomplished.

The algorithm to process phase 2 with interleaved movement of data in preparation for phase 3 is as follows. Node n $(0 \leq n < 4096)$ starts executing Y-direction FFTs in tile column $(n \gg 6)$, processing tile columns in increasing order, modulo $(n \gg 6)$, until they're all done. This unusual ordering ensures that later data movement is evenly spread across the nodes, and doesn't overwhelm certain nodes with a huge amount of incoming data while other nodes aren't receiving any data at all. It also helps ensure minimal memory waste, if your system has less than 33 or 34 GB of memory per node. As a tile column is completed, its tiles are immediately scheduled for transfer to the proper node for Z-direction processing. The tile whose (tile-row, tile-column) value is (r, c) will be transfered to node $(c \ll 6) + r$. The memory layout for the Z-direction should be thought of in terms of 4096 tiles, where each tile corresponds to Z depth. So, tile t $(0 \leq t < 4096)$ contains z values $[4 * t, 4 * t + 3]$.

The destination memory location depends on how much memory you have in each node. If you happen to have at least 33 or 34 GB of memory per node, you can use plan A. Otherwise, use plan B.

Plan A: Having sufficient room for this luxury, the memory space for the Z data is allocated completely disjoint from the memory space where the data were originally loaded for X and Y processing. Each source node has exactly one tile destined for any given destination node. That one tile should be written into the memory location in the destination node whose index matches the node number of the source node. Synchronize just once at the end, before beginning phase 3 processing.

Plan B: A space corresponding to 64 large tiles is reserved somewhere in the destination node's memory, disjoint from the data originally loaded for X and Y processing. This space occupies 0.25 GB, and thus the large arrays in any node can be held in 16.25 GB of memory, total. It is assumed that the operating system and all software necessary to accomplish this can reside in the remaining 0.75 GB of our *17 GB system*. If not, then bump it up to 18 GB. All nodes compute their first column of tiles in the Y-direction, and the tiles in that column will be written into that reserved space, where the index into that space is just the least significant six bits of the node number of the source node. As it happens, exactly one of these 64 tiles in this first column ends up being a local copy within the same node. Subsequent columns will be written into the destination memory in a location within its original data area. When source node N $(0 \leq N < 4096)$ completes processing of its phase 2 tile column C $(0 \leq C < 64)$ each of the 64 tiles in that column index by

R $(0 \leq R < 64)$ is transferred to destination node $64 * C + R$ at tile offset $N - 64 \bmod 4096$. All nodes synchronize after each Y-direction tile column is processed, to guarantee that the destination memory is available.

After this point, the processing of the third phase can begin. If plan B was used, the code for computing the Z-direction FFTs will have to be aware of the unusual data layout.

9.8   Conclusion

The techniques presented in this chapter for FFTs are applicable to other multicore applications. It is established that developing software for parallel systems of varying designs is a non-trivial problem. In order to achieve a near-optimal implementation, the designers have to adopt a systematic approach that takes into account the attributes of both the application and target system.

A performance upper bound must first be established to determine an optimization metric. The upper bound is useful in identifying performance bottlenecks so programmers may make algorithmic adjustments that exploit additional parallel resources and ultimately remove the system bottlenecks.

The new parallel resources of the multicore architectures introduce a new programming challenge that has not been well supported by today's traditional programming languages. New language constructs are required for developers to specify performance design knowledge. Such knowledge is critical for a multicore compiler to map an algorithm and schedule operations effectively on the available computational resources. A successful implementation depends upon a deep understanding of the data access patterns, computation properties and available hardware resources. Applications like FFTs with regular data access and computation patterns can readily take advantage of generalized performance planning techniques to produce successful implementations across a wide variety of multicore architectures.

Bibliography

[1] D.H. Bailey. A high-performance FFT algorithm for vector supercomputers. *International Journal of Supercomputer Applications*, vol. 2, no. 1, 1988.

[2] D.H. Bailey. FFTs in external or hierarchical memory. *Journal of Supercomputing*, vol. 4, no. 1, 1990, pages 23–35.

[3] A.L. Belady. A study of replacement algorithms for virtual storage computers. *IBM Systems Journal*, vol. 5, no. 2, 1966, page 78.

[4] S. Chellappa, F. Franchetti, and M. Pueschel. Computer generation of fast Fourier transforms for the Cell Broadband Engine. *International Conference on Supercomputing*, 2009, pages 26–35.

[5] S. Cheon, D. Kim, and B.P. Zeigler. DEVS model composition by system entity structure. *Information Reuse and Integration*, 2008, pages 479–484.

[6] A.C. Chow. A multi-facet description of the Fast Fourier Transform algorithms in System Entity Structure. Paper presented at the High Performance Computing and Simulation System (HPC 2009).

[7] A.C. Chow, G. Fossum, and D. Brokenshire. A Programming Example: Large FFT on the Cell Broadband Engine. *Proceedings of GSPx 2005 Conference*.

[8] J.W. Cooley and J.W. Tukey. An algorithm for the machine calculation of complex Fourier series. *Math. Comput.*, vol. 19, no. 2, April 1965, pages 297–301.

[9] P. Duhamel. Implementation of split-radix FFT algorithms for complex, real, and real-symmetric data. *IEEE Transactions on ASSP*, vol. 34, April 1986, pages 285–295.

[10] P. Duhamel and H. Hollmann. Split-radix FFT algorithm. *Electronic Letters*, vol. 20, no. 1, 1984, pages 14–16.

[11] P. Duhamel and M. Vetterli. Fast Fourier transforms: A tutorial review and a state of the art. *Signal Processing*, 19, 1990, pages 259–299.

[12] K. Fatahalian, T.J. Knight, M. Houston, M. Erez, D.R. Horn, L. Leem, J. Park, M. Ren, A. Aiken, W.J. Dally, and P. Hanrahan. Sequoia: Programming the Memory Hierarchy. *Proceedings of the 2006 ACM/IEEE conference on Supercomputing*, pages 4.

[13] M. Frigo, 2003. Register Allocation in Kernel Generators. Presentation at Workshop on Automatic Tuning for Petascale Systems, 2007.

[14] M. Frigo and S.G. Johnson. FFTW: An adaptive software architecture for the FFT. *Proc. ICASSP 1998*, 3, page 1381.

[15] W.M. Gentleman and G. Sande. Fast Fourier transforms for fun and profit. *Proc. AFIPS*, vol. 29, 1966, pages 563–578.

[16] M. Hegland. A self-sorting in-place fast Fourier transform algorithm suitable for vector and parallel processing. *Numerische Mathematik*, vol. 68, 1994.

[17] M.T. Heideman, D.H. Johnson, and C.S. Burrus. Gauss and the History of the Fast Fourier Transform. *IEEE ASSP Magazine*, October 1984.

[18] IBM Corporation. 2007. Cell Broadband Engine Architecture specification. www.ibm.com/chips/techlib/techlib.nsf/products/ Cell_Broadband_Engine

[19] S.G. Johnson and M. Frigo. A modified split-radix FFT with fewer arithmetic operations. *IEEE Transactions Signal Processing*, vol. 55, no. 1, 2007, pages 111–119.

[20] D. Jones. Decimation-in-Frequency (DIF) Radix-2 FFT. http://cnx.org/ content/m12018/latest, Sept 17, 2006.

[21] H. Karner, M. Auer, and C.W. Ueberhuber. Optimum Complexity FFT Algorithms for RISC Processors. *Technology Report AURORA TR1998-03*, Institute for Applied and Numerical Mathematics, Vienna University of Technology, 1998.

[22] Y. Li, L. Zhao, H. Lin, A.C. Chow, and J.R. Diamond. A performance model for fast Fourier transform. *Parallel and Distributed Processing Symposium*, vol. 0, 2009, pages 1–11.

[23] R. Motwani, K.V. Palen, V. Sarkar, and S. Reyen. Combining Register Allocation and Instruction Scheduling. Technical Report, 1995.

[24] A. Saidi. Decimation-in-Time-Frequency FFT Algorithm. *Proceedings of the IEEE International Conference on Acoustics, Speech, and Signal Processing*, IEEE ICASSP-94, Adelaide, Australia, April 19–22 1994, p. III:453–456.

[25] R.C. Singleton. On computing the fast Fourier transform. *Communications of the ACM*, vol. 10, 1967, pages 647–654.

[26] B.P. Zeigler, H. Praehofer, and T.G. Kim. *Theory of Modeling and Simulation*. Academic Press, 2000, 2nd Edition.

Part V

Combinatorial Algorithms

Chapter 10

Combinatorial Algorithm Design on the Cell/B.E. Processor

David A. Bader

Georgia Institute of Technology, Atlanta, GA

Virat Agarwal

IBM T.J. Watson Research Center, Yorktown Heights, NY

Kamesh Madduri

Lawrence Berkeley National Laboratory, Berkeley, CA

Fabrizio Petrini

IBM T.J. Watson Research Center, Yorktown Heights, NY

10.1 Introduction

Combinatorial algorithms play a subtle, yet important, role in traditional scientific computing. Perhaps the most well-known example is the graph partitioning formulation for load-balanced parallelization of scientific simulations. Partitioning algorithms are typically composed of several combinatorial kernels such as graph coloring, matching, sorting, and permutations. Combinatorial algorithms also appear in auxiliary roles for efficient parallelization of linear algebra, computational physics, and numerical optimization computations. In the last decade or so, the paradigm of data-intensive scientific discovery has significantly altered the landscape of computing. Combinatorial approaches are now at the heart of massive data analysis routines, systems biology, and in general, the study of natural phenomena involving networks and complex interactions.

Combinatorial kernels are computationally quite different from floating-point intensive scientific routines involving matrices and vectors. String and graph algorithms are typically highly memory-intensive; make heavy use of data structures such as lists, sets, queues, and hash tables; and exhibit a combination of data and task-level parallelism. Due to power constraints and the *memory wall*, we are rapidly converging towards widespread use of multi- and many-core chips and accelerators for speeding up computation. Unfortunately, several known parallel algorithms for combinatorial problems do not easily map onto current multicore architectures. The mismatch arises because current architectures lean towards efficient execution of regular computations with low memory footprints and working sets, and heavily penalize memory-intensive codes with irregular memory accesses; parallel algorithms in the past were mostly designed assuming an underlying, well-balanced compute-memory platform.

List ranking [7, 13, 17] is a fundamental combinatorial kernel, and stands in stark contrast to regular scientific computations. Given an arbitrary linked list that is stored in a contiguous area of memory, the objective of list ranking problem is to determine the distance from each node to the head of the list. This problem specification is a simpler instance of generalized prefix computations on linked lists. List ranking is a key subroutine in parallel graph algorithms for tree contraction and expression evaluation [1], minimum spanning forest [2] evaluation, and ear decomposition.

The serial algorithm for list ranking is straightforward and based on pointer jumping. It is easy to see that the performance is heavily dependent on the ordering of the linked list nodes in memory. There is very poor locality and no inherent data parallelism in the pointer-jumping serial algorithm if the nodes are randomly ordered. On cache-based architectures and for problem instances where the linked list does not fit in cache, the serial algorithm would incur the penalty of accessing main memory for every pointer traversal. Thus, the performance is largely dependent on the latency to main memory, and the algorithm hardly utilizes or requires any of the on-chip compute-resources.

In this chapter, we present the design and optimization of a parallel list ranking algorithm for the heterogenous Cell Broadband Engine (or the Cell/B.E.) architecture. The Cell/B.E. [6] is a novel architectural design by Sony, Toshiba, and IBM (STI), primarily targeting high-performance multimedia and gaming applications. It consists of a traditional microprocessor (called the PPE) that controls eight single instruction multiple data (SIMD) co-processing units called synergistic processor elements (SPEs), a high-speed memory controller, and a high-bandwidth bus interface (termed the element interconnect bus, or EIB), all integrated on a single chip. The Cell/B.E. chip is a computational workhorse, and offers a theoretical peak single-precision floating-point performance of 204.8 GFlop/s. The eight SPEs can each execute two instructions per clock cycle. Further, the SPE supports both scalar as well as SIMD computations [12]. The on-chip interconnection network elements have been specially designed to cater for high performance on bandwidth-intensive applications (such as those in gaming and multimedia). Cell/B.E. is the central processing unit of the Sony PlayStation 3 video game console [8], and the accelerator unit of the Roadrunner supercomputer [4]. A detailed description of this architecture has been presented in earlier chapters.

An interesting aspect of the Cell/B.E. is that it gives the programmer a lot of control over data movement. Unlike cache-based architectures, the programmer has to explicitly fetch data from main memory to the on-chip scratch space close to the SPEs. Williams et al. [20] analyzed the performance of Cell/B.E. for key scientific kernels such as dense matrix multiply, sparse matrix vector multiply, and 1D and 2D fast Fourier transforms. They demonstrate that the Cell/B.E. performs impressively for applications with predictable memory access patterns, and that communication and computation can be overlapped more effectively on the Cell/B.E. than on conventional cache-based systems. Scarpazza et al. [18] present an implementation of the breadth first search algorithm on the Cell processor.

In this chapter, we study the performance of list ranking, a unique representative from the class of combinatorial algorithms. Our work challenges the general perception that the Cell/B.E. architecture is not suited for problems that involve fine-grained and irregular memory accesses, and where there is insufficient computation to hide memory latency. By applying new techniques for latency-tolerance and load-balancing with *software-managed threads* (SM-Threads), our list ranking implementation on the Cell/B.E. processor achieves significant parallel speedup. We conduct an extensive experimental study comparing our code on IBM QS20 Cell blade with implementations on current cache-based architectures, and large shared memory symmetric multiprocessor and multithreaded architectures. Our main results are summarized here:

- Our latency-hiding technique boosts Cell/B.E. performance by a factor of about 4.1 for both random and ordered lists.

- By tuning just one algorithm parameter, our list ranking implementation is load-balanced across the SPEs with high probability, even for random lists.

- The Cell/B.E. achieves an average speedup of 8 over the performance on current cache-based microprocessors (for input instances that do not fit into the L2 cache).

- On a random list of 1 million nodes, we obtain a speedup of 8.34 compared to a single-threaded PPE-only implementation. For an ordered list (with stride-1 accesses only), the speedup over a PPE-only implementation is 1.56.

We discuss list ranking in Section 10.3, and introduce SM-Threads and our latency-hiding technique in Section 10.3.3.

The heterogeneous processors, limited on-chip memory, and multiple avenues for parallelism on the Cell/B.E. processor make algorithm design and implementation a new challenge, with potentially high payoffs in terms of performance. Analyzing algorithms using traditional sequential complexity models like the random access memory (RAM) model fail to account for several Cell/B.E. architectural intricacies. There is currently no simple and accepted model of computation for the Cell/B.E., and in general, for multicore architectures. We use a simple complexity model for the design and analysis of parallel algorithms on the Cell/B.E. architecture. We express the algorithm complexity on the Cell/B.E. processor using the triplet $\langle T_C, T_D, T_B \rangle$, where T_C denotes the computational complexity, T_D the number of direct memory access (DMA) requests, and T_B the number of branching instructions, all expressed in terms of the problem size. We explain the rationale behind the choice of these three parameters in Section 10.2. We then present a *systematic methodology* for analyzing algorithms using this complexity model, and illustrate this with an example of matrix multiplication.

10.2   Algorithm Design and Analysis on the Cell/B.E.

10.2.1   A Complexity Model

There are several architectural features of the Cell/B.E. processor that can be exploited for performance.

- The SPEs are designed as compute-intensive co-processors, while the PowerPC unit (the PPE) orchestrates the control flow. So it is necessary to partition the computation among the SPEs, and an efficient SPE implementation should also exploit the SIMD instruction set.

- The SPEs operate on a limited on-chip memory (256 KB local store [LS]) that stores both instructions and data required by the program. Unlike the PPE, the SPE cannot access memory directly, but has to transfer data and instructions using asynchronous coherent DMA commands. Algorithm design must account for DMA transfers (i.e., the latency of DMA transfers, as well as their frequency), which may be a significant cost.

- The SPE also differs from conventional microprocessors in the way branches are handled. The SPE does not support dynamic branch prediction, but instead relies on compiler-generated branch hints to improve instruction prefetching. Thus, there is a significant penalty associated with branch misprediction, and branching instructions should be minimized for designing an efficient implementation.

We present a complexity model to simplify the design of parallel algorithms on the Cell/B.E. Let n denote the problem size. We model the execution time using the triplet $\langle T_C, T_D, T_B \rangle$, where T_C denotes the computational complexity, T_D the number of DMA requests, and T_B the number of branching instructions. We consider the computation on the SPEs ($T_{C,SPE}$) and PPE ($T_{C,PPE}$) separately, and T_C denotes the sum of these terms. $T_{C,SPE}$ is the maximum of $T_{C,SPE(i)}$ for $1 \leq i \leq p$, where p is number of SPEs. In addition, we have T_D, an upper bound on the number of DMA requests made by a single SPE. This is an important parameter, as the latency due to a large number of DMA requests might dominate over the actual computation. In cases when the complexity of $T_{C,SPE}$ dominates over T_D, we can ignore the overhead due to DMA requests. Similarly, branch mispredictions constitute a significant overhead. Since it may be difficult to compute the actual percentage of mispredictions, we just report the asymptotic number of branches in our algorithm. For algorithms in which the misprediction probability is low, we can ignore the effects of branching.

Our model is similar to the Helman-JáJá model for symmetric multiprocessors (SMPs) [10] in that we try to estimate memory latency in addition to computational complexity. Also, our model is more tailored to heterogeneous multicore systems than general-purpose parallel computing models such as LogP [9], BSP [19], and QSM [15]. The execution time is dominated by the SPE that does the maximum amount of work. We note that exploiting the SIMD features results in only a constant factor improvement in the performance, and does not affect the asymptotic analysis. This model does not take into account synchronization mechanisms such as on-chip mailboxes, and synergistic processing unit (SPU) operation under the isolated mode. Also, our model does not consider the effect of floating-point precision on the performance of numerical algorithms, which can be quite significant [20].

10.2.2 Analyzing Algorithms

We now discuss a systematic procedure for analyzing algorithms using the above model:

- We compute the computational complexity $T_{C,SPE}$.

- Next, we determine the complexity of DMA requests T_D in terms of the input parameters.

  - If the DMA request complexity is a constant, then typically computation would dominate over memory accesses and we can ignore the latency due to DMA transfers.

  - Otherwise, we need to further analyze the algorithm, taking into consideration the size of DMA transfers, as well as the computational granularity.

- It is possible to issue non-blocking DMA requests on the SPE, and so we can keep the SPE busy with computation while waiting for a DMA request to be completed. However, if there is insufficient computation in the algorithm between a DMA request and its completion, the SPE will be idle. We analyze this effect by computing the *computational complexity in the average case* between a DMA request and its completion. If this term is a function of the input parameters, this implies that memory latency can be hidden by computation.

- Finally, we compute the number of branching instructions T_B in the algorithm. These should be minimized as much as possible in order to design an efficient algorithm.

We present a simple example to illustrate the use of our model, as well as the above algorithm analysis procedure.

Matrix multiplication ($C = A * B$, $c_{ij} = \sum_{k=1}^{n} a_{ik} * b_{kj}$, for $1 \leq i, j \leq n$) on the Cell/B.E. is analyzed as follows:

- We partition computation among the eight SPEs by assigning each SPE the Matrix B and $\frac{n}{p}$ rows of Matrix A.

- Let us assume that each SPE can obtain b rows of A and b columns of B in a single DMA transfer. Thus, we would require $O(\frac{n^2}{b^2})$ DMA transfers, and T_D is $O(\frac{n^2}{pb^2})$.

Chen et al. [6] describe their Cell/B.E. implementation of this algorithm. The algorithmic complexity is given by $T_C = O(\frac{n^3}{p})$, $T_D = O(\frac{n^2}{pb^2})$ and $T_B = O(n^2)$. Using the analysis procedure, we note that the DMA request complexity is not a constant. Following step 3, we compute the average case of the computational complexity between the DMA request and its completion, assuming non-blocking DMAs and double buffering. This is given by $O(nb^2)$ (as the complexity of computing b^2 elements in C is $O(n)$). Thus, we can ignore the constant DMA latency for each transfer, and the algorithm running time is dominated by computation for sufficiently large n. However, note that we have $O(n^2)$ branches due to the absence of a branch predictor on the SPE, which might degrade performance if they result in mispredicts. Using SIMD and dual-issue features within the SPE, it is possible to achieve a peak cycles per instruction (CPI) *of* 0.5. Chen et al. in fact obtain a CPI of 0.508 for their implementation of the above algorithm, incorporating optimizations such as SIMD, double buffering, and software pipelining.

Algorithm design and analysis is more complex for irregular memory-intensive applications, and problems exhibiting poor locality. *List ranking* is representative of this class of problems, and is a fundamental technique for the design of several combinatorial and graph-theoretic applications on parallel processors. After a brief introduction to list ranking in the next section, we describe our design of an efficient algorithm for the Cell/B.E.

10.3   List Ranking

Given an arbitrary linked list that is stored in a contiguous area of memory, the list ranking problem determines the distance of each node to the head of the list. For a random list, the memory access patterns are highly irregular, and this makes list ranking a challenging problem to solve efficiently on parallel architectures. Implementations that yield parallel speedup on shared memory systems exist [3, 10], yet none are known for distributed memory systems. In another work, Rehman et al. [16] present their design of parallel list ranking on the GPU.

10.3.1   A Parallelization Strategy

List ranking is an instance of the more general prefix problem [3]. Let X be an array of n elements stored in arbitrary order. For each element i, let $X(i).value$ denote its value and $X(i).next$ the index of its successor. Then for any binary associative operator $\oplus$, compute $X(i).prefix$ such that $X(head).prefix = X(head).value$ and $X(i).prefix = X(i).value \oplus X(predecessor).prefix$, where $head$ is the first element of the list, i is not equal to $head$, and $predecessor$ is the node preceding i in the list. If all values are 1 and the associative operation is addition, then prefix reduces to list ranking. We assume that we know the location of the head h of the list, otherwise we can easily locate it. The parallel algorithm for a canonical parallel computer with p processors is as follows:

- Partition the input list into s sublists by randomly choosing one node from each memory block of $n/(s-1)$ nodes, where s is $\Omega(p \log n)$. Create the array *Sublists* of size s.

- Traverse each sublist computing the prefix sum of each node within the sublists. Each node records its sublist index. The input value of a node in the *Sublists* array is the sublist prefix sum of the last node in the previous *Sublists*.

- The prefix sums of the records in the *Sublists* array are then calculated.

- Each node adds its current prefix sum value (value of a node within a sublist) and the prefix sum of its corresponding *Sublists* record to get its final prefix sums value. This prefix sum value is the required label of the leaves.

Ordered List

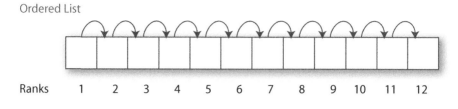

Ranks   1   2   3   4   5   6   7   8   9   10   11   12

Random List

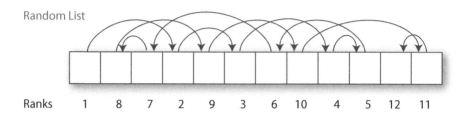

Ranks   1   8   7   2   9   3   6   10   4   5   12   11

FIGURE 10.1: List ranking for an ordered (top) and random (bottom) list. From D.A. Bader, V. Agarwal, and K. Madduri, "On the Design and Analysis of Irregular Algorithms on the Cell Processor: A case study on list ranking," *21th IEEE International Parallel and Distributed Processing Symposium* (IPDPS), Long Beach, CA, March 2007; and D.A. Bader, V. Agarwal, K. Madduri, and S. Kang, "High Performance Combinatorial Algorithm Design on the Cell Broadband Engine Processor," *Parallel Computing*, 33(10–11):720–740, 2007. With permission.

We map this to Cell/B.E. and analyze it as follows. Assume that we start with eight sublists, one per SPE. Using DMA fetches, the SPEs continue obtaining the successor elements until they reach a sublist end, or the end of the list.

10.3.2   Complexity Analysis

Analyzing the complexity of this algorithm using our model, we have $T_C = O(\frac{n}{p})$, $T_D = O(\frac{n}{p})$, and $T_B = O(1)$. From step 2 of the procedure, since the complexity of DMA fetches is a function of n, we analyze the computational complexity in the average case between a DMA request and its completion. This is clearly $O(1)$, since we do not perform any significant computation while waiting for the DMA request to complete. This may lead to processor stalls, and since the number of DMA requests is $O(n)$, stall cycles might dominate the optimal $O(n)$ work required for list ranking. Our asymptotic analysis offers only a limited insight into the algorithm, and we have to inspect the algorithm at the instruction level and design alternative approaches to hide DMA latency.

10.3.3   A Novel Latency-Hiding Technique for Irregular Applications

Due to the limited local store (256 KB) within an SPE, memory-intensive applications that have irregular memory access patterns require frequent DMA transfers to fetch the data. The relatively high latency of a DMA transfer creates a bottleneck in achieving performance for these applications. Several combinatorial problems, such as the ones that arise in graph theory, belong to this class of problems. Formulating a general strategy that helps overcome the latency overhead will provide direction to the design and optimization of irregular applications on Cell/B.E.

Since the Cell/B.E. supports non-blocking memory transfers, memory transfer latency will not be a problem if we have sufficient computation between a request and completion. However, if we do not have enough computation in this period (for instance, the Helman-JáJá list ranking algorithm), the SPE will stall for the request to be completed. A generic solution to this problem would be to restructure the algorithm such that the SPE does useful computation until the memory request is completed. This essentially requires identification of an additional level of parallelism/concurrency within each SPE. Note that if the computation can be decomposed into several independent tasks, we can overcome latency by exploiting concurrency in the problem.

Our technique is analogous to the concept of tolerating latency in architectures such as the Cray MTA-2 and NVIDIA G80 that use massive multithreading to hide latency. The SPE does not have support for hardware multithreading, and so we manage the computation through SM-Threads. The SPE computation is distributed to a set of SM-Threads and at any instant, one thread per SPE is active. We keep switching software contexts so that we do computation between a DMA request and its completion. We use a round-robin schedule for the threads.

Through instruction-level profiling, it is possible to determine the minimum number of SM-Threads that are needed to hide the memory latency. Note that utilizing more SM-Threads than required incurs a significant overhead. Each SM-Thread introduces additional computation and also requires memory on the limited LS. Thus, we have a trade-off between the number of SM-Threads and latency due to DMA stalls. In the next section, we will use this technique to efficiently implement list ranking on Cell/B.E.

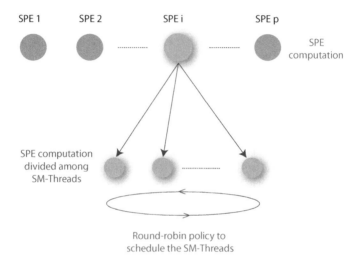

FIGURE 10.2: Illustration of the latency-tolerance technique employed by our parallel implementation. From D.A. Bader, V. Agarwal, and K. Madduri, "On the Design and Analysis of Irregular Algorithms on the Cell Processor: A case study on list ranking," *21th IEEE International Parallel and Distributed Processing Symposium* (IPDPS), Long Beach, CA, March 2007; and D.A. Bader, V. Agarwal, K. Madduri, and S. Kang, "High Performance Combinatorial Algorithm Design on the Cell Broadband Engine Processor," *Parallel Computing*, 33(10–11):720–740, 2007. With permission.

10.3.4   Cell/B.E. Implementation

Our Cell/B.E. implementation (described in high level in the following four steps) is similar to the Helman-JáJá algorithm. Let us assume p SPEs in the analysis.

- We uniformly pick s *head nodes* in the list and assign them to the SPEs. So, each SPE will traverse s/p sublists.

- Using these s/p sublists as independent SM-Threads, we adopt the latency-hiding technique. We divide the s/p sublists into b DMA list transfers. Using one DMA list transfer, we fetch the next elements for a set of s/pb lists. After issuing a DMA list transfer request, we move on to the next set of sublists and so forth, thus keeping the SPU busy until this DMA transfer is complete. Figure 10.3 illustrates step 3 of this algorithm.

  We maintain temporary structures in the LS for these sublists, so that the LS can create a contiguous sublist out of these randomly scattered sublists, by creating a chain of next elements for the sublists.

  After one complete round, we manually revive this SM-Thread and wait for the DMA transfer to complete. Note that there will be no stall if we have sufficient number of SM-Threads (we determine this number in Section 10.3.5) to hide the latency. We store the elements that are fetched into the temporary structures, initiate a new DMA list transfer request for fetching the successors of these newly fetched elements, and move on to the next set of sublists.

  When these temporary structures get full, we initiate a new DMA list transfer request to transfer back these elements to the main memory.

  At the end of step 2, we have the prefix sum of each node within the sublist for each sublist within the SPU. Also, we have the randomly scattered sublists stored into a contiguous area of memory.

- Compute the rank of each sublist head node using the power processing unit (PPU).

The running time for step 2 of the algorithm dominates over the rest of algorithm by an order of magnitude. In the asymptotic notation, this step is $O(n)$. It consists of an outer loop of $O(s)$ and an inner loop of $O(length\ of\ the\ sublist)$. Since the lengths of the sublists are different, the amount of work performed by each SM-Thread differs. For a large number of threads, we get sufficient computation for the SPE to hide DMA latency even when the load is imbalanced. Helman and JáJá [10, 11] established that with high probability, no processor would traverse more $\alpha(s)\frac{n}{p}$ elements for $\alpha(s) \geq 2.62$. Thus, the load is balanced among various SPEs under this constraint.

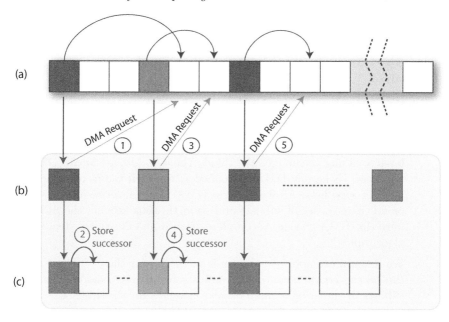

FIGURE 10.3: Step 2 of list ranking on Cell/B.E. (a) The input list. Colored nodes are allocated to SPE(i). (b) View from SPE(i)—it has s/p sublist head nodes to traverse concurrently. (c) A temporary buffer to store sublists in a contiguous area of memory. When this buffer becomes full, we transfer it to the main memory. From D.A. Bader, V. Agarwal, and K. Madduri, "On the Design and Analysis of Irregular Algorithms on the Cell Processor: A case study on list ranking," *21th IEEE International Parallel and Distributed Processing Symposium* (IPDPS), Long Beach, CA, March 2007; and D.A. Bader, V. Agarwal, K. Madduri, and S. Kang, "High Performance Combinatorial Algorithm Design on the Cell Broadband Engine Processor," *Parallel Computing*, 33(10–11):720–740, 2007. With permission.

In our implementation, we incorporate recommended software strategies [5] and techniques to exploit the architectural features of Cell/B.E. For instance, we use manual loop unrolling, branch hints, and design our implementation for a limited LS.

10.3.5   Performance Results

We report our performance results from actual runs on an IBM BladeCenter QS20, with two 3.2 GHz Cell/B.E. processors, 512 KB Level 2 cache per processor, and 1 GB memory (512 MB per processor). We use one processor for measuring performance and compile the code using the gcc compiler provided with Cell/B.E. SDK 2.0, with level 3 optimization.

Similar to [3, 10] we use two classes of lists to test our code, *Ordered* and *Random*. An ordered list representation places each node in the list according to its rank. Thus node i is placed at position i, and its successor is at position $i+1$. A random list representation places successive elements randomly in the array.

Our significant contribution in this chapter is a generic work partitioning technique to hide memory latency. We demonstrate the results of this technique for list ranking: we use SM-Threads to vary the number of outstanding DMA requests on each SPE, as well as partition the problem and allocate more sublists to each SPE. Figure 10.4 shows the performance boost we obtain as we tune the DMA parameter.

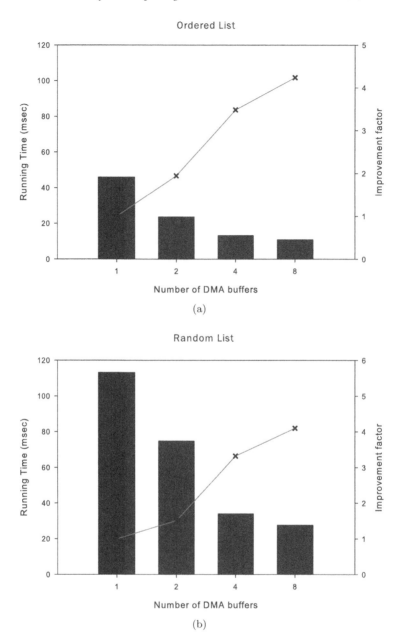

FIGURE 10.4: Achieving latency tolerance through DMA parameter tuning for ordered (a) and random (b) lists of size 2^{20}. From D.A. Bader, V. Agarwal, and K. Madduri, "On the Design and Analysis of Irregular Algorithms on the Cell Processor: A case study on list ranking," *21th IEEE International Parallel and Distributed Processing Symposium* (IPDPS), Long Beach, CA, March 2007; and D.A. Bader, V. Agarwal, K. Madduri, and S. Kang, "High Performance Combinatorial Algorithm Design on the Cell Broadband Engine Processor," *Parallel Computing*, 33(10–11):720–740, 2007. With permission.

From instruction-level profiling of our code we determine that the exact number of computational clock cycles between a DMA transfer request and its completion are 75. Comparing this with the DMA transfer latency (90ns, i.e. about 270 clock cycles) suggests that four outstanding DMA requests should be sufficient for hiding the DMA latency. Our results confirm this analysis and we obtain an improvement factor of 4.1 using eight DMA buffers.

In Figure 10.5 we present the results for load balancing among the eight SPEs, as the number of sublists are varied. For ordered lists, we allocate equal chunks to each SPE. Thus, load is balanced among the SPEs in this case. For random lists, since the length of each sublist varies, the work performed by each SPE varies. We achieve a better load-balancing by increasing the number of sublists. Figure 10.5 illustrates this: load-balancing is better for 64 sublists than the case of 8 sublists.

We present a performance comparison of our implementation of list ranking on Cell/B.E. with other single-processor and parallel architectures. We consider both random and ordered lists with 8 million nodes.

Figure 10.6 shows the running time of our Cell/B.E. implementation compared with efficient implementations of list ranking on the following architectures:

Intel_x86: 3.2 GHz Intel Xeon processor, 1 MB L2 cache, Intel C compiler v9.1.

Intel_i686: 2.8 GHz Intel Xeon processor, 2 MB L2 cache, Intel C compiler v9.1.

Intel_ia64: 900 MHz Intel Itanium 2 processor, 256 KB L2 cache, Intel C compiler v9.1.

SunUS_III: 900 MHz UltraSparc-III processor, Sun C compiler v5.8.

Intel_WC: 2.67 GHz Intel Dual-Core Xeon 5150 processor (Woodcrest), Intel C compiler v9.1.

MTA-[1,2,8]: 220 MHz Cray MTA-2 processor, no data cache. We report results for 1, 2, and 8 processors.

SunUS-[1,2,8]: 400 MHz UltraSparc II Symmetric Multi-processor system (Sun E4500), 4 MB L2 cache, Sun C compiler. We report results for 1, 2, and 8 processors.

For Intel Xeon 5150 (Woodcrest) we use a parallel implementation [3] running on two threads. Table 10.2 gives the comparison of running time of various steps in the list ranking algorithm. The table shows that step 2 dominates the entire running time of the algorithm. Finally, we demonstrate a substantial speedup of our Cell/B.E. implementation over a sequential implementation using the PPE only. We compare the best sequential approach that uses pointer-chasing to our algorithm using different problem instances. Figure 10.7 shows

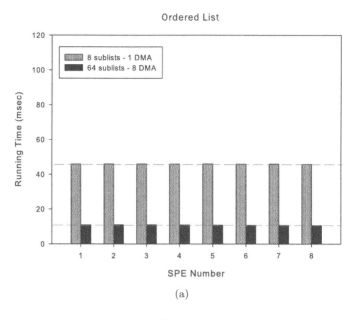

(a)

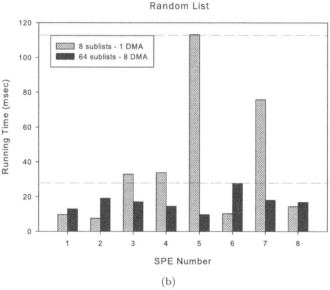

(b)

FIGURE 10.5: Load-balancing among SPEs for ordered (a) and random (b) lists of size 2^{20}. From D.A. Bader, V. Agarwal, and K. Madduri, "On the Design and Analysis of Irregular Algorithms on the Cell Processor: A case study on list ranking," *21th IEEE International Parallel and Distributed Processing Symposium* (IPDPS), Long Beach, CA, March 2007; and D.A. Bader, V. Agarwal, K. Madduri, and S. Kang, "High Performance Combinatorial Algorithm Design on the Cell Broadband Engine Processor," *Parallel Computing*, 33(10–11):720–740, 2007. With permission.

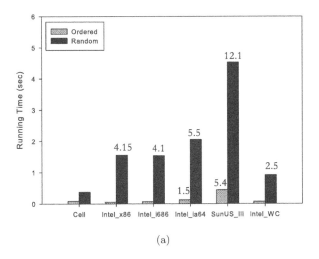

(a)

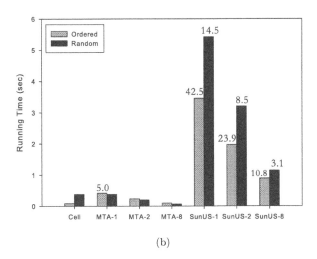

(b)

FIGURE 10.6: Performance of list ranking on Cell/B.E. compared to other architectures, for an input list of size 8 million nodes. The speedup of the Cell/B.E. implementation over the architectures is indicated above the respective bars. From D.A. Bader, V. Agarwal, and K. Madduri, "On the Design and Analysis of Irregular Algorithms on the Cell Processor: A case study on list ranking," *21th IEEE International Parallel and Distributed Processing Symposium* (IPDPS), Long Beach, CA, March 2007; and D.A. Bader, V. Agarwal, K. Madduri, and S. Kang, "High Performance Combinatorial Algorithm Design on the Cell Broadband Engine Processor," *Parallel Computing*, 33(10–11):720–740, 2007. With permission.

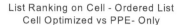

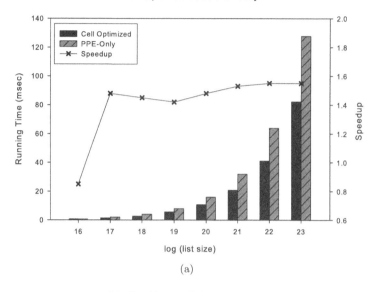

(a)

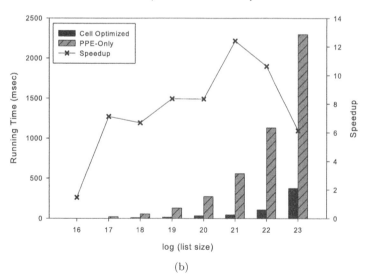

(b)

FIGURE 10.7: Performance comparison of the sequential implementation on the PPE to our parallel implementation for ordered (a) and random (b) lists.

TABLE 10.1: Speedup achieved by our Cell/B.E. implementation over implementations on other architectures.

Architecture	Speedup (ordered)	Speedup (random)
Intel_x86	0.8	4.1
Intel_i686	0.9	4.1
Intel_ia64	1.5	5.5
SunUS_III	5.4	12.1
Intel_WC	0.8	2.5
MTA-1	5.0	1.0
SunUS_II-1	42.5	14.5
SunUS_II-2	23.9	8.5
SunUS_II-8	10.8	3.1

TABLE 10.2: Running time comparison of the various steps of list ranking on Cell/B.E. for ordered and random lists of size 8 million nodes.

Task	Ordered	Random
Step 1: Identification of sublist head nodes	0.09%	0.02%
Step 2: Traversal of sublists	99.84%	99.96%
Step 3: Computing the ranks	0.07%	0.02%

that for random lists we get an overall speedup of 8.34 (1 million vertices), and even for ordered lists we get a speedup of 1.5.

10.4   Conclusions

In summary, we present a complexity model to simplify the design of algorithms on Cell/B.E. multicore architecture, and a systematic procedure to evaluate their performance. To estimate the execution time of an algorithm, we consider the computational complexity, memory access patterns (DMA transfer sizes and latency), and the complexity of branching instructions. This model also helps identify potential bottlenecks in the algorithm. We also present a generic work partitioning technique to hide memory latency on Cell/B.E. This technique can be applied to many combinatorial graph algorithms having exhibiting irregular memory access patterns. Using this technique, we develop a fast parallel implementation of the list ranking algorithm for the Cell/B.E. processor. We confirm the efficacy of our technique by demonstrating an improvement factor of 4.1 as we tune the DMA parameter. We also demonstrate an overall speedup of 8.34 of our implementation over an efficient

PPE-only sequential implementation. We also show substantial speedups by comparing the performance of our list ranking implementation with several single-processor and parallel architectures.

Acknowledgments

This work was supported in part by NSF Grants CNS-0614915, CAREER CCF-0611589, and an IBM Shared University Research (SUR) grant. We would like to acknowledge the support of the Center for Adaptive Supercomputing Software for Multithreaded Architectures (CASS-MT), led by Pacific Northwest National Laboratory, and thank Sidney Manning (IBM Corporation) and Vipin Sachdeva (IBM Research) for providing valuable input during the course of our research.

Bibliography

[1] D.A. Bader and G. Cong. A fast, parallel spanning tree algorithm for symmetric multiprocessors (SMPs). In *Proc. Int'l Parallel and Distributed Processing Symp. (IPDPS 2004)*, Santa Fe, NM, April 2004.

[2] D.A. Bader and G. Cong. Fast shared-memory algorithms for computing the minimum spanning forest of sparse graphs. In *Proc. Int'l Parallel and Distributed Processing Symp. (IPDPS 2004)*, Santa Fe, NM, April 2004.

[3] D.A. Bader, G. Cong, and J. Feo. On the architectural requirements for efficient execution of graph algorithms. In *Proc. 34th Int'l Conf. on Parallel Processing (ICPP 2005)*, Oslo, Norway, June 2005.

[4] K.J. Barker, K. Davis, A. Hoisie, D.J. Kerbyson, M. Lang, S. Pakin, and J.C. Sancho. Entering the petaflop era: The architecture and performance of Roadrunner. In *Proc. of the 2008 ACM/IEEE Conference on Supercomputing (SC08)*, pages 1–11, Piscataway, NJ, 2008. IEEE Press.

[5] D.A. Brokenshire. Maximizing the power of the Cell Broadband Engine Processor: 25 tips to optimal application performance. IBM developerWorks technical article, June 2006.

[6] T. Chen, R. Raghavan, J. Dale, and E. Iwata. Cell Broadband Engine Architecture and its first implementation. IBM developerWorks technical article, November 2005.

[7] R. Cole and U. Vishkin. Faster optimal prefix sums and list ranking. *Information and Computation*, 81(3):344–352, 1989.

[8] Sony Corporation. Sony PlayStation 3. `http://www.us.playstation.com/`.

[9] D.E. Culler, R.M. Karp, D.A. Patterson, A. Sahay, K.E. Schauser, E. Santos, R. Subramonian, and T. von Eicken. LogP: Towards a realistic model of parallel computation. In *Proc. 4th ACM SIGPLAN Symp. on Principles and Practice of Parallel Programming (PPOPP 1993)*, pages 1–12, San Diego, CA, May 1993.

[10] D.R. Helman and J. JáJá. Designing practical efficient algorithms for symmetric multiprocessors. In *Proc. 1st Intl. Workshop on Algorithm Engineering and Experimentation (ALENEX 1999)*, volume 1619 of *Lecture Notes in Computer Science*, pages 37–56, Baltimore, MD, January 1999. Springer-Verlag.

[11] D.R. Helman and J. JáJá. Prefix computations on symmetric multiprocessors. *J. Parallel & Distributed Comput.*, 61(2):265–278, 2001.

[12] C. Jacobi, H.-J. Oh, K.D. Tran, S.R. Cottier, B.W. Michael, H. Nishikawa, Y. Totsuka, T. Namatame, and N. Yano. The vector floating-point unit in a synergistic processor element of a Cell processor. In *Proc. 17th IEEE Symp. on Computer Arithmetic (ARITH 2005)*, pages 59–67, Washington, DC, 2005.

[13] J. JáJá. *An Introduction to Parallel Algorithms*. Addison-Wesley Publishing Company, New York, 1992.

[14] J.A. Kahle, M.N. Day, H.P. Hofstee, C.R. Johns, T.R. Maeurer, and D. Shippy. Introduction to the Cell multiprocessor. *IBM J. Res. Dev.*, 49(4/5):589–604, 2005.

[15] V. Ramachandran. A general-purpose shared-memory model for parallel computation. In M.T. Heath, A. Ranade, and R. S. Schreiber, editors, *Algorithms for Parallel Processing*, volume 105, pages 1–18. Springer-Verlag, New York, 1999.

[16] M.S. Rehman, K. Kothapalli, and P.J. Narayanan. Fast and scalable list ranking on the GPU. In *ICS '09: Proceedings of the 23rd International Conference on Supercomputing*, pages 235–243, New York, NY, 2009. ACM.

[17] M. Reid-Miller. List ranking and list scan on the Cray C-90. *J. Comput. Syst. Sci.*, 53(3):344–356, December 1996.

[18] D.P. Scarpazza, O. Villa, and F. Petrini. Efficient breadth-first search on the Cell/B.E. processor. *Parallel and Distributed Systems, IEEE Transactions on*, 19(10):1381–1395, October 2008.

[19] L.G. Valiant. A bridging model for parallel computation. *Communications of the ACM*, 33(8):103–111, 1990.

[20] S. Williams, J. Shalf, L. Oliker, S. Kamil, P. Husbands, and K. Yelick. The potential of the Cell processor for scientific computing. In *Proc. 3rd Conf. on Computing Frontiers (CF 2006)*, pages 9–20, Ischia, Italy, 2006.

This chapter is based on the authors' previously published work:

- D.A. Bader, V. Agarwal, and K. Madduri, "On the Design and Analysis of Irregular Algorithms on the Cell Processor: A case study on list ranking," *21th IEEE International Parallel and Distributed Processing Symposium* (IPDPS), Long Beach, CA, March 2007.

- D.A. Bader, V. Agarwal, K. Madduri, and S. Kang, "High Performance Combinatorial Algorithm Design on the Cell Broadband Engine Processor," *Parallel Computing*, 33(10–11):720–740, 2007.

Part VI

Stencil Algorithms

Chapter 11

Auto-Tuning Stencil Computations on Multicore and Accelerators

Kaushik Datta

University of California, Berkeley

Samuel Williams

Lawrence Berkeley National Laboratory

Vasily Volkov

University of California, Berkeley

Jonathan Carter

Lawrence Berkeley National Laboratory

Leonid Oliker

Lawrence Berkeley National Laboratory

John Shalf

Lawrence Berkeley National Laboratory

Katherine Yelick

Lawrence Berkeley National Laboratory

11.1   Introduction

The recent transformation from an environment where gains in computational performance came from increasing clock frequency and other hardware engineering innovations, to an environment where gains are realized through the deployment of ever increasing numbers of modest performance cores has profoundly changed the landscape of scientific application programming. This exponential increase in core count represents both an opportunity and a challenge: access to petascale simulation capabilities and beyond will require that this concurrency be efficiently exploited. The problem for application programmers is further compounded by the diversity of multicore architectures that are now emerging [4]. From relatively complex out-of-order CPUs with complex cache structures, to relatively simple cores that support hardware multithreading, to chips that require explicit use of software controlled memory, designing optimal code for these different platforms represents a serious impediment. An emerging solution to this problem is auto-tuning: the automatic generation of many versions of a code kernel that incorporate various tuning strategies, and the benchmarking of these to select the highest performing version. Typical tuning strategies might include: maximizing in-core performance with loop unrolling and restructuring; maximizing memory bandwidth by exploiting non-uniform memory access (NUMA), engaging prefetch by directives; and minimizing memory traffic by cache blocking or array padding. Often a key parameter is associated with each tuning strategy (e.g., the amount of loop unrolling or the cache blocking factor), and these parameters must be explored in addition to the layering of the basic strategies themselves.

This study focuses on the key numerical technique of stencil computations, used in many different scientific disciplines, and illustrates how auto-tuning can be used to produce very efficient implementations across a diverse set of current multicore architectures. In Section 11.2, we give an overview of the two stencils studied, followed by a description of the multicore architectures that form our testbed in Section 11.3. This is followed, in Section 11.4, by our performance expectations across the testbed based on the computational characteristics of the stencil kernels coupled with a Roofline model analy-

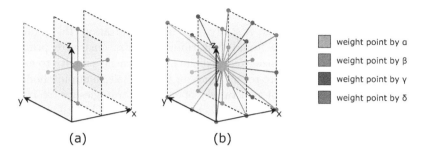

FIGURE 11.1: Visualization of the two stencils used in this work. (a) 7-point stencil. (b) 27-point stencil. Note: color represents the weighting factor for each point in the linear combination stencils.

sis. We summarize the applied optimizations and the parameter search in Sections 11.5 and 11.6. Finally, we present performance results followed by conclusions in Sections 11.7 and 11.8.

11.2   Stencil Overview

Partial differential equation (PDE) solvers are employed by a large fraction of scientific applications in such diverse areas as heat diffusion, electromagnetics, and fluid dynamics. These applications are often implemented using iterative finite-difference or similar techniques that sweep over a spatial grid, performing nearest neighbor computations called *stencils*. In the simplest stencil operations, each point in the grid is updated with a linear combination of its neighbors in both time and space. These operations are then used to build solvers that range from simple Jacobi iterations to complex multigrid and adaptive mesh refinement methods [5].

Stencil calculations perform global sweeps through data structures that are typically much larger than the capacity of the available data caches. In addition, the degree of data reuse is limited to the number of points in a stencil, typically fewer than 30. As a result, these computations generally achieve a low fraction of theoretical peak performance on modern microprocessors as data cannot be transferred from main memory fast enough to avoid stalling the computational units. Reorganizing these stencil calculations to take full advantage of memory hierarchies has been the subject of much investigation over the years. These have principally focused on tiling optimizations [19, 22, 23] that attempt to exploit locality by performing operations on cache-sized blocks of data before moving on to the next block. As seen in Chapter 11 it is possible to block explicit stencil methods in the time dimension in addition to the three spatial dimensions. Such strategies can dramatically improve performance by increasing arithmetic intensity. A study of stencil optimization [16] on (single-core) cache-based platforms found that tiling optimizations were primarily effective when the problem size exceeded the on-chip cache's ability to exploit temporal recurrences. A more recent study of lattice-Boltzmann methods [28] employed auto-tuners to explore a variety of effective strategies for refactoring lattice-based problems for multicore processing platforms. This study expands on prior work by developing new optimization techniques and applying them to a broader selection of processing platforms including accelerators like the Cell Broadband Engine and graphics processors units (GPUs).

In this work, we explore the performance of a single Jacobi (out-of-place) iteration of the 3D 7-point and 27-point stencils. In the Jacobi method, we maintain two separate double-precision (DP) 3D arrays. In a given iteration, one array is only read from and the second is only written to. This avoids all data dependencies and maximizes parallelism.

The 7-point stencil, shown in Figure 11.1(a), weights the center point by some constant α and the sum of its six neighbors (two in each dimension) by a second constant β. Naïvely, a 7-point stencil sweep can be expressed as a triply nested ijk loop over the following computation:

$$
\begin{aligned}
B_{i,j,k} \;=\; & \alpha A_{i,j,k} + \\
& \beta(A_{i-1,j,k} + A_{i,j-1,k} + A_{i,j,k-1} + A_{i+1,j,k} + A_{i,j+1,k} + A_{i,j,k+1})
\end{aligned}
\tag{11.1}
$$

where each subscript represents the 3D index into array A or B. The 27-point 3D stencil, as shown in Figure 11.1(b), is similar to the 7-point stencil, but with additional points to include the edge and corner points of a 3×3×3 cube surrounding the center grid point. It also introduces two additional constants — γ, to weight the sum of the edge points, and δ, to weight the sum of the corner points. Across all machines and experiments we preserve the stencil functionality and mandate a common, albeit parameterized, data structure.

11.3   Experimental Testbed

Although all multicore processor computers implement a cache-based shared memory architecture, their microarchitectural implementations are extremely diverse. As such, the work required to attain good performance and efficiency varies dramatically among them. Programming heterogeneous, accelerator-based computers compounds this challenge as there is no consensus on the broader question of memory hierarchy. To fully appreciate and understand the effects of architectural decisions and to demonstrate the ability of our auto-tuner to provide performance portability, we use one of the broadest sets of computers imaginable. Although the node architectures are diverse, they represent the building-blocks of current and future ultra-scale supercomputing systems. The key architectural features of these systems appears in Table 11.1. The details of these machines can be found in the literature [1, 2, 8–11, 13–15, 20, 21, 24].

In addition to using the recent evolution of commodity x86 processor architectures (Intel's Core2, AMD's Barcelona, Intel's Nehalem) which represent the addition of integrated memory controllers and multithreading, we also examine IBM's Blue Gene/P compute node, as well as the Sun's chip multithreaded (CMT) dual-socket Niagara2 (Victoria Falls). Although the in-order BGP will deliver rather low per-node performance, because it is optimized for power efficiency, its value in ultra-scale systems is immense. Unlike the other architectures, Niagara2 uses CMT to solve the instruction- and memory-level parallelism challenges with a single programming paradigm.

To explore the advantages of using accelerators, we explored both performance and efficiency for both the QS22 PowerXCell 8i enhanced DP Cell blade as well the GTX280 GPU. Cell uses a shared memory programming model in which all threads, regardless of the core they run on, may access a common pool of dynamic random access memory (DRAM). Cell exploits heterogeneity (essentially productivity vs. efficiency) in that the PowerPC cores use caches, but the synergistic processor elements (SPEs) use disjoint, DMA-filled, local stores.

Conversely, the GPU employs a partitioned DRAM memory. GPU cores may directly address the GPU DRAM, but may not address CPU DRAM. Such an architecture allows specialization in the form of optimization for capacity or performance; the GPU DRAM bandwidth is far greater than any other architecture. We obviate the complexity of this architecture (shuffling data between the CPU and GPU) by assuming the problems of interest fit within GPU DRAM. The GPU cores use a local store architecture similar to Cell, but fill it via multithreaded vector loads rather than DMA.

11.4   Performance Expectation

Before discussing potential optimizations or performance results, we first perform some basic kernel analysis to set realistic performance expectations. We commence with some analysis of the two kernels (and a substantive optimization to the latter) and proceed to analyze their performance on each of our seven architectures of interest.

TABLE 11.1: Architectural summary of evaluated platforms. [1]Superscalar Out-of-Order (OoO). [2]Concurrent *CUDA thread blocks* (max 8) per SM. [3]16 KB local-store partitioned among concurrent thread blocks. [4]System power was measured with a digital power meter while under full computational load. [5]Power running Linpack averaged per blade. (www.top500.org) [6]GTX280 system power shown for the entire system under load (450W) and GTX280 card itself (236W).

Core	Intel Nehalem	AMD Barcelona	Intel Core2	IBM PPC450d	Sun Niagara2	STI Cell eDP SPE	NVIDIA GT200 SM
Clock (GHz)	2.66	2.30	2.66	0.85	1.16	3.20	1.30
type	OoO[1]	OoO	OoO	in-order	in-order	in-order	vector
threads per core	2	1	1	1	8	1	8[2]
DP (GFlop/s)	10.7	9.2	10.7	3.4	1.16	12.8	2.6
local store	—	—	—	—	—	256KB	16KB[3]
L1 D\$	32KB	64KB	32KB	32KB	8KB	—	—
private L2\$	256KB	512KB	—	—	—	—	—

Socket	Intel Nehalem	AMD Barcelona	Intel Core2	IBM BGP chip	Sun Niagara2	STI Cell Processor	NVIDIA GT200
cores per socket	4	4	4 (MCM)	4	8	8 SPEs + 1 PPE	30
shared LL\$	8MB	2MB	2×4MB (2 cores/\$)	8MB	4MB	—	—
memory parallelism	HW prefetch	HW prefetch	HW prefetch	HW prefetch	MT	DMA	MT w/ coalescing

Node	Xeon X5550 Nehalem	Opteron 2356 Barcelona	Xeon E5355 Clovertown	BGP Compute Node	UltraSparc T5140 VF	QS22 Cell Blade	GeForce GTX280
sockets per SMP	2	2	2	1	2	2	1 GPU (+CPU)
Peak DP (GFlop/s)	85.3	73.6	85.3	13.6	18.7	204.8	78.0
DRAM Pin BW (GB/s)	51.2	21.33	21.33(Rd) 10.66(Wr)	13.6	42.66(Rd) 21.33(Wr)	51.2	141
Flop:Byte	1.66	3.45	2.66	1.00	0.29	4.00	0.55
DRAM size (GB)	16	16	12	2	32	32	1 (device) 4 (host)
DRAM type	DDR3-1066	DDR2-800	FBDIMM-667	DDR2-425	FBDIMM-667	DDR2-800	GDDR3-1100
System Power (W)[4]	375	350	530	31[5]	610	265[5]	450 (236)[6]
Threading	POSIX Threads	POSIX Threads	POSIX Threads	POSIX Threads	POSIX Threads	libspe 2.1	CUDA 2.0
Compiler	icc 10.0	icc 10.0	icc 10.0	xlc 9.0	gcc 4.0.4	gcc 4.1.1	nvcc 0.2.1221
STREAM (GB/s)	35.3	15.2	7.16	12.8	24.9	37	127

11.4.1   Stencil Characteristics

Consider the 7-point stencil. We observe that in the naïve implementation each stencil presents 7 reads and 1 write to the memory subsystem. In the 27-point stencil, the number of reads per stencil increases to 27, but the number of writes remains 1. However, when one considers adjacent stencils, we observe substantial reuse. Thus, to attain good performance, a cache (if present) must filter the requests and present only the two compulsory (in 3C's parlance) requests per stencil to DRAM [12]. There are two compulsory requests per stencil because every point in the grid must be read once and written once. One should be mindful that many caches are *write allocate*. That is, on a write miss, they first load the target line into the cache. Such an approach implies that writes generate twice the memory traffic as reads even if those addresses are written but never read. The two most common approaches to avoiding this superfluous memory traffic are *write through* caches or cache bypass stores.

Table 11.2 shows the per stencil average characteristics for both the 7- and 27-point stencils as well as the highly optimized common subexpression elimination (CSE) version of the 27-point stencil (discussed in detail in Section 11.5.4). Observe that all three stencils perform dramatically different numbers of floating-point operations and loads. Although an ideal cache would distill these loads and stores into 8 bytes of compulsory DRAM read traffic and 8 bytes of compulsory DRAM write traffic, caches are typically not write through, infinite in capacity, or fully associative. As naïve codes are not cache blocked, we expect an additional 8 bytes of DRAM write allocate traffic, and another 16 bytes of capacity miss traffic (based on the caches found in superscalar processors and the reuse pattern of these stencils)—a 2.5× increase in memory traffic. Auto-tuners for structured grids will actively or passively attempt to elicit better cache behavior and less memory traffic on the belief that reducing memory traffic and exposed latency will improve performance. If the auto-tuner can eliminate all cache misses, we can improve performance by 1.65×, but if the auto-tuner also eliminates all write allocate traffic, then it may improve performance by 2.5×.

Arithmetic intensity is a particularly useful term in bounding performance expectations. For our purposes, we define arithmetic intensity as the ratio of floating-point operations to DRAM bytes transferred (*i.e.* the memory traffic not filtered by the cache). High arithmetic intensities suggest high temporal locality and thus a propensity to achieve high performance. Low arithmetic intensities imply very little computation per memory transaction and thus performance limited by memory bandwidth. In the latter case, performance is bounded by the product of arithmetic intensity and DRAM bandwidth.

11.4.2   A Brief Introduction to the Roofline Model

The Roofline model [27, 29, 30] provides a visual assessment of potential performance and impediments to performance constructed using bound and

TABLE 11.2: Average stencil characteristics. Arithmetic Intensity is $\frac{Total\ Flops}{Total\ DRAM\ Bytes}$. WB is an abbreviation for write back. Numbers in parentheses assume exploitation of cache bypass. Capacity misses are estimated based on capturing only the temporal recurrence within a plane. The potential benefit from auto-tuning (elimination of write allocations and capacity misses) is about 1.65× and 2.5× using cached and cache-bypass stores, respectively.

Type	(per stencil) flops	$ refs	Memory Traffic (Bytes) Compulsory Reads	Write WB's	Capacity Allocate	Misses	Arithmetic Intensity Naïve	Optimized
7-pt	8	8	8	8	8	16	0.20	0.33 (0.50)
27-pt	30	28	8	8	8	16	0.75	1.25 (1.88)
27-pt (CSE)	18	10	8	8	8	16	0.45	0.75 (1.13)

bottleneck analysis [18]. Each model is constructed using a communication-computation abstraction where data are moved from a memory to computational units. This "memory" could be registers, L1, L2, or L3, but is typically DRAM. Computation for our purposes will be the floating-point datapaths. We use arithmetic intensity as a means of expressing the balance between computation and communication. Often a first order model (*e.g.* DRAM–FP) is sufficient for a given architecture for a range of similar kernels. However, for certain kernels, depending on the degree of optimization, bandwidth from the L3 could be the actual bottleneck. For purposes of this paper, we will only use a DRAM–FP Roofline model.

The Roofline model defines three types of potential bottlenecks: computation, communication, and locality (arithmetic intensity). Evocative of the roofline analogy, these are labeled as ceilings and walls. The in-core ceilings (or computation bounds) are perhaps the easiest to understand. To achieve peak performance, a number of architectural features must be exploited— thread-level parallelism (TLP) (*e.g.* multicore), instruction-level parallelism (ILP) (*e.g.* keep functional units busy by unrolling and jamming loops), data-level parallelism (DLP) (*e.g.* single instruction multiple data [SIMD]), and proper instruction mix (*e.g.* balance between multiplies and adds or total use of fused multiply add. If one fails to exploit one of these (either a failing of the compiler or programmer), the performance is diminished. We define ceilings as impenetrable impediments to improved performance without the corresponding optimization. Bandwidth ceilings are similar but are derived from incomplete expression and exploitation of memory-level parallelism. As such we often define ceilings such as no NUMA, or no prefetching. Finally, locality walls represent the balance between computation and communication. For many kernels the numerator of this ratio is fixed (*i.e.* the number of floating-point operations is fixed), but the denominator varies as compulsory misses are augmented with capacity or, conflict misses, as well as speculative or write allocation traffic. As these terms are progressively added they define a new arithmetic intensity and thus a new locality wall. Moreover, for each of these terms there is a corresponding optimization which must be applied (*e.g.* cache blocking for capacity misses, array padding for conflict misses, or cache bypass for write allocations) to remove this potential impediment to performance. It should be noted that the ordering of ceilings is based on the perceived abilities of compilers. Those ceilings least likely to be addressed by a compiler are placed at the top.

One may use the Roofline model to identify potential bottlenecks for each architecture. Given an arithmetic intensity, one may simply scan upward from the x-axis. Performance may not exceed a ceiling until the corresponding optimization has been implemented. For example, with cache bypass, the 27-point stencil on Nehalem requires full ILP including unroll and jam, full DLP using streaming SIMD extensions (SSE) instructions (SIMDization), and NUMA-aware allocation to have any hope of achieving peak performance. Conversely, without cache bypass, the 7-point will not even require full TLP, that is using all cores, to achieve peak performance.

11.4.3 Roofline Model-Based Performance Expectations

Figure 11.2 presents a Roofline model for each of our seven computers. The table in the legend describes the parallelism mix for each in-core ceiling and defines a circular symbol (A, B, C, D) for cross-referencing the figure. The in-core ceilings are based on theoretical architectural capabilities. Similarly, we define a series of bandwidth ceilings based on empirical performance obtained via an optimized version of the STREAM benchmark that can be configured not to exploit NUMA or cache-bypass [6]. These are denoted by diamonds X, Y, and Z. Hashed bars represent the expected per-architecture range in ideal arithmetic intensity for the 7- and 27-point (non-CSE) stencils differentiated by use of cache-bypass instructions (see Section 11.5.3) or a local store architecture. Although cache-bypass behavior can improve arithmetic intensity by 50%, depending on the implemented optimizations and the underlying architecture, commensurate improvements in performance may not be attainable.

Please note, there is a range in arithmetic intensity due to explicit array padding, speculative loads from hardware prefetchers, implicit ghost zones arising from cache blocking, and potential conflict misses. Although on a local store architecture like Cell, DMA provides us the ability to more precisely set a tighter upper limit to arithmetic intensity, the obscurities of memory coalescing and minimum memory quanta (opaque microarchitectural issues) on the GTX280 result in a somewhat larger range in arithmetic intensity. Moreover, only a CSE version of the 27-point stencil was implemented on

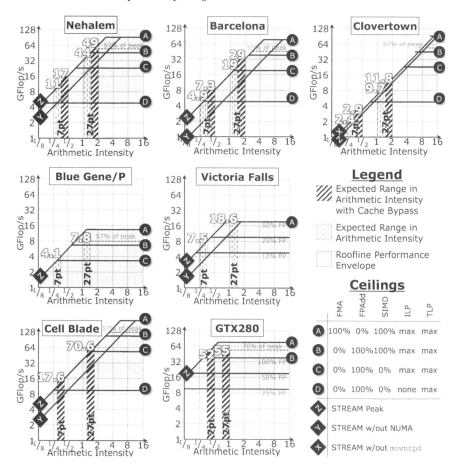

FIGURE 11.2: Roofline model–predicted performance (in GFlop/s) for both the 7- and the (non-CSE) 27-point stencils with and without write allocate. Notes: The table in the legend provides a symbol and definition for each type of ceiling. The ceilings for Victoria Falls and the GTX280 are the fraction of the dynamic instruction mix that is floating-point. Victoria Falls and Blue Gene/P do not exploit cache bypass (they always use write allocate).

the GTX280. Finally, for clarity we express Roofline-predicted performance in GFlop/s. To convert to GStencil/s, divide the 7-point stencil by 8, and the 27-point (non-CSE) stencil by 30.

Using the Roofline models for the architectures in Figure 11.2, combined with the knowledge of each kernel's arithmetic intensity and instruction mix, we may not only bound ultimate performance for each architecture, but also broadly enumerate the optimizations required to achieve it.

Nehalem: For the 7-point stencil, Nehalem will ultimately be memory-bound with or without cache-bypass. Given the STREAM bandwidth and ideal arithmetic intensity, 17 GFlop/s (2.1 GStencil/s) is a reasonable performance bound. To achieve it, some ILP or DLP coupled with correct NUMA allocation is required. However, as we move to the 27-point stencil, we observe that Nehalem will likely become compute limited; likely achieving between 40 and 49 GFlop/s (1.3–1.6 GStencil/s). There is some uncertainty here due to the confluence of a broad range in arithmetic intensity at a point on the Roofline where computation is nearly balanced with communication. The transition from memory-bound to compute-bound implies that the benefits of cache-bypass will be significantly diminished as one moves from the 7-point to the 27-point stencil.

Barcelona: When one considers Barcelona, a processor architecturally similar to Nehalem but built on a previous generation's technology, we see that although it exhibits similar computational capability, its substantially lower memory bandwidth mandates that all stencil kernels be memory-bound. Barcelona should be limited to 7.3 GFlop/s (0.9 GStencil/s) and 29 GFlop/s (0.98 GStencil/s) for the 7- and 27-point stencils, respectively. However, to achieve high performance on the latter, SIMD (DLP), substantial unrolling (ILP), and proper NUMA allocation will be required.

Clovertown: Our third x86 architecture is Intel's Clovertown; an even older, front side bus (FSB) based architecture. It too has similar computational capabilites to Nehalem and Barcelona, but has even lower memory bandwidth. As such, all kernels will be memory-bound. Interestingly, although the STREAM benchmark time-to-solution is superior using the cache-bypass store (*movntpd* instruction), the observed STREAM bandwidth (based on total bytes including those from write allocations) is substantially lower than the standard bandwidth (*movpd* instruction). As such, the benefit of exploiting cache bypass on stencil operations is muted to perhaps 20% instead of the ideal 50%. Clovertown will be so heavily memory-bound that simple parallelization should be enough to achieve peak performance on the 7-point, where only moderate unrolling is sufficient on the 27-point. We expect Clovertown performance to be limited to 2.9 GFlop/s (0.36 GStencil/s) and 12 GFlop/s (0.39 GStencil/s) for the 7- and 27-point stencils, respectively

Blue Gene/P: Architectures like the chip used in Blue Gene/P are much more balanced, dedicating a larger fraction of their power and design budget to DRAM performance. This is not to say they have higher absolute memory bandwidth, but rather the design is more balanced given the low-frequency quad core processors. As we were not able to exploit cache bypass we see substantially lower arithmetic intensity than the x86 architectures. This simple, first order model suggests that if we were able to perfectly SIMDize and unroll the code, we would expect the 7-point stencil to be memory-bound, yielding 4.1 GFlop/s or 0.5 GStencil/s, but the 27-point to be compute-bound limited by the relatively small fraction of multiplies in the code to 7.8 GFlop/s or 0.26 GStencil/s. This may be difficult to achieve given the limited issue-width and in-order PPC450 architecture.

Victoria Falls: Although Victoria Falls uses a dramatically different mechanism for expression of memory-level parallelism (massive TLP), its performance characteristics should be similar to Blue Gene/P in that it will be memory-bound for the 7-point stencil, and compute-bound for the 27-point with performances of 7.5 GFlop/s (0.94 GStencil/s) and 18.6 GFlop/s (0.6 GStencil/s), respectively. We did not exploit any form of cache-bypass on Victoria Falls. As multithreading provides an attractive solution to avoiding the ILP pitfall, our primary concern after proper NUMA allocations is that floating-point instructions dominate the instruction mix on the 27-point stencil. As such, the Victoria Falls Roofline is shown with computational ceilings corresponding to various ratios of floating-point instructions.

Cell: Although Cell is a local store–based architecture, we may seamlessly analyze it using the Roofline model. DMA obviates the need for write allocate behavior, but comes with a severe alignment penalty. With proper array padding and blocking for a rather small local store, we may achieve 78% of the ideal x86 arithmetic intensity. More succinctly, Cell will certainly generate more compulsory and capacity memory traffic than its x86 counterparts. Unfortunately, this is a rather unattractive situation given the QS22's DDR2-based memory bandwidth is now no higher than the top-of-the-line x86 Nehalem processor. As such, we expect Cell to be memory-bound on both stencils, delivering up to 17.6 GFlop/s (2.2 GStencil/s) and 70.6 GFlop/s (2.35 GStencil/s). NUMA and appropriate loop unrolling will be required for both, and SIMDization for the latter.

GTX280: Although one could ostensibly classify the GTX280 as a local store–based architecture, the lack of documentation on the memory subsystem behavior results in a wide range of possible arithmetic intensities. Overall, the shape of the roofline for the GTX280 is much more in line with Blue Gene/P (albeit a much higher roofline) than with Cell. Unfortunately, for the 7-point stencil to achieve peak performance (52 GFlop/s or 6.5 GStencil/s), we would require 100% of the dynamic instructions to be floating-point. As a large fraction will need to be loads and address calculations, we expect performance to be significantly lower. It should be noted that just as cache-based microarchitectures must access DRAM in full cache lines, the GPU microarchitecture must access DRAM in units of the memory quanta. This complexity is hidden from programmers, but the performance impacts may come as a surprise. That is, if the memory quanta were large, arithmetic intensity would be so depressed (lack of spatial locality on certain accesses) that the 7-point stencil might become memory bound. Due to the complexity of GPU programming, only the CSE version of the 27-point stencil was implemented. Given user and compiler elimination of floating-point operations, the number of operations per stencil was reduced to about 17. Although this pushes the kernel toward the memory-bound region, it can significantly improve performance to about 3.2 GStencil/s. Of course, this also assumes the unrealistic 100% floating-point instruction mix.

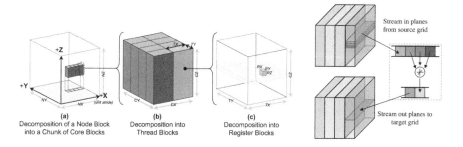

FIGURE 11.3: (LEFT) Four-level problem decomposition: In (a), a *node block* (the full grid) is broken into smaller *chunks*. One core block from the chunk in (a) is magnified in (b). A single thread block from the core block in (b) is then magnified in (c) and decomposed into register blocks. (RIGHT) Circular queue optimization: planes are streamed into a queue containing the current time step, processed, written to out queue, and streamed back.

11.5   Stencil Optimizations

Compilers utterly fail to achieve satisfactory stencil code performance (even with manual pthread parallelization) because implementations optimal for one microarchitecture may deliver suboptimal performance on another (an artifact perhaps eventually mitigated by incorporating auto-tuning into compilers). Moreover, their ability to infer legal domain-specific transformations, given the freedoms of the C language, is limited and permanent. To that end, we discuss a number of optimizations at the source level to improve stencil performance, including: NUMA-aware data allocation, array padding, multilevel blocking, loop unrolling and reordering, CSE, as well as prefetching for cache-based architectures and DMA for local–store based architectures. Additionally, we present two novel stencil optimizations: circular queue and thread blocking. The optimizations can roughly be divided into five categories: problem decomposition (parallelization), data allocation, bandwidth optimizations, in-core optimizations, and algorithmic transformations. In this section, we discuss each of these techniques in greater detail. Then, in Section 11.6, we cover our overall auto-tuning strategy, including architecture-specific exceptions. These optimizations are an extension of those used in our previous work [7].

11.5.1 Parallelization and Problem Decomposition

In this work, we examine parallelization through threading and geometric problem decomposition. Across all architectures, we applied a four-level geometric decomposition strategy, visualized in Figure 11.3(left), that simultaneously implements parallelization, cache blocking, and loop unrolling. This multilevel decomposition encompasses three separate optimizations: *core blocking*, *thread blocking*, and *register blocking*. Note that the nature of out-of-place iterations implies that all blocks are independent and can be computed in any order. This greatly facilitates parallelization.

We now discuss the decomposition strategy from the largest structures to the finest. First, a *node block* (the entire problem) of size $NX \times NY \times NZ$ is partitioned in all three dimensions into smaller *core blocks* of size $CX \times CY \times CZ$, where X is the unit stride dimension. This first step is designed to avoid last level cache capacity misses by effectively cache blocking the problem. Each core block is further partitioned into a series of *thread blocks* of size $TX \times TY \times CZ$. Core blocks and thread blocks are the same size in the Z (least unit stride) dimension, so when $TX = CX$ and $TY = CY$, there is only one thread per core block. This second decomposition is designed to exploit the common locality threads may have within a shared cache or local memory. Then, our third decomposition partitions each thread block into *register blocks* of size $RX \times RY \times RZ$. The dimensions of the register block indicate how many times the inner loop has been unrolled in each of the three dimensions. This allows us to explicitly express DLP and ILP rather that assuming the compiler may discover it.

To facilitate NUMA allocation, core blocks are grouped together into *chunks* of size *ChunkSize* and assigned in bulk to an individual core. The number of threads in a core block ($Threads_{core}$) is simply $\frac{CX}{TX} \times \frac{CY}{TY}$, so we then assign these chunks to groups of $Threads_{core}$ threads in a round-robin fashion (similar to the *schedule* clause in OpenMP's *parallel for* directive). Note that all the core blocks in a chunk are processed by the same subset of threads. When *ChunkSize* is large, concurrent core blocks may map to the same set in cache, causing conflict misses. However, we do gain a benefit from diminished NUMA effects. In contrast, when *ChunkSize* is small, concurrent core blocks are mapped to contiguous set addresses in a cache, reducing conflict misses. This comes at the price of magnified NUMA effects. We therefore tune *ChunkSize* to find the best tradeoff of these two competing effects. In general, this decomposition scheme allows us to explain shared cache locality, cache blocking, register blocking, and NUMA-aware allocation within a single formalism.

11.5.2   Data Allocation

The layout of our data array can significantly affect performance. As a result, we implemented a *NUMA-aware allocation* to minimize inter-socket communication and *array padding* to minimize intra-thread conflict misses.

Our stencil code implementation allocates the source and destination grids as separate large arrays. On NUMA systems that implement a "first touch" page mapping policy, a memory page will be mapped to the socket where it is initialized. Naïvely, if we let a single thread fully initialize both arrays, then all the memory pages containing those arrays will be mapped to that particular socket. Then, if we used threads across multiple sockets to perform array computations, they would perform expensive inter-socket communication to retrieve their needed data.

Since our decomposition strategy has deterministically specified which thread will update each array point, a better alternative is to let each thread initialize the points that it will later be processing. This *NUMA-aware allocation* correctly pins data to the socket tasked to update them. This optimization is only expected to help when we scale from one socket to multiple sockets, but without it, performance on memory-bound architectures could easily be cut in half.

The second data allocation optimization that we utilized is *array padding*. Some architectures have relatively low associativity shared caches, at least when compared to the product of threads and cache lines required by the stencil. On such computers, conflict misses can significantly impair performance. In other cases, some architectures prefer certain alignments for coalesced memory accesses; failing to do so can greatly reduce memory bandwidth. To avoid these pitfalls, we pad the unit-stride dimension ($NX \leftarrow NX + pad$).

11.5.3   Bandwidth Optimizations

For stencils with low arithmetic intensities, the 7-point stencil being an obvious example, memory bandwidth is a valuable resource that needs to be managed effectively. As a result, we introduce three bandwidth optimizations: *software prefetching* to hide memory latency and thereby increase effective memory bandwidth, *circular queue* to minimize conflict misses, and the *cache bypass* instruction to dramatically reduce overall memory traffic.

The architectures used in this paper employ four principal mechanisms for hiding memory latency: hardware prefetching, software prefetching, DMA, and multithreading. The x86 architectures use hardware stream prefetchers that can recognize unit-stride and strided memory access patterns. When such a pattern is detected successive cache lines are prefetched without first being demand requested. Hardware prefetchers will not cross TLB boundaries (only 512 consecutive doubles), and can be easily halted by either spurious memory requests or discontinuities in the address stream. The former demands the hardware prefetcher be continually prodded to prefetch more. The latter may arise when $CX < NX$, that is, when core blocking results in stanza access patterns and jumps in the address stream. Although this is not an issue on multithreaded architectures, they may not be able to completely cover all cache and memory latency. In contrast, *software prefetching*, which is available on all cache-based processors, does not suffer from either limitation. However, it can only express a cache line's worth of memory-level parallelism. In addition, unlike a hardware prefetcher (where the prefetch distance is implemented in hardware), software prefetching must specify the appropriate distance to effectively hide memory latency. DMA is only implemented on Cell, but can easily express the stanza memory access patterns. DMA operations are decoupled from execution and are implemented as double-buffered reads of core block planes.

The *circular queue* implementation, visualized in Figure 11.3(right), is a technique that allows efficient parallelization, eliminates conflict misses, and allows efficient expression of memory-level parallelism. This approach allocates a shadow copy of the planes of a core block in local memory or registers. The 7-point stencil requires three read planes to be allocated, which are then populated through loads or DMAs. However, it can often be beneficial to allocate an output plane and double-buffer reads and writes as well (six cache blocked planes). The advantage of the circular queue is the potential avoidance of lethal conflict misses. We currently explore this technique only on the local-store architectures but note that future work will extend this to the cache-based architectures.

So far we have discussed optimizations designed to hide memory latency to improve memory bandwidth, but we can extend this discussion to optimizations that minimize memory traffic. As described in Section 11.4.1 we may eliminate 33% of the memory traffic and thus increase arithmetic intensity by 50% by bypassing any write-allocate cache. If bandwidth bound, this can also increase performance by 50%. This benefit is clearly implicit on the cache-less Cell and GT200 architectures. However, this optimization is not supported on either Blue Gene/P or Victoria Falls.

11.5.4   In-Core Optimizations

For stencils with higher arithmetic intensities, the 27-point stencil being a good example, computation can often become a bottleneck. To address this issue, we perform *register blocking* to effectively utilize a given platform's registers and functional units.

Although superficially simple, there are innumerable ways of optimizing the execution of a 7-point or 27-point stencil. After tuning for bandwidth and memory traffic, it often helps to explore the space of inner-loop transformations to find the fastest possible code. To this end, we wrote a code generator that could generate any unrolled, jammed, and reordered version of the stencil. Register blocking is, in essence, unroll and jam in X, Y, and Z. This creates small $RX \times RY \times RZ$ blocks that sweep through each thread block. Larger

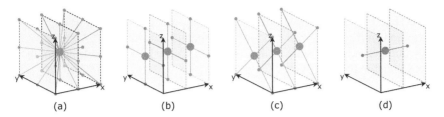

FIGURE 11.4: Visualization of common subexpression elimination. (a) Reference 27-point stencil. (b)-(d) decomposition into seven simpler stencils. As one loops through x, two of the stencils from both (b) and (c) will be reused for $x + 1$.

register blocks have better surface-to-volume ratios and thus reduce the demands for L1 cache bandwidth while simultaneously expressing ILP and DLP. However, in doing so, they may significantly increase register pressure.

Although the standard code generator produces portable C code, compilers often fail to effectively SIMDize the resultant code. As such, we created several instruction set architecture (ISA) specific variants that produce *explicitly SIMDized* code for x86, Blue Gene/P, and Cell using intrinsics. These versions will deliver much better in-core performance than a compiler. However, as one might expect, this may have a limited benefit on memory-bound stencils like the 7-point.

11.5.5   Algorithmic Transformations

Our final optimization involves identifying and eliminating common expressions across several points. This type of CSE can be considered to be an algorithmic transformation because of two reasons: the flop count is being reduced, and the flops actually being performed may be performed in a different order than our original implementation. Due to the non-associativity of floating-point operations, this may well produce results that are not bit-wise equivalent to those from the original implementation.

For the 7-point stencil, there was very little opportunity to identify and eliminate common subexpressions. Hence, this optimization was not performed, and 8 flops are always performed for every point. The 27-point stencil, however, presents such an opportunity. Consider Figure 11.4. If one were to perform the reference 27-point stencil for successive points in x, we perform 30 flops per stencil. However, as we loop through x, we may dynamically create several temporaries (unweighted reductions)—Figure 11.4(b) and (c). For 27-point stencils at x and $x+1$, there is substantial reuse of these temporaries. On fused multiply-add (FMA) based architectures, we may implement the 27-point stencil by creating these temporaries and performing a linear combination using two temporaries from Figure 11.4(b), two from Figure 11.4(c), and

TABLE 11.3: Attempted optimizations and the associated parameter spaces explored by the auto-tuner for a 256^3 stencil problem ($NX, NY, NZ = 256$). All numbers are in terms of doubles. †On Cell, the 7-point stencil only used 2×8 register blocks.

Category	Optimization Parameter	Name	parameter tuning range by architecture				
			x86	BGP	VF	Cell	GTX280
Data Alloc	NUMA Aware		✓	N/A	✓	✓	N/A
	Pad (max):		32	32	32	15	15
	Pad (multiple of):		1	1	1	16	16
Domain Decomp	Core Block Size	CX	NX	NX	{8...NX}	{64...NX}	{16...32}
		CY	{4...NY}	{4...NY}	{4...NY}	{8...NY}	CX
		CZ	{4...NZ}	{4...NZ}	{4...NZ}	{128...NZ}	64
	Thread Block Size	TX	CX	CX	{8...CX}	CX	1
		TY	CY	CY	{8...CY}	CY	CY/4
	Chunk Size		$\{1...\frac{NX \times NY \times NZ}{CX \times CY \times CZ \times NThreads}\}$				N/A
Low Level	Register Block Size	RX	{1...8}	{1...8}	{1...8}	{1...16}†	TX
		RY	{1...4}	{1...4}	{1...4}	{1...8}†	TY
		RZ	{1...4}	{1...4}	{1...4}	1	1
	(SIMDized)		✓	✓	N/A	✓	N/A
	Prefetch Distance		{0...64}	{0...64}	{0...64}	N/A	N/A
	DMA Size		N/A	N/A	N/A	CX×CY	N/A
	Cache Bypass		✓	—	N/A	implicit	implicit
	Circular Queue		—	—	—	✓	✓
Tuning	Search Strategy		Iterative Greedy			Exhaustive	Hand
	Data-aware		✓	✓	✓	✓	N/A

the stencil shown in Figure 11.4(c). This method requires about 15 instructions. On the x86 architectures, we create a second group of temporaries by weighting the first set. With enough loop unrollings in the inner loop, the CSE code has a lower bound of 18 flops/point. Disappointingly, neither the gcc nor icc compilers were able to apply this optimization automatically. However, on the GTX280, a combination of a 24-flop hand-coded CSE implementation and the nvcc compiler was able to produce a 17-flop implementation.

11.6   Auto-tuning Methodology

Thus far, we have described our applied optimizations in general terms. In order to take full advantage of the optimizations mentioned in Section 11.5, we developed an auto-tuning environment [7] similar to that exemplified by libraries like ATLAS [26] and OSKI [25]. To that end, we first wrote a Perl code generator that produces multithreaded C code variants encompassing our stencil optimizations. This approach allows us to evaluate a large optimization space while preserving performance portability across significantly varying architectural configurations.

The parameter space for each optimization individually, shown in Table 11.3, is certainly tractable—but the parameter space generated by combining these optimizations results in a combinatorial explosion. Moreover, these optimizations are not independent of one another; they can often interact in subtle ways that vary from platform to platform. Hence, the second component of our auto-tuner is the search strategy used to find a high-performing parameter configuration. For this study, we afforded ourselves the luxury of spending many hours tuning a single node, since large-scale stencil applications may be scaled to thousands of nodes and run many times. At this level of parallelism, it is vital to ensure that the software is as efficient as possible.

To find the best configuration parameters, we employed an iterative "greedy" search on the cache-based machines. First, we fixed the order of optimizations. Generally, they were ordered by their level of complexity, but there was some expert knowledge employed as well. This ordering is shown in the legends of Figures 11.5 and 11.6; the relevant optimizations were applied in order from bottom to top. Within each individual optimization, we performed an exhaustive search to find the best performing parameter(s). These values were then fixed and used for all later optimizations. We consider this to be an iterative greedy search. If all applied optimizations were independent of one another, this search method would find the global performance maxima. However, due to subtle interactions between certain optimizations, this usually will not be the case. Nonetheless, we expect that it will find a good-performing set of parameters after doing a full sweep through all applicable optimizations.

In order to judge the quality of the final configuration parameters, two metrics can be used. The more useful metric is the Roofline model, which provides an upper bound on kernel performance. If our fully tuned implementation approaches this bound, then further tuning will not be productive. The second metric is the performance improvement obtained from doing a second pass through our greedy iterative search. This is represented by the topmost color in the legends of Figures 11.5 and 11.6. If this second pass improves performance substantially, then our initial greedy search obviously was not effective.

For the local-store architectures, two other search strategies were used. On the Cell, the size of the local store sufficiently restricted the search space so that an exhaustive search could be performed. For the GTX280 GPU, a CUDA code generator was not written; instead, since the code was only being deployed on a single architecture, it was hand-tuned (manual-search) by a knowledgeable programmer.

11.6.1   Architecture-Specific Exceptions

Due to limited potential benefit and architectural characteristics, not all architectures implement all optimizations or explore the same parameter spaces. Table 11.3 details the range of values for each optimization parameter by architecture. In this section, we explain the reasoning behind these exceptions to the full auto-tuning methodology. To make the auto-tuning search space tractable, we typically explored parameters in powers of two.

The x86 architectures, Barcelona and Clovertown, rely on hardware stream prefetching as their primary means for hiding memory latency. The Nehalem architecture adds multithreading as another mechanism to tolerate memory latency and improve instruction scheduling. As previous work [17] has shown that short stanza lengths severely impair memory bandwidth, we prohibit core blocking in the unit-stride (X) dimension, so $CX = NX$. Thus, we expect the hardware stream prefetchers to remain engaged and effective. Although we utilized both Nehalem threads for computation, we did not attempt to perform thread blocking on this architecture. Further, neither of the two other x86 architectures support multithreading. Thus, the thread blocking search space was restricted so that $TX = CX$, and $TY = CY$. As x86 processors implement SSE2, we implemented a special SSE SIMD code generator for the x86 ISA that would produce both explicit SSE SIMD intrinsics for computation as well as the option of using a non-temporal store *movntpd* to bypass the cache. On these computers, the threading model was Pthreads.

For the BG/P architecture, the optimization and parameter space restrictions are mostly the same as those for x86. Hardware stream prefetch is implemented on this processor, so again we prohibit blocking in the unit-stride dimension. We also implemented a BG/P-specific SIMD code generator, but did not use any cache bypass feature.

Although Victoria Falls is also a cache-coherent architecture, its multi-threading approach to hiding memory latency is very different from out-of-order execution coupled with hardware prefetching. As such, we allow core blocking in the unit-stride dimension. Moreover, we allow each core block to contain either one or eight thread blocks. In essence, this allows us to conceptualize Victoria Falls as either a 128-core machine or a 16-core machine with eight threads per core. In addition, there are no supported SIMD or cache bypass instrinsics, so only the portable pthreads C code was run.

Unlike the previous four computers, Cell uses a cache-less local-store architecture. Moreover, instead of prefetching or multithreading, DMA is the architectural paradigm utilized to express memory-level parallelism and hide memory latency. This has a secondary advantage in that it also eliminates superfluous memory traffic from the cache line fill on a write miss. The Cell code generator produces both C and SIMDized code. However, our use of SDK 2.1 resulted in poor DP code scheduling as the compiler was scheduling for a QS20 rather than a QS22. Unlike the cache-based architectures, we implement the dual circular queue approach on each SPE. Moreover, we double-buffer both reads and writes. For optimal performance, DMA must be 128 byte (16 doubles) aligned. As such, we pad the unit-stride (X) dimension of the problem so that $NX + 2$ is a multiple of 16. For expediency, we also restrict the minimum unit-stride core blocking dimension (CX) to be 64. The threading model was IBM's libspe.

The GT200 has architectural similarities to both Victoria Falls (multi-threading) and Cell (local-store based). However, it differs from all other architectures in that the device DRAM is disjoint from the host DRAM. Unlike the other architectures, the restrictions of the CUDA programming model constrained optimization to a very limited number of cases. First, we explore only two core block sizes: 32×32 and 16×16. We depend on CUDA to implement the threading model and use thread blocking as part of the tuning strategy. The thread blocks for the two core block sizes are restricted to 1×8 and 1×4, respectively. Since the GT200 contains no automatically-managed caches, we use the circular queue approach that was employed in the Cell stencil code. However, the register file is four times larger than the local memory, so we chose register blocks to be the size of thread blocks ($RX = TX, RY = TY, RZ = 1$) and chose to keep some of the planes in the register file rather than shared memory.

11.7   Results and Analysis

To understand the performance of the 7-point and 27-point stencils detailed in Section 11.2, we apply one out-of-place stencil sweep at a time to a 256^3 grid. The reference stencil code uses only two large flat 3D scalar arrays as data structures, and that is maintained through all subsequent tuning. We do increase the size of these arrays with an array padding optimization, but this does not introduce any new data structures nor change the array ordering. In addition, in order to get accurate measurements, we report the average of at least five timings for each data point, and there was typically little variation among these readings.

Below we present and analyze the results from auto-tuning the two stencils on each of the seven architectures. Please note, in all figures we present performance as GStencil/s (10^9 stencils per second), to allow a meaningful comparison between CSE and non-CSE kernels, and we order threads to first exploit all the threads on a core, then populate all cores within a socket, and finally use multiple sockets. We stack bars to represent the performance as the auto-tuning algorithm progresses through the greedy search, *i.e.* subsequent optimizations are built on best configuration of the previous optimization.

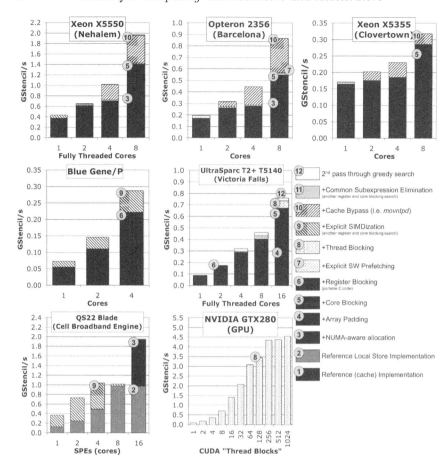

FIGURE 11.5: 7-point stencil performance. In all of the graphs above, "GStencil/s" can be converted to "GFlop/s" by multiplying by 8 flops/stencil.

11.7.1 Nehalem Performance

Figure 11.5 shows 7-point stencil performance, and we observe several interesting features. First, if we only examine the reference implementation, performance is fairly constant *regardless of core count*. This is discouraging news for programmers; the compiler, even with all optimization flags set, cannot take advantage of the extra resources provided by more cores.

The first optimization we applied was using a NUMA-aware data allocation. By correctly mapping memory pages to the socket where the data will be processed, this optimization provides a speedup of 2.5× when using all eight cores of the SMP. Subsequently, core blocking and cache bypass also produced performance improvements of 74% and 37%, respectively. Both of these optimizations attempt to reduce memory traffic (capacity misses), suggesting

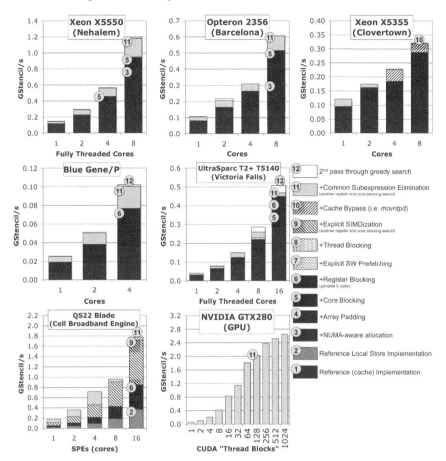

FIGURE 11.6: 27-point stencil performance. Before the *Common Subexpression Elimination* optimization is applied, "GStencil/s" can be converted to "GFlop/s" by multiplying by 30 flops/stencil. Note, explicitly SIMDized BG/P performance was slower than the scalar form both with and without CSE. As such, it is not shown.

that performance was bandwidth-bound at high core counts. By looking at the Roofline model for the Nehalem (shown in Figure 11.2), we see that this is indeed the case. The model predicts that if the stencil calculation can achieve STREAM bandwidth, while minimizing non-compulsory cache misses, then the 7-point stencil will attain a maximum of 2.1 GStencil/s (16.8 GFlop/s). In actuality, we achieve 2.0 GStencil/s (15.8 GFlop/s). As this is acceptably good performance, we can stop tuning. Overall, auto-tuning produced a speedup of 4.5× at full concurrency and also showed an improvement of 4.5× when scaling from one to eight cores.

Observe that register blocking and software prefetching ostensibly had little performance benefit—a testament to `icc` and hardware prefetchers. Remember, the auto-tuning methodology explores a large number of optimizations in the hope that they may be useful on a given architecture–compiler combination. As it is difficult to predict this beforehand, it is still important to try each relevant optimization.

The 27-point stencil performs 3.8× more flops per grid point than the 7-point stencil (before application of CSE), so a very different set of bottlenecks may ultimately limit performance. We see in Figure 11.6 that the performance of the reference implementation improves by 3.3× when scaling from one to four cores, but then drops slightly when we use all eight cores across both sockets. This performance quirk is eliminated when we apply the NUMA-aware optimization.

There are several indicators that strongly suggest that it is compute-bound—core blocking shows less benefit than for the 7-point stencil, cache bypass does not show any benefit, performance scales linearly with the number of cores, and the CSE optimization is successful across all core counts. Again, this is correctly predicted by the Roofline model; however, due to the arithmetic intensity uncertainty factors mentioned in Section 11.4.3, the model is overly optimistic in predicting performance as 1.6 GStencil/s (49.1 GFlop/s) when 0.95 GStencil/s (28.5 GFlop/s) is actually achieved. Nonetheless, after tuning, we see a 3.6× speedup when using all eight cores. Moreover, we also see parallel scaling of 8.1× when going from one to eight cores—the ideal multicore scaling.

11.7.2 Barcelona Performance

In many ways, the performance of the 7-point stencil on Barcelona is very similar to that of Nehalem (no surprise given the very similar architecture). In Figure 11.5, we again see that the reference implementation shows no parallel scaling at all. However, the NUMA-aware version increased performance by 115% when both sockets were engaged.

Like the 7-point stencil on Nehalem, the optimizations that made the biggest impact are cache bypass and core blocking. The cache bypass (streaming store) intrinsic provides an additional improvement of 55% when using all eight cores—indicative of its importance when the machine is memory-bound.

Unlike the 27-point stencil on Nehalem, the sub-linear scaling of Barcelona in Figure 11.6 seems to indicate that the kernel is constrained by memory bandwidth. However, the fact that cache bypass did not improve performance, while the CSE optimization improves performance by 18% at maximum concurrency, hints that it is close to being compute-bound. Overall, auto-tuning was able to produce a 4.1× speedup using all eight cores. In addition, scaling from one to eight cores now produces a speedup of 5.7×.

The Roofline model for Barcelona, shown in Figure 11.2, again predicts that both the 7-point and 27-point stencils will be bandwidth-bound.

For the 7-point stencil (shown in Figure 11.5), this is likely true, as the Roofline predicts 0.91 GStencil/s (7.3 GFlop/s), which reconciles well with the 0.86 GStencil/s (6.9 GFlop/s) we actually attained. The fact that the cache bypass instruction produced speedups commensurate with the reduction in memory traffic further corroborates this idea.

Similar to the Nehalem predictions, the 27-point stencil (without CSE) predictions for Barcelona are looser than for the 7-point stencil. The Roofline model predicts an upper bound of about 0.98 GStencil/s (29.3 GFlop/s), but our attained performance, as seen in Figure 11.6, is 0.51 GStencil/s (15.4 GFlop/s). This suggests that as one approaches a compute-bound state, one should employ multiple Roofline models per architecture. On might infer that the 27-point stencil is likely compute-bound on Barcelona as cache bypass was not beneficial where CSE optimization was.

11.7.3   Clovertown Performance

Unlike the Nehalem and the Barcelona computers, the Clovertown is a Uniform Memory Access (UMA) machine with an older FSB architecture. This implies that the NUMA-aware optimization will not be useful and that both stencil kernels will likely be bandwidth-constrained. If we look at Figure 11.5, both these predictions are true for the 7-point stencil. Only memory optimizations like core blocking and cache bypass seem to be of any use. After full tuning, we attain a performance of 0.32 GStencil/s (2.54 GFlop/s), which aligns well with the Roofline upper bound of 0.36 GStencil/s (2.9 GFlop/s) shown in Figure 11.2. Due to the severe bandwidth limitations on this machine, auto-tuning had diminished effectiveness; using all eight cores, performance improves by only 1.9×. In addition, Clovertown's single-core performance of 0.17 GStencil/s (1.37 GFlop/s) grows only by 1.9× when using all eight cores, resulting in aggregate node performance of only 0.32 GStencil/s (2.54 GFlop/s).

Clovertown's poor multicore scaling indicates that the system rapidly becomes memory-bound. Given the snoopy coherency protocol overhead, it is not too surprising that the performance only improves by 38% between the four-core and eight-core experiments (when both FSBs are engaged), despite the doubling of the peak aggregate FSB bandwidth.

For the 27-point stencil, shown in Figure 11.6, memory bandwidth is again an issue at the higher core counts. When we run on one or two cores, cache bypass is not helpful, while the CSE optimization produces speedups of at least 30%, implying that the lower core counts are compute-bound. However, as we scale to four and eight cores, we observe a transition to being memory-bound. The cache bypass instruction improves performance by at least 10%, while the effects of CSE are negligible. This behavior is well explained by the different streaming bandwidth ceilings on Clovertown's Roofline model. The Roofline model also predicts a performance upper bound of approximately 0.39 GStencil/s (11.8 GFlop/s), while we actually attained 0.32 GStencil/s

(9.7 GFlop/s)—hence, further tuning will have diminishing returns. All in all, full tuning for the 27-point stencil resulted in a 1.9× improvement using all eight cores, as well as a 2.7× speedup when scaling from one to eight cores.

11.7.4   Blue Gene/P Performance

Unlike the three previous architectures, the IBM Blue Gene/P implements the PowerPC ISA. In addition, the xlc compiler does not generate or support cache bypass at this time. As a result, the best arithmetic intensity we can achieve is 0.33 for the 7-point stencil and 1.25 for the 27-point stencil. The performance of a Blue Gene/P node is an interesting departure from the bandwidth-limited x86 architectures, as it seems to be compute-bound both for the 7-point and 27-point stencils. As seen in Figures 11.5 and 11.6, in neither case do memory optimizations like padding, core blocking, or software prefetching make any noticeable difference. The only optimizations that help performance are computation related, like register blocking, SIMDization, and CSE. After full tuning, both stencil kernels show perfect multicore scaling.

Interestingly, when we modified our stencil code generator to produce SIMD intrinsics we observed very different results on the 7- and 27-point stencils. We observe nearly a 30% increase in performance when using SIMD intrinsics on the 7-point, but a 10% decrease in performance on the 27-point CSE implementation using SIMD. One should note that unlike x86, Blue Gene does not support an unaligned SIMD load. As such, to load an unaligned stream of elements (and write to aligned), one must perform permutations and asymptotically require two instructions for every two elements. Clearly this is no better than a scalar implementation of one load per element.

If we look at the Roofline model in Figure 11.2, we would conclude both stencils are likely compute limited, although further optimization will quickly make the 7-point memory-limited. After full tuning of the 7-point stencil, we see an improvement of 4.4× at full concurrency. Similarly, for the 27-point stencil, performance improves by a factor of 2.9× at full concurrency.

11.7.5   Victoria Falls Performance

Like the Blue Gene/P, the Victoria Falls does not exploit cache bypass. Moreover, it is a highly multithreaded architecture with low-associativity caches. Initially, if we look at the performance of our reference 7-point stencil implementation in Figure 11.5, we see that the we attain 0.16 GStencil/s (1.29 GFlop/s) at four cores, but only 0.09 GStencil/s (0.70 GFlop/s) using all 16 cores! Clearly the machine's resources are not being utilized properly. Now, as we begin to optimize, we find that properly-tuned padding improves performance by 4.8× using eight cores and 3.4× when employing all 16 cores. The padding optimization produces much larger speedups on Victoria Falls than for all previous architectures, primarily due to the low associativity of its caches.The highly multithreaded nature of the architecture results in each thread receiving only 64 KB of L2 cache. Consequently, core blocking also becomes vital, and, as expected, produces large gains across all core counts.

A new optimization that we introduced specifically for Victoria Falls was thread blocking. In the original implementation of the stencil code, each core block is processed by only one thread. When the code is thread blocked, threads are clustered into groups of eight; these groups work collectively on one core block at a time. When thread blocked, we see a 3% performance improvement with eight cores and a 12% improvement when using all 16 cores. However, the automated search to identify the best parameters was relatively lengthy, since the parameter space is larger than conventional threading optimizations.

Finally, we also saw a small improvement when we performed a second pass through our greedy algorithm. For the higher core counts, this improved performance by about 0.025 GStencil/s (0.20 GFlop/s). Overall, the tuning for the 7-point stencil resulted in an $8.7\times$ speedup at maximum concurrency and an $8.6\times$ parallel speedup as we scale to 16 cores.

For the 27-point stencil, shown in Figure 11.6, the reference implementation scales well. Nonetheless, auto-tuning was still able to achieve significantly better results than the reference implementation alone. Many optimizations combined together to improve performance, including array padding, core blocking, CSE, and a second sweep of the greedy algorithm. After full tuning, performance improved by $1.8\times$ when using all 16 cores, and we also see parallel scaling of $13.1\times$ when scaling to 16 cores. The fact that we almost achieve linear scaling strongly hints that it is compute-bound. This is confirmed by examining the Roofline model in Figure 11.2.

The Victoria Falls performance results are even more impressive considering that one must regiment 128 threads to perform one operation; this is $8\times$ as many as the Nehalem, $16\times$ more than either the Barcelona or Clovertown, and $32\times$ more than the Blue Gene/P.

11.7.6   Cell Performance

The Cell blade is the first of two local-store architectures discussed in this study. Recall that generic microprocessor-targeted source code cannot be naïvely compiled and executed on the SPE's software controlled memory hierarchy. Therefore, we use a local-store implementation as the baseline performance for our analysis. It should be noted this baseline implementation naïvely blocks the code for the local store. Our Cell-optimized version utilizes an auto-tuned circular queue algorithm that searches for the optimal local store and register blockings.

For both local-store architectures, data movement to and from DRAM is explicitly controlled through DMAs, so write allocation is neither needed nor supported. Therefore, the ideal arithmetic intensity for the local-store architectures is 0.50 for the 7-point stencil and 1.88 for the 27-point stencil.

In practice, however, the Cell's arithmetic intensity is significantly below this ideal due to extra padding for the DMA operations.

Examining the behavior of the 7-point stencil on Cell (shown in Figure 11.5) reveals that the system is clearly computationally bound for the baseline stencil calculation when using one or two cores. In this region, there is a significant performance advantage in using hand-optimized SIMD code. However, at concurrencies greater than four cores, there is essentially no advantage—the machine is clearly bandwidth-limited. The only useful optimization is NUMA-aware data placement. Exhaustively searching for the optimal core blocking provided no appreciable speedup over the naïve approach of maximizing local-store utilization. Although the resultant performance of 15.6 GFlop/s is a low fraction of peak performance, it achieves about 90% of the streaming memory bandwidth, as evidenced in Roofline model in Figure 11.2.

Unlike for the 7-point stencil, register blocking becomes useful for the 27-point stencil (shown in Figure 11.6). As evidenced by the linear scaling across SPEs, this kernel is limited by computation until CSE is applied. After CSE is applied, it is compute-bound from one to four cores, but likely bandwidth-bound when utilizing eight cores on a single socket or all 16 cores across the SMP. Clearly, like the x86 version of `gcc`, the Cell version of `gcc` is incapable of SIMDization or CSE. Overall, full tuning allowed us to achieve a 4.8× speedup at full concurrency and an improvement of 9.9× when scaling from 1 to 16 SPEs.

Although this Cell blade does not provide a significant performance advantage over the previous incarnation for memory intensive codes, it provides a tremendous productivity advantage by ensuring DP performance is never the bottleneck—one only needs to focus on DMA and blocking.

11.7.7 GTX280 Performance

The NVIDIA GT200 GPU (GeForce GTX280) is our other local-store architecture. As an accelerator, the GTX280 links its address space to the disjoint CPU address space via the PCIExpress bus. However, for this study, we assume that the grids being operated on are already present in the local GPU memory, thus ignoring the data transfer time from CPU memory. In addition, our GTX280 27-point implementation only exploits a 24-flop CSE implementation. The compiler further reduced the flop count to approximately 17 per stencil, so this platform is performing slightly more than half of the 30 flops typically performed by the non-CSE kernel on other machines.

For the 7-point and 27-point stencils, presented in Figures 11.5 and Figure 11.6, we manually select the appropriate decomposition and number of threads. Unfortunately, the problem decomposes into a power-of-two number of *CUDA thread blocks* which we must run on 30 streaming multiprocessors. Clearly, when the number of *CUDA thread blocks* is fewer than 30, there is a linear mapping without load imbalance. However, at 32 thread blocks the load imbalance is maximal (two cores are tasked with twice as many blocks

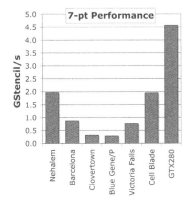 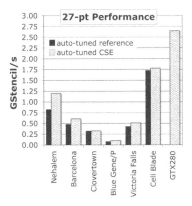

FIGURE 11.7: A performance comparison for all architectures at maximum concurrency after full tuning. The left graph shows auto-tuned 7-point stencil performance, while the right graph displays performance for the auto-tuned 27-point stencil with and without common subexpression elimination.

as others). As concurrency increases beyond 32 thread blocks, load imbalance diminishes and performance saturates at 4.56 GStencil/s (36.5 GFlop/s).

Our 27-point stencil implementation, shown in Figure 11.6, only exploits CSE. The performance profile still looks similar to the 7-point stencil, however—performance monotonically increases as we scale from 1 to 1024 *CUDA thread blocks*, and at maximum concurrency we peak at 2.64 GStencil/s (45.5 GFlop/s)—a compute-bound result.

11.7.8 Cross-Platform Performance and Power Comparison

At ultra-scale, power has become a severe impediment to increased performance. Thus, in this section not only do we normalize performance comparisons by looking at entire nodes rather than cores, we also normalize performance with power utilization. To that end, we use a power efficiency metric defined as the ratio of sustained performance to sustained system power — GStencil/s/kW. This is essentially the number of stencil operations one can perform per joule of energy.

Although manufactured by different companies, the evolution of x86 multicore chips from the Intel Clovertown, through the AMD Barcelona, and finally to the Intel Nehalem is an intriguing one. The Clovertown is a UMA architecture that uses an older FSB architecture and supports only a single hardware thread per core. In terms of DRAM, it employs fully buffered dual in-line memory module (FBDIMMs) running at a relatively slow 667 MHz. Consequently, it is not surprising to see in Figure 11.7 that the Clovertown is the slowest x86 architecture for either stencil. In addition, due in part to the use of power-hungry FBDIMMs, it is also the least power efficient

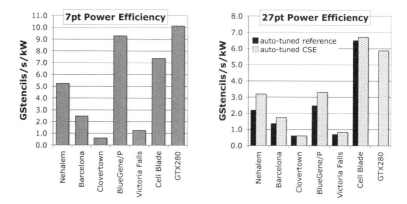

FIGURE 11.8: A power efficiency comparison for all architectures at maximum concurrency after full tuning. The left graph shows auto-tuned 7-point stencil power efficiency, while the right graph displays power efficiency for the auto-tuned 27-point stencil with and without common subexpression elimination.

x86 platform (as evidenced in Figure 11.8). The AMD Barcelona has several significant upgrades over the Clovertown. It employs a modern multisocket architecture, meaning that it is NUMA with integrated on-chip memory controllers and an inter-chip network. It also uses standard DDR2 DIMMs (half the power, two thirds the bandwidth). These features allow for noticeably better effective memory bandwidth, resulting in a 2.7× speedup for the 7-point stencil over Clovertown. The 27-point stencil, which is less likely to be constrained by memory, still produces a 1.9× speedup over the Clovertown (with CSE). As previously mentioned, Intel's new Nehalem improves on previous x86 architectures in several ways. Notably, Nehalem features an integrated on-chip memory controller, the QuickPath inter-chip network, and simultaneous multithreading (SMT). It also uses three channels of DDR3 DIMMs running at 1066 MHz. These enhancements are reflected in the bandwidth-intensive 7-point stencil performance, which is 6.2× better than Clovertown *and* 2.3× better than Barcelona. On the compute-intensive 27-point stencil (with CSE), we still see a 3.7× improvement over Clovertown and a 2.0× speedup over Barcelona.

The IBM Blue Gene/P was designed for large-scale parallelism, and one consequence is that it is tailored for power efficiency rather than performance. This trend is starkly laid out in Figures 11.7 and 11.8. For both stencil kernels, the Blue Gene/P delivered the lowest performance per SMP among all architectures. Despite this, it attained the *best* power efficiency among the cache-based processors, and even bested Cell on the 7-point. It should be noted that Blue Gene/P is two process technology generations behind Nehalem.

Victoria Falls' CMT mandates one exploit 128-way parallelism. We see that

Victoria Falls achieves performance close to that of Barcelona, but certainly better than either Clovertown or Blue Gene/P. However, in terms of power efficiency, it is second to last, besting only Clovertown—no surprise given they both use power-inefficient FBDIMMs.

Finally, the STI Cell blade and NVIDIA's GTX280 GPU are the two local-store architectures that we studied. They are different from the cache-based architectures in two important ways. First, they are both heterogeneous; the Cell has a PowerPC core as well as eight SIMD SPE units per socket, while the GTX280 is an accelerator that links to a disjoint CPU. For this study, we did not exploit heterogenity. All the computation for the Cell was conducted on the SPEs, and all the computation on the GTX280 system was performed on the GPU. Second, they employ different programming models that render our portable C code useless; the Cell's SPEs require DMA operations to move data between main memory and each SPE's local store, while the GTX280 uses the CUDA programming language. Nonetheless, the potential productivity loss may be justified by the performance and power efficiency numbers attained using these two architectures. We see that Cell's performance is at least as good as any of the cache-based processors, while the GTX280's performance is at least *twice* as good. Moreover, the power efficiency of both these architectures is significantly better than any of the cache-based architectures on the 27-point stencil. Nevertheless, it is important to note that neither the data transfer time between CPU and GPU, nor any impacts from Amdahl's law [3] were included in our performance results.

11.8 Conclusions

In this work, we examined the application of auto-tuning to the 7- and 27-point stencils on the widest range of multicore architectures explored in the literature. The chip multiprocessors examined in our study lie at the extremes of a spectrum of design tradeoffs that range from replication of existing core technology (multicore) to employing large numbers of simpler cores (manycore) and novel memory hierarchies (streaming and local-store). Results demonstrate that parallelism discovery is only a small part of the performance challenge. Of equal importance is selecting from various forms of hardware parallelism and enabling memory hierarchy optimizations, made more challenging by the separate address spaces, software-managed memory local stores, and NUMA features that appear in multicore systems today.

Our work leverages the use of auto-tuners to enable portable, effective optimization across a broad variety of chip multiprocessor architectures, and successfully achieves the fastest multicore stencil performance to date. Analysis shows that every optimization was useful on at least one architecture (Figures 11.5 and 11.6), highlighting the importance of optimization within an auto-tuning framework. A key contribution to our study is the Roofline model, which effectively provides a visual assessment of potential performance and bottlenecks for each architectural design. Using this model allowed us to access the impact of our auto-tuning methodology, and determine that overall performance was generally close to its practical limit. Clearly, among the cache-based architectures, auto-tuning was essential in providing substantial speedups for both numerical kernels regardless of whether the computational balance ultimately became memory- or compute-bound; on the other hand, the reference implementation often showed no (or even negative) scalability.

Overall results show substantial benefit in raw performance and power efficiency for novel architectural designs, which use a larger number of simpler cores and employ software controlled memories (Cell and GTX280). However, the software control of local-store architectures results in a difficult tradeoff, since it gains performance and power efficiency at a significant cost to programming productivity. Conversely the cache-based CPU architectures offer a well-understood and more productive programming paradigm. Results show that Nehalem delivered the best performance of any of the cache-based systems, achieving more than 2× improvement versus Barcelona, and more than a 6× speedup compared the previous generation Intel Clovertown—due, in-part, to the elimination of the FSB in favor of on-chip memory controllers. However, the low-power BG/P design offered one of the most attractive power efficiencies in our study, despite its poor single node performance; this highlights the importance of considering these design tradeoffs in an ultra-scale, power-intensive environment. Due to the complexity of reuse patterns endemic to stencil calculations coupled with relatively small per-thread cache capacities, Victoria Falls was perhaps the most difficult machine to optimize—it needed virtually every optimization.

Now that power has become the primary impediment to future processor performance improvements, the definition of architectural efficiency is migrating from a notion of "sustained performance" towards a notion of "sustained performance per watt." Furthermore, the shift to multicore design reflects a more general trend in which software is increasingly responsible for performance as hardware becomes more diverse. As a result, architectural comparisons should combine performance, algorithmic variations, productivity (at least measured by code generation and optimization challenges), and power considerations. We believe that our work represents a template of the kind of architectural evaluations that are necessary to gain insight into the tradeoffs of current and future multicore designs.

Acknowledgments

The authors acknowledge Georgia Institute of Technology, its Sony-Toshiba-IBM Center of Competence, and the National Science Foundation for the use of Cell resources. We would like to express our gratitude to Sun and NVIDIA for their machine donations. We also thank the Argonne Leadership Computing Facility for use of their Blue Gene/P cluster. ANL is supported by the Office of Science of the U.S. Department of Energy under contract DE-AC02-06CH11357. This work and its authors are supported by the Director, Office of Science, of the U.S. Department of Energy under contract number DE-AC02-05CH11231 and by NSF contract CNS-0325873. Finally, we express our gratitude to Microsoft, Intel, and U.C. Discovery for providing funding (under Awards #024263, #024894, and #DIG07-10227, respectively) and for the Nehalem computer used in this study.

Bibliography

[1] Software Optimization Guide for AMD Family 10h Processors, May 2007.

[2] AMD64 Architecture Programmers Manual Volume 2: System Programming, September 2007.

[3] Gene M. Amdahl. Validity of the single processor approach to achieving large scale computing capabilities. In *AFIPS '67 (Spring): Proceedings of the April 18-20, 1967, Spring Joint Computer Conference*, pages 483–485, New York, NY, USA, 1967. ACM.

[4] Kevin Barker, Kei Davis, Adolfy Hoisie, Darren Kerbyson, Michael Lang, Scott Pakin, and Jose Carlos Sancho. Entering the petaflop era: The architecture and performance of Roadrunner. In *SC '08: Proceedings of the 2008 ACM/IEEE conference on Supercomputing*, Piscataway, NJ, USA, 2008. IEEE Press.

[5] M. Berger and J. Oliger. Adaptive mesh refinement for hyperbolic partial differential equations. *Journal of Computational Physics*, 53:484–512, 1984.

[6] Kaushik Datta. *Auto-tuning Stencil Codes for Cache-Based Multicore Platforms*. PhD thesis, EECS Department, University of California, Berkeley, December 2009.

[7] Kaushik Datta, Mark Murphy, Vasily Volkov, Samuel Williams, Jonathan Carter, Leonid Oliker, David Patterson, John Shalf, and Katherine Yelick.

Stencil computation optimization and auto-tuning on state-of-the-art multicore architectures. In *SC '08: Proceedings of the 2008 ACM/IEEE Conference on Supercomputing*, pages 1–12, Piscataway, NJ, USA, 2008. IEEE Press.

[8] J. Doweck. Inside intel core microarchitecture. In *HotChips 18*, 2006.

[9] B. Flachs, S. Asano, S. H. Dhong, et al. A streaming processor unit for a cell processor. *ISSCC Dig. Tech. Papers*, pages 134–135, February 2005.

[10] M. Gschwind. Chip multiprocessing and the Cell Broadband Engine. In *CF '06: Proceedings of the 3rd Conference on Computing Frontiers*, pages 1–8, New York, NY, USA, 2006.

[11] M. Gschwind, H. P. Hofstee, B. K. Flachs, M. Hopkins, Y. Watanabe, and T. Yamazaki. Synergistic processing in Cell's multicore architecture. *IEEE Micro*, 26(2):10–24, 2006.

[12] M. D. Hill and A. J. Smith. Evaluating associativity in CPU caches. *IEEE Trans. Comput.*, 38(12):1612–1630, 1989.

[13] Intel64 and IA-32 Architectures Optimization Reference Manual, May 2007.

[14] Intel 64 and IA-32 Architectures Software Developer's Manual, September 2008.

[15] J. A. Kahle, M. N. Day, H. P. Hofstee, C. R. Johns, T. R. Maeurer, and D. Shippy. Introduction to the cell multiprocessor. *IBM J. Res. Dev.*, 49(4/5):589–604, 2005.

[16] S. Kamil, K. Datta, S. Williams, L. Oliker, J. Shalf, and K. Yelick. Implicit and explicit optimizations for stencil computations. In *ACM SIGPLAN Workshop Memory Systems Performance and Correctness*, San Jose, CA, USA, 2006.

[17] S. Kamil, P. Husbands, L. Oliker, J. Shalf, and K. Yelick. Impact of modern memory subsystems on cache optimizations for stencil computations. In *3rd Annual ACM SIGPLAN Workshop on Memory Systems Performance*, Chicago, IL, USA, 2005.

[18] Edward D. Lazowska, John Zahorjan, G. Scott Graham, and Kenneth C. Sevcik. *Quantitative System Performance: Computer System Analysis using Queueing Network Models*. Prentice-Hall, Inc., Upper Saddle River, NJ, USA, 1984.

[19] A. Lim, S. Liao, and M. Lam. Blocking and array contraction across arbitrarily nested loops using affine partitioning. In *Proceedings of the ACM SIGPLAN Symposium on Principles and Practice of Parallel Programming*, June 2001.

[20] D. Pham, S. Asano, M. Bollier, et al. The design and implementation of a first-generation cell processor. *ISSCC Dig. Tech. Papers*, pages 184–185, February 2005.

[21] S. Phillips. Victoria falls: Scaling highly-threaded processor cores. In *HotChips 19*, 2007.

[22] G. Rivera and C. Tseng. Tiling optimizations for 3D scientific computations. In *Proceedings of SC'00*, Dallas, TX, USA, November 2000. Supercomputing 2000.

[23] S. Sellappa and S. Chatterjee. Cache-efficient multigrid algorithms. *International Journal of High Performance Computing Applications*, 18(1):115–133, 2004.

[24] The SPARC Architecture Manual Version 9, 1994.

[25] R. Vuduc, J. Demmel, and K. Yelick. OSKI: A library of automatically tuned sparse matrix kernels. In *Proc. of SciDAC 2005, J. of Physics: Conference Series*. Institute of Physics Publishing, June 2005.

[26] R. C. Whaley, A. Petitet, and J. Dongarra. Automated empirical optimization of software and the ATLAS project. *Parallel Computing*, 27(1-2):3–35, 2001.

[27] S. Williams. *Auto-tuning Performance on Multicore Computers*. PhD thesis, EECS Department, University of California, Berkeley, December 2008.

[28] S. Williams, J. Carter, L. Oliker, J. Shalf, and K. Yelick. Lattice Boltzmann simulation optimization on leading multicore platforms. In *Interational Conference on Parallel and Distributed Computing Systems (IPDPS)*, Miami, FL, USA, 2008.

[29] S. Williams, D. Patterson, L. Oliker, J. Shalf, and K. Yelick. The roofline model: A pedagogical tool for auto-tuning kernels on multicore architectures. In *IEEE HotChips Symposium on High-Performance Chips (HotChips 2008)*, August 2008.

[30] S. Williams, A. Watterman, and D. Patterson. Roofline: An insightful visual performance model for floating-point programs and multicore architectures. *Communications of the ACM*, April 2009.

Chapter 12

Manycore Stencil Computations in Hyperthermia Applications

Matthias Christen

Computer Science Department, University of Basel, Switzerland

Olaf Schenk

Computer Science Department, University of Basel, Switzerland

Esra Neufeld

Foundation for Research on Information Technologies in Society (IT'IS), ETH Zurich, Switzerland

Maarten Paulides

Erasmus MC–Daniel den Hoed Cancer Center, Rotterdam, The Netherlands

Helmar Burkhart

Computer Science Department, University of Basel, Switzerland

12.1 Introduction

Multi- and manycore as well as heterogeneous microarchitecture today play a major role in the hardware landscape. Specialized hardware, such as commodity graphics processing units, have proven to be compute accelerators that are capable of solving specific scientific problems orders of magnitude faster than conventional CPUs.

In this chapter, we study optimizations of a computational kernel appearing within a biomedical application, hyperthermia cancer treatment, on some of the latest microarchitectures, including Intel's Xeon Nehalem and AMD's Opteron Barcelona multicore processors, the Cell Broadband Engine Architecture (Cell BE), NVIDIA's graphic processing units (GPUs), and two cluster computers: a "traditional" CPU and a Cell BE cluster.

Hyperthermia is a relatively new treatment modality that is used as a complementary therapy to radio- or chemo-therapies. Clinical studies have shown that the effect of both radio- and chemo-therapies can be substantially enhanced by combining them with hyperthermia. The computationally demanding part of treatment planning consists of solving a large-scale nonlinear, nonconvex partial differential equation (PDE)-constrained optimization problem as well as the forward problem, which is discussed here and which can be used to solve the inverse problem.

The optimizations discussed in this chapter concern bandwidth-saving algorithmic transformations and implementations on the architectures mentioned above.

12.2   Hyperthermia Applications

Hyperthermia cancer treatment, i.e., application of moderate heat to the body, is a promising modality in oncology that is used for a wide variety of cancer types (including pelvic, breast, cervical, uterine, bladder, and rectal cancers, as well as melanoma and lymphoma). Both animal and clinical studies have proven that hyperthermia intensifies both radio- and chemo-therapies by factors of 1.2 up to 10, depending on heating quality, treatment modality combination, and type of cancer [8, 27, 31]. Hyperthermia is therefore applied in conjunction with both radio- and chemo-therapies.

An effect of hyperthermia treatment is apoptosis of tumor cells, which have a chaotic vascular structure resulting in poorly perfused regions in which cells are very sensitive to heat. Hyperthermia further makes the tumor cells more susceptible to both radiation and certain cancer drugs. There are a variety of reasons for this. Among others, heat naturally increases blood flow and therefore increases drug delivery to the tumor cells, and also increases the toxicity of some drugs. The effect of radiation is amplified as a result of improved oxygenation due to increases in blood flow. In our setting, we are dealing with local hyperthermia where the aim is to focus the energy noninvasively only at the tumor location. This is done by creating a constructive interference at the tumor location using nonionizing electromagnetic radiation (microwaves) and thereby aiming at heating the tumor to $42 - 43°C$, but even lower temperatures can be beneficial.

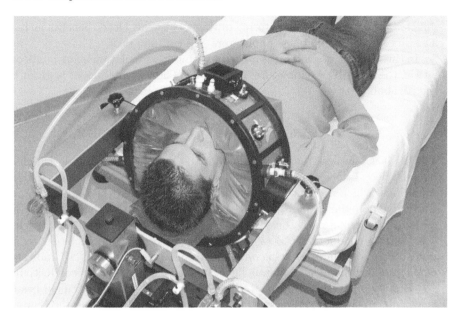

FIGURE 12.1: Head and neck hyperthermia applicator "HYPERcollar" developed at the Hyperthermia Group of Erasmus MC (Daniel den Hoed Cancer Center, Rotterdam, the Netherlands).

Cylindrical applicators, such as the one shown in Fig. 12.1, which shows the "HYPERcollar" applicator developed for head and neck hyperthermia treatment at the Hyperthermia Group of Erasmus MC (Daniel den Hoed Cancer Center) in Rotterdam, the Netherlands [18, 19], feature a number of dipole antennae arranged on the circumference of the applicator, whose amplitudes and phase shifts can be controlled to create the desired electric field inducing heat. Here the aim is to avoid cold spots within the tumor and hot spots in healthy tissue to maximize tumor heating and limit pain and tissue damage. The water bolus between the patient's skin and the applicator prevents heating of the skin and more importantly, from an engineering point of view, it is an efficient transfer medium for the electromagnetic waves into the tissue: it reduce reflections and allows for smaller antenna sizes. In treatment planning, it is therefore highly relevant to determine the therapeutically optimal antenna settings, given the patient geometry. This leads to a large-scale nonlinear, nonconvex PDE-constrained optimization problem, the PDE being the thermal model shown in Eq. (12.1), which is known as Pennes's bioheat equation [20], or a variant thereof. Another important aspect of treatment planning is simulating the temperature distribution within the human body given the antenna parameters, which has been successfully demonstrated to accurately predict phenomena occurring during treatment [17,26]. Simulations are helpful in order to determine the correct doses and help to overcome the difficulty of temperature monitoring during treatment. Other benefits include assistance in developing new applicators and training staff.

In this chapter, we will focus on accelerating the simulation using multicore CPUs, the Cell BE Architecture, and GPUs. The thermal model is given by the parabolic PDE

$$\rho C_{\mathrm{p}} \frac{\partial u}{\partial t} = \nabla \cdot (k \nabla u) - \rho_{\mathrm{b}} W(u) C_{\mathrm{b}} (u - T_{\mathrm{b}}) + \rho Q + \frac{\sigma}{2} \|\mathbf{E}\|^2, \qquad (12.1)$$

which is the simplest thermal model. On the boundary we impose Dirichlet boundary conditions to model constant skin temperature, Neumann boundary conditions, which account for constant heat flux through the skin surface, or convective boundary conditions [16]. In Eq. (12.1), u is the temperature field, for which the equation is solved, ρ is density, C_{p} is the specific heat capacity, k is thermal conductivity, $W(u)$ is the temperature-dependent blood perfusion rate, T_{b} is the arterial blood temperature (the subscript "b" indicates blood properties), Q is a metabolic heat source, σ is electric conductivity, and $\mathbf{E}$ is the electric field generated by the antenna. The electric field has to be calculated (by solving Maxwell's equations) before computing the thermal distribution. More elaborate models include, e.g., temperature-dependent material parameters or tensorial heat conductivity to account for the directivity of blood flow [28].

We use a finite volume method for discretizing Eq. (12.1). The blood perfusion is modeled as a piecewise linear function of the temperature. As a discretized version of Eq. (12.1) we obtain the expression

$$
\begin{aligned}
u_{i,j,k}^{(n+1)} =\ & u_{i,j,k}^{(n)} \left(a_{i,j,k} u_{i,j,k}^{(n)} + b_{i,j,k} \right) + c_{i,j,k} \qquad (12.2) \\
& + d_{i,j,k} u_{i-1,j,k}^{(n)} + e_{i,j,k} u_{i+1,j,k}^{(n)} \\
& + f_{i,j,k} u_{i,j-1,k}^{(n)} + g_{i,j,k} u_{i,j+1,k}^{(n)} \\
& + h_{i,j,k} u_{i,j,k-1}^{(n)} + l_{i,j,k} u_{i,j,k+1}^{(n)}.
\end{aligned}
$$

The lower indices i, j, k denote grid coordinates; the expression, also called a "stencil," is evaluated for each point in the discretized domain. The upper index n is the timestep. By letting the stencil sweep over the grid multiple times ($n = 0, 1, 2, \ldots$), PDE Eq. (12.1) is solved with a simple explicit Euler time integration scheme. Each evaluation of the stencil expression amounts to 16 floating-point operations. Note that the coefficients $a, \ldots, l$ depend on their location in space. Therefore, naïvely, we need 16 load and 1 store operations per stencil call, which, for single precision, results in a rather low arithmetic intensity of $16/68$ Flops/Byte. In certain cases it is also desired to make them time-dependent to account for changes of physical properties in time or to mimic modifying antenna parameters during treatment.

12.3 Bandwidth-Saving Stencil Computations

A stencil is a geometric structure on a structured grid consisting of a particular arrangement of nodes around a center node. In a stencil computation, the values of these nodes are used to update the value of the center nodes. Data types of the node values are not restricted to scalar values; they can be any data types for which meaningful operations can be defined.

Stencils are often used to solve PDEs; they arise when a numerical method is chosen that involves discretization of differential operators, e.g., in finite difference of finite volume methods. Stencil computations also occur in multigrid methods (smoothing, restriction, and prolongation operators are essentially stencil computations), and in image processing (e.g., blur filters).

Stencil computations typically only perform a very limited number of operations per node, and the number of operations is constant regardless of the problem size. Hence the performance of stencil computations is typically limited by the available bandwidth. Therefore it is crucial to minimize memory transfers. The grid traversal can, in fact, be designed in such a way that memory transfers are reduced with respect to the naïve grid traversal (i.e., a d-fold nested loop, d being the dimensionality of the domain).

Another aspect of bandwidth saving is nonuniform memory access (NUMA) awareness. Typically, NUMA optimizations give rise to substantial performance benefits. The goal is to minimize accesses to distant memory, i.e., memory to which a core does not have affinity. On systems using the "first touch" policy, the thread that first accesses a memory page will "own" that memory page. Therefore the correct initialization of data is crucial: each thread should initialize the data on which it will work later on.

Note that thread affinity plays an important role. Letting threads migrate among cores of a CPU or even among CPUs (the default behavior of the scheduler) can annihilate any performance benefits achieved through blocking techniques or NUMA awareness because of cache thrashing.

12.3.1 Spatial Blocking and Parallelization

The key idea in reducing memory transfers from and to slower memory in the memory hierarchy is to reuse the data that have already been transferred to the faster and closer memory (the L2 cache on a CPU system with two levels of cache, the local store of the synergistic processor units (SPUs) on the Cell BE Architecture, shared memory on NVIDIA GPUs, etc.).

Assume the 7-point stencil calculation is dependent on its immediate neighbors along the axes:

$$u'_{x,y,z} = \varphi(u_{x,y,z}, u_{x-1,y,z}, u_{x+1,y,z}, u_{x,y-1,z}, u_{x,y+1,z}, u_{x,y,z-1}, u_{x,y,z+1}).$$

Then, if the values for a fixed $z = z_0$ have been loaded into the fast memory and they still reside in fast memory when moving to $z = z_0 + 1$, all of these data can be reused for the computation of the stencil on the points $(x, y, z_0 + 1)$, which depend on the data points (x, y, z_0). Hence, for our hyperthermia stencil we can assume that we only need to load one value of u (instead of seven), and the arithmetic intensity increases to 16/44 Flops/Byte. If domain sizes are large, it is likely that the points (x, y, z_0) have been evicted from the fast memory before they can be reused in iteration $z = z_0 + 1$, and they have to be transferred another time.

This can be prevented by only working on small blocks of the domain at a time, with the block sizes chosen such that a subplane of data can be kept in fast memory until it is reused.

Usually it is beneficial to apply this idea recursively with decreasing block sizes to account for the multiple layers in the memory hierarchy (L2 cache, which is possibly shared among multiple cores, per core L1 cache, registers).

The idea also lends itself to parallelization. In a shared memory environment that introduces a layer of spatial blocking in which each thread is assigned a distinct subset of blocks, the stencil sweep can be executed in parallel without any synchronization points within the sweep, since the stencil computation on one point is completely independent of the calculation of other points.

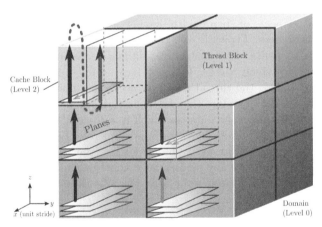

FIGURE 12.2: Levels of blocking: decomposition of the domain into thread and cache blocks for parallelization and data locality optimization.

Fig. 12.2 shows a possible hierarchical decomposition of a cube-shaped domain into blocks. In level 1, each thread is assigned a "thread block." Thread blocks are executed in parallel. Each thread block is decomposed into "cache blocks" to optimize for highest level cache data locality. When choosing the cache block sizes, there are a number of facts that have to be kept in mind. Typically, the highest level cache is shared among multiple threads (e.g., on Intel Xeon X5550 (Nehalem) the four cores on a socket share an 8 MB L3 cache, and each dual-threaded core has its own unified 256 KB L2 cache and 32 KB L1 data cache; the caches of AMD's Opteron 2356 (Barcelona) are organized similarly). In order to benefit from the hardware prefetchers on modern CPUs, the thread block should not be cut in x (the unit stride) direction, as shown in Fig. 12.2. However, it is not necessary that the cache holds all of the data in the cache block. It suffices that the neighboring planes of the plane being calculated fit into the cache. To avoid loading a plane twice, thread blocks should not be cut in z direction either, if it can be avoided.

In a distributed memory environment, usually a ghost layer is added at the block borders that replicates the nodes of the adjacent blocks in order to account for the data dependencies. If multiple sweeps have to be run on the data, the ghost nodes need to be updated after each sweep. The simplest parallelization scheme simply synchronizes the execution after the sweep and exchanges the values on the artificial boundaries. A more elaborate scheme that allows overlapping the computation and the communication of the boundary values (and thus scales if the block sizes are large enough) is described in Fig. 12.3. It is assumed that each process is assigned exactly one block.

```
in each process do
   for t = 0...t_max
      copy boundary data to send buffers
      initiate asynchronous transfers of the send buffers to the
         corresponding neighbors into receive buffers
      compute
      wait for the data transfers from the neighbors to complete
      update the bndry nodes with the data in the receive buffers
      apply boundary conditions
   end for
end do
```

FIGURE 12.3: Scalable distributed memory parallelization.

The update step in line 8 requires that the stencil expression can be separated into single components depending only on neighboring nodes in one direction. This is given, e.g., if the stencil expression is a weighted sum over the values of the neighboring nodes. Note that there might be stencils for which the separation is not possible, and hence the algorithm is not applicable.

12.3.2 Temporal Blocking

In certain cases, the stencil kernel does multiple sweeps over the entire grid. If intermediate results are of no interest, we can think of blocking the stencil computation not only in space, but also in the time dimension, i.e., blocking the outermost loop and thereby saving intermediate write-backs and loads.

As illustrated by Fig. 12.4, temporal blocking increases the number of data dependencies with increasing time block size. It has the effect of applying the stencil to each "input node" of the stencil, i.e., the nodes in the neighborhood of the center point that is required for the stencil computation. Hence, for constant output node set size the input node set increases as the time block size increases.

A lot of research has been done on how to block the time loop of stencil computations, or, more generally how to block imperfectly nested loops, so that data dependencies are respected [12, 24], or how to transform the loops so that the data dependencies are respected [9].

To make the computation amenable to parallelization, we choose a simple blocking strategy known as the "circular queue" strategy [3–5]. The circular queue method loads and computes the block boundaries redundantly as shown in Fig. 12.4(b). Let us assume we have a 1-dimensional domain and a stencil that depends only on its left and right neighbors, and we want to compute two timesteps on the input data, as shown in Fig. 12.4(b). Then, in order to compute the values on an n node block, $n + 2 \cdot 2$ nodes are required for the computation. When going to the next block, the circular queue method simply loads four node values redundantly (in this example, as indicated in the figure by the dotted line) and computes two node values redundantly. The benefit is that now both blocks can be computed completely independently, i.e., in parallel.

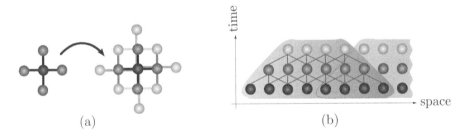

(a)   (b)

FIGURE 12.4: (a) Effect of temporal blocking on data dependencies. (b) The number of the computable nodes decreases with each sweep due to data dependencies.

Note that the number of timesteps applicable on the data block is limited by two factors. The size of the planes directly limits the number of timesteps that can be performed simultaneously on the data, since the size of the intermediate planes decreases with each timestep. Therefore, planes should be as large as possible. On the other hand, with each additional timestep the number of intermediate planes that need to be kept in fast memory increases. The memory size limits the size of the planes. Hence a trade-off must be found. In Section 12.3.2.2 this is discussed in detail for the stencil used in the thermal simulation for the hyperthermia application. The arithmetic intensity is now increased to approximately $16b/44$ Flops/Byte since we need to load one set of coefficients and load and store one temperature field per b timesteps. Note, however, that this number is (not a sharp) upper bound, and the actual number depends on the plane sizes we can store in the local memory.

In the shared memory environment, no explicit treatment of the artificial block boundary has to be carried out. We simply have to make sure that the overlaps of the input data of adjacent blocks are large enough so that the result planes are layed out seamlessly when written back to main memory.

Distributed memory environments again require the artificial boundaries to be exchanged. The scalable algorithm described above (Fig. 12.3) can still be used (with the same restrictions), but the boundary node update step (line 8) becomes more involved and, unfortunately, larger and more computationally intense. The width of the boundaries that are exchanged is multiplied by the number of timesteps that are performed on the data, and in the update step the time-blocked code is executed on the boundary strip before incorporating the values into the actual domain nodes.

Boundary conditions are to be applied after each timestep on the blocks that contain boundaries. Dirichlet boundary conditions (boundaries whose values do not change in time) can be easily implemented by simply copying the boundary values on the input plane forward to the planes belonging to the intermediate time levels. Note that, e.g., discretized Neumann or convective boundary conditions can be rewritten as Dirichlet boundary conditions, a process that involves only modifications of the stencil coefficients on the boundary [16].

12.3.2.1 Temporally Blocking the Hyperthermia Stencil

We assume that the coefficient fields $a, \ldots, l$ in Eq. (12.1) are constant in time, or at least constant within a time block, which enables the use of a time-blocking scheme to solve the PDE. As the coefficients are not constant in space, they put additional pressure on the local memory.

Fig. 12.5 shows the layout of the planes u in memory for time block size $b = 2$. It shows the most compact memory layout that can be explicitly programmed on architectures with a software-managed cache (e.g., the Cell BE). The lower indices of u denote the spatial indices and the upper indices are the indices in time. The figure illustrates that the spaces in memory are filled as time advances (from top to bottom in the diagram): u_0^0, u_1^0, u_2^0, u_3^0 are loaded in succession (the white items symbolize loading of data into fast memory). In the fourth row, u_0^0, u_1^0, and u_2^0 are available, and u_1^1 can be computed. The memory locations that receive calculated data are drawn in black. In the next step, u_0^0 is no longer needed and can be overwritten to contain data of u_4^0, while u_2^1 is calculated from u_1^0, u_2^0, and u_3^0. In the sixth row, after calculating u_3^1, the input planes to compute the first plane of the next timestep, u_2^2, are available. As we want to overlap computation and communication, u_2^2 is written back from fast to slower memory only a step later (this can, of course, only be controlled in architectures which support explicit memory transfers).

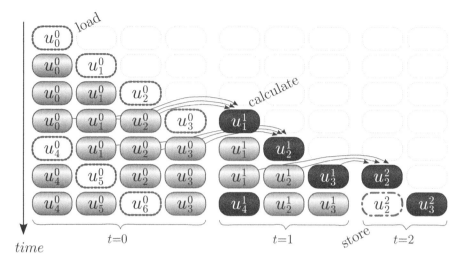

FIGURE 12.5: Memory layout of planes in the time-blocking scheme.

For a given time in computation (i.e., on a specific row in the diagram), the spatial indices of the planes being calculated are skewed. For example, in the last row of the diagram, plane 4 of timestep 1 (u_4^1) and plane 3 of timestep 2 (u_3^2) are calculated. As a consequence, since the coefficients can be reused in time but not in space, a set of coefficients has to be kept in memory for each timestep.

12.3.2.2 Speedup for the Hyperthermia Stencil

Let b be the time block size, n_x, n_y, n_z the dimensions of a write-back block, and N_m the number of points in a plane with boundary width m; i.e., $N_m = (n_x + 2m)(n_y + 2m)$. The data elements that have to be kept in local memory are a solution plane with boundary width b for preloading; a "truncated pyramid" of solution planes with boundary widths $b, b-1, \ldots 0$; a write-back solution plane with no additional boundary; nine coefficient planes for each of the b timesteps; and a set of preloading coefficient planes with boundary width $b-1$. The boundary width of the coefficient planes is chosen to be constant throughout the steps in a time block to avoid fragmenting local memory or copying data in local memory. Hence, the total number of data elements that needs to fit into local memory is given by

$$N_b + 3 \sum_{j=0}^{b} N_{b-j} + N_0 + 9(b+1)N_{b-1} = \alpha n_x n_y + \beta(n_x + n_y) + \gamma,$$

with $\alpha = 12b + 14$, $\beta = 21b^2 + 5b - 18$, and $\gamma = 40b^3 - 26b^2 - 34b + 36$.

For a fixed number n_x of data elements in the unit stride direction x and

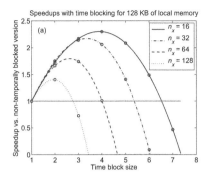
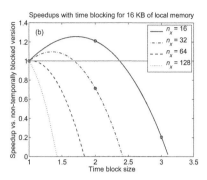

FIGURE 12.6: Theoretical maximum speedup numbers for temporal blocking. (a) 128 KB of usable local memory (Cell BE), (b) 16 KB local memory (NVIDIA G80 and G200 series GPUs).

a given size K of the local memory, the number of lines in a plane n_y as a function of the timestep size b is therefore limited by

$$n_y(b) \leq \max\left\{\left\lfloor \frac{K - \alpha - \beta n_x}{\beta + \gamma n_x} \right\rfloor, 0\right\}.$$

The number of lines depends on the time block size, since the more timesteps are performed on a piece of data, the more intermediate data (both solution and coefficient data) must be kept in the local memory.

The data volume V that needs to be transferred per block per single timestep is

$$V(b) = \tfrac{1}{b}\left(N_b(b) + 9N_{b-1}(b) + N_0(b)\right)\left\lceil \tfrac{n_y(1)}{n_y(b)} \right\rceil,$$

$N_m(b) = (n_x + 2m)(n_y(b) + 2m)$ being the number of data points in a plane with boundary width m and $n_y(b)$ the number of lines in a plane that fit into local memory for a fixed number of points n_x in unit stride direction and fixed timestep size b, as determined above.

We need to transfer a solution plane with boundary width b from main memory, as well as 9 $b - 1$ boundary-wide coefficient planes, and we need to write back one solution plane. We need to do this only every bth timestep (hence the fraction $\tfrac{1}{b}$), but as b increases the maximum number of lines in a plane will decrease because of the limited local memory space. Hence, more loads and stores will be necessary; specifically, $\left\lceil \tfrac{n_y(1)}{n_y(b)} \right\rceil$ more as compared to the non-time-blocked case.

Since the kernel is still bandwidth limited, the speedup S of a b-time-blocked version is $S(b) = \text{time}(1)/\text{time}(b) = V(1)/V(b)$.

The graphs in Fig. 12.6 depict the theoretically achievable speedups for two different local memory sizes, 128 KB (a) and 16 KB (b), assuming that all data elements are single-precision 4-byte floating-point numbers. The 128 KB case represents the Cell BE implementation, since there are 256 KB of local store memory available per SPU, which has to contain both the program code and all of the data. In our implementation, 128 KB is a good estimate of what memory is left for holding the solution and coefficient data. The 16 KB represents the NVIDIA GPU, where each of the streaming multiprocessors is equipped with 16 KB of shared memory.

The speedup numbers are plotted on the vertical axis as a function of the time block size b on the horizontal axis. The four curves show the speedup numbers for varying line lengths $n_x \in \{16, 32, 64, 128\}$.

The figures show that on the Cell BE the best speedups are obtained for time block sizes between 2 and 4, depending on the line length, which yields speedups between $1.5\times$ to $2.5\times$. In our prototype code, we did not decompose the domain along the unit stride direction, so the line length is predetermined by the problem size. Decomposing the domain in unit stride direction has several consequences. The data need to be gathered from main memory into the local store instead of simply being streamed in. Also, because of alignment restrictions and code SIMDization, there will be overlaps at the artificial boundaries, which will have an impact on performance. If cutting along the unit stride direction is performed, there is a trade-off between optimal speedup from temporal blocking and performance impact by using too short line lengths.

Fig. 12.6(b) demonstrates that temporal blocking on the GPU equipped with only a very small amount of local memory is not beneficial for the stencil kernel, hence temporal blocking was not done on the GPU. However, in NVIDIA's latest GPU architecture ("Fermi"), which features 64 KB of local per-multiprocessor memory, which can be configured as 48 KB shared memory and 16 KB L1 cache (using the whole memory as shared memory is not supported), temporal blocking could again prove beneficial. The theoretical model projects speedups of up to $1.6\times$ on this architecture.

12.4   Experimental Performance Results

All the performance benchmarks have been performed in single precision on two CPU systems with an Intel Xeon X5550 (Nehalem) and an AMD Opteron 2356 (Barcelona) processor, an IBM QS22 Blade Center, an NVIDIA Tesla S870 GPU system, and an NVIDIA GeForce GTX 285 GPU. Some architectural features are summarized in Table 12.1. The bandwidths shown in the table have been measured using the STREAM benchmark [14] for the CPU systems, NVIDIA's `bandwithTest` coming with the CUDA SDK. For the Cell BE bandwidth, we relied on [10].

TABLE 12.1: Characteristics of the architectures on which the benchmarks were carried out and parallelization paradigm.

Processor	Intel Xeon X5550	AMD Opteron 2356	IBM QS22 Cell BE	NVIDIA Tesla S870 NVIDIA GTX 285
Type	CPU	CPU	Cell BE	GPU
Clock	2.67 GHz	2.3 GHz	3.2 GHz	1.35 GHz (S870) 1.48 GHz (GTX285)
Memory Hierarchy	L1: 32KB/core L2: 256KB/core L3: 8MB/socket DRAM: 16GB	L1: 64KB/core L2: 512KB/core L3: 2MB/socket DRAM: 16GB	LS: 256KB/SPU L1 (PPU): 32KB L2 (PPU): 512KB DRAM: 16GB	shmem: 16KB/SM DRAM: 1.5GB/GPU (S870) 1GB (GTX285)
DRAM Bandwidth	32GB/s	13.5 GB/s	35 GB/s	64.5 GB/s (device, S870) 1GB/s (dev. ↔ host, S870) 125 GB/s / 4GB/s (GTX285)
Concurrency	2 sockets 4 cores/socket 2 threads/core	2 sockets 4 cores/socket	2 sockets 1 PPU/socket 8 SPUs/socket	4 GPUs 16 SMs/GPU (S870) 30 SMs/GPU (GTX285) 8 processors/SM
Parallelization	OpenMP	OpenMP	pthreads/libspe	CUDA

12.4.1   Kernel Benchmarks

Before focusing on the entire application, we would like to give some results on the actual compute kernel. The benchmarks presented here have been done in single precision, since this is the default mode of the thermal solver integrated in the simulation package "SEMCAD X" [25] for which this code has been developed. If the timestep is sufficiently large, simulation results in single precision are adequate.

The CPU codes were compiled using Intel's Fortran compiler 10.1 (Fortran was chosen because the compiler was able to generate up to 20% faster code than Intel's C compiler). The parallelization was done with OpenMP. On the Cell BE, the GNU C compiler was used (gcc 4.3), and NVIDIA's nvcc on the GPU systems (gcc 4.4.1 for the host code) together with CUDA 2.3. The Cell BE implementation follows the low-level SDK approach using pthreads and the libspe for parallelization on the synergistic processing element (SPEs).

In Fig. 12.7(a) through (e), the performance and the impact of optimization techniques on the hyperthermia stencil are shown for three different problem sizes ($128 \times 128 \times 64$, $256 \times 256 \times 64$, $512 \times 512 \times 32$) on the different architectures. Fig 12.7(f) shows a performance benchmark (weak scaling) on an Intel Xeon X5355 and a Cell BE cluster. The distributed memory parallelization has been done with message passing interface (MPI).

On the Intel Xeon X5550 (Nehalem) (Fig. 12.7(a)), doing a NUMA-aware implementation and pinning threads to cores were the most beneficial optimizations. Blockings (spatial, which is not included in the graph because no real performance improvement arose from it, and temporal, which had some effect when only two threads were used) only added marginally to the overall GFlop/s rate. As a thread affinity policy a compact scheme has been used (which, for Intel's OpenMP implementation, can be done by setting the KMP_AFFINITY environment variable to compact), which places thread $j + 1$ as close as possible on the hardware to thread j, i.e., cores get filled before sockets. Threads were allowed to migrate within a single core. Affinity causes performance degradation with fewer than 16 threads (it seems that the Linux scheduler does a good job for this kernel). The benefit if threads are pinned if all 16 threads are used is substantial. Note that these observations are only valid for this particular kernel. Other types of stencils will react differently to affinity.

Intel does not support affinity on AMD processors, hence there is no such optimization in Fig. 12.7(b). The performance graphs suggest that the threads were pinned by default, i.e., on both Intel and AMD platforms the same affinity mapping was used resulting in a fair comparison. In a naïve implementation, the memory bandwidth is obviously exhausted when only two threads are used. Adding another socket (going from four to eight threads) doubles bandwidth (each socket has its own L3 cache and memory interface) and hence performance. Spatial blockings and array padding have a larger effect on the AMD architecture; time blocking, however, did not have any beneficial effect.

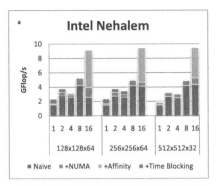

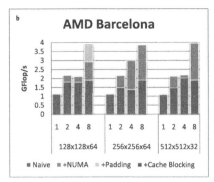

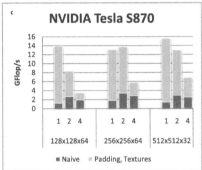

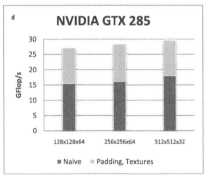

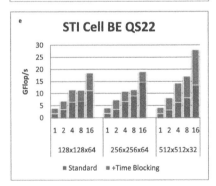

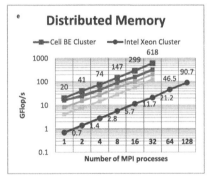

FIGURE 12.7: Performances of the hyperthermia stencil and impact of different optimization techniques for three different problem sizes on different architectures. (a) Intel Xeon Nehalem X5550 (two sockets, four cores per socket, two threads per core), (b) AMD Opteron Barcelona 2356 (two sockets, four cores per socket), (c) NVIDIA's Tesla S870 computing solution (four GPUs of the G80 series), (d) NVIDIA GTX 285, and (e) IBM Cell BE QS22 blade. (f) Performance of an MPI parallelization on Intel Xeon X5355 (Clovertown) and a Cell BE cluster on a log-log scale. Note that this is weak scaling.

Fig. 12.7(c) shows the performance of the NVIDIA Tesla S870, which essentially consists of four first-generation Tesla GPUs. At first sight, astonishingly, the performance drops when adding more GPUs. The CUDA kernel itself, which is executed on the GPUs, experiences a linear speedup. The problem is that after each stencil sweep the execution is synchronized and the artificial boundaries have to be copied to the device memory of the neighboring GPU. Exchanging boundary data has to be done via the host (and hence over the PCIe bus), which is a serious bottleneck. A distributed memory style time blocking has not yet been implemented, and could prove beneficial. Also, computation and data transfers should be overlapped, but this is not supported by the hardware (newer hardware, however, supports this feature).

Comparing Fig. 12.7(c) to Fig. 12.7(d) shows that the padding and texture optimizations, i.e., using the GPU's specialized cached texture memory in the unit stride direction, are not as significant on newer NVIDIA GPU generations (here the GTX 285) as on older ones (G80 series). On GPUs of the G80 series, applying padding to the unit stride dimension in order to enforce coalesced memory accesses (i.e., respect alignment restrictions and a uniform memory access pattern) and making use of cached texture memory is crucial to performance: it results in a factor of $10\times$ in speedup relative to the naïve implementation. On the GTX 285 these optimizations give rise to merely a factor of about 1.7. Furthermore, because of the increased bandwidth on the GTX 285 (125 GB/s vs. 64 GB/s on the Tesla G80), the absolute overall performance is higher.

In Fig. 12.7(e) we compare the performances of the Cell BE implementation without temporal blocking and with the time-blocking optimization. As predicted in Section 12.3.2.2 we obtain a speedup of about $2\times$. The code scales up to four SPEs, and then the bandwidth limit is reached. With 16 SPEs, i.e., using both processors on the blade, the performance is about doubled because a NUMA-aware implementation takes into consideration that each of the chips has its separate interface to the memory subsystem. That is, the available bandwidth is doubled.

The reason for the better performance of the $512 \times 512 \times 32$ problem lies in the fact that we did not cut the domain in unit stride dimension (here the z dimension). Therefore, with only 32 data points in unit stride direction, larger planes can be stored in the local store (cf. Section 12.3.2.1), and thus fewer redundant data transfers and calculations are performed.

Fig. 12.7(f) shows the results of a benchmark on a CPU cluster (the nodes were equipped with Intel Xeon Clovertown X5355 CPUs) and a Cell BE cluster. Note that the figure is in log-log scale and that the figure shows weak scaling. The parallelization was done with MPI. It is simply a pure MPI implementation; i.e., OpenMP has not been used for fine-grain thread-level parallelization. The horizontal axis shows the number of MPI processes. We ran the benchmark on up to 32 nodes on both the CPU and the Cell BE cluster; since the Intel Xeon X5355 has twice four cores, up to 128 MPI processes were run on the CPU cluster. The circles mark the performance of the CPU

FIGURE 12.8: (a) Model of a lucite cone applicator with patient model and (b) simulated thermal distribution after 3600 seconds simulation time. (c) The heating of a boy model induced by the "HYPERcollar" head-and-neck applicator.

cluster, the squares that of the Cell BE cluster. The gray lines with square markers are the performance numbers for one, two, four, and eight SPEs per Cell BE blade.

We observe linear speedups on both the CPU and the Cell BE cluster, which means that the MPI communication was ideally overlapped with computation.

12.4.2 Application Benchmarks

To test the speed advantage offered by the Cell BE processor and GPU implementations, a realistic problem was simulated: the treatment of a superficial tumor using hyperthermia.

Real medical image data were segmented to obtain a patient model. A model of a lucite cone applicator (LCA, [7, 22]) was positioned above the tumor and a water bolus inserted between the LCA and the patient was used to reduce reflections at interfaces and to influence the surface temperature; cf. Fig. 12.8. An EM simulation (harmonic voltage, 433 MHz) based on the FDTD software SEMCAD X [25] and a subsequent thermal simulation of the induced temperature increase after 1 hour of exposure were performed using a graded mesh with high resolution at relevant structures with 3.1 million cells (minimal step: 0.6 mm). Convective boundary conditions were applied at the patient/air interface (convective coefficient $h = 4\,\mathrm{Wm}^{-2}\mathrm{K}^{-1}$, $T_{\mathrm{air}} = 25°\mathrm{C}$) and the patient/water bolus interface ($h = 20\,\mathrm{Wm}^{-2}\mathrm{K}^{-1}$, $T_{\mathrm{water}} = 38°\mathrm{C}$).

Table 12.2 lists the timings that were required for the simulation on the individual architectures for two models, the previously discussed LCA model and a model ("boy") using the "HYPERcollar" applicator to heat the tumor in the esophagus by constructive interference; cf. Fig. 12.8(c). (The boy model has been taken from the Virtual Family [2] and is not a real patient.)

TABLE 12.2: Timings and performance results on CPU, Cell BE, and GPU architectures for the thermal simulation.

Architecture	LCA 195 × 196 × 198 voxels 11083 iterations		Boy 160 × 169 × 123 voxels 2606 iterations	
	Simulation Time [s]	Performance [MCells/s]	Simulation Time [s]	Performance [MCells/s]
Intel Xeon X5550				
Original code	1069	79	134	68
1 thread	864	98	96	97
2 threads	914	92	104	89
4 threads	472	179	57	173
8 threads	281	303	38	286
16 threads	140	617	23	576
STI Cell BE QS22				
PPE	13335	6	4347	2
1 SPE	488	184	154	92
2 SPEs	271	349	110	175
4 SPEs	166	617	87	316
8 SPEs	140	764	75	578
16 SPEs	97	1258	70	952
NVIDIA GTX 285	52	1655	13	1434

The table shows the total simulation time (left column) and the performance in million cells (voxels) per second (right column).

On the Intel Nehalem (X5550) there is about a 4-second overhead to read the input files for the LCA model, and about 8 seconds to read the inputs for the boy model, since the electric fields need to be loaded for each of the 12 antennae. The overhead is included in the execution time in Table 12.2. Note that the performance is worse when using two threads on the X5550 instead of one. The reason is that the affinity scheme that was used places both threads on a single core and therefore they have to share resources (e.g., the L1 cache, the memory controller).

The overheads are much larger on the Cell BE, since the files are read in the PPE program. They are approximately 30 and 60 seconds, respectively, hence the high numbers in the time column. They are a result of the PPE being a weak processor compared to current Intel or AMD CPUs. If the entire simulation is run on the PPE only, it achieves only a fraction of the performance (6 and 2 MCells/s, respectively) that is reached when the actual calculation is done by an SPE.

On the GPU system (a single GTX 285 GPU), the overheads are about 2 and 7 seconds, respectively. The data are loaded into host memory by the CPU (an Intel i7 920, 2.67 GHz), then transferred to device memory and kept there until the end of the calculation. For the simulation, the advantage of the GPU system over the Cell BE is that a powerful CPU is available that can do the preprocessing, which reduces the overall time significantly. The simulations also take advantage of the fact that the coefficient fields do not change in time and hence no host-to-device data transfers are necessary during the computation.

12.5   Related Work

Temporal blocking and tiling algorithms for stencil computations have been investigated and described in [11, 13, 21, 23, 30]. Frigo and Strumpen [9] describe an interesting approach of cache-oblivious blockings, which can be beneficial to support a broad range of architectures by finding optimal block sizes algorithmically at runtime. Another approach is to use autotuning to determine the best block sizes offline [6]. More general work on blocking/parallelization of imperfectly nested loops can be found in [12, 24].

Stencil performance on the Cell BE processor has been benchmarked in [5, 29]. Using temporal blocking, linear speedups and performances of up to 65 GFlop/s for single-precision stencil computations are reported. The stencil considered in this work is a stencil with constant coefficients, i.e., does not suffer from bandwidth limitations if time blocking is applied.

In [1] GPUs are explored as accelerators for a computational fluid dynamics simulation. The method used to compute the simulation translates to stencil computations. Speedups of 29× (in 2D) and 16× (in 3D) over the Fortran reference implementation running on an Intel Core 2 Duo 2.33 GHz CPU are reported. More recent work [15] shows nice scaling results of 3D stencil computations on a newer Tesla 10-series system, which supports concurrent data transfer and kernel execution.

12.6   Conclusion

Stencil-based computations constitute an important class of numerical methods, ranging from simple PDE solvers to constituent kernels in multigrid and adaptive mesh refinement methods as well as image processing applications. In this chapter, we have considered a type of stencil arising in a biomedical simulation (hyperthermia cancer treatment planning) and studied the performances of implementations on various architectures, including the latest Intel and AMD CPUs, the STI Cell Broadband Engine, and NVIDIA GPUs. As stencil computations are bandwidth-limited due to their low arithmetic intensity, typically only a fraction of the hardware's peak performance can be reached. It is therefore important to minimize bandwidth consumption by considering spatial and temporal data locality. It has been discussed how spatial and temporal blocking of the stencil loop can help achieve this goal.

The broad variety of stencil types make an approach that incorporates the entire class of stencils into a single code unfeasible and undesirable. Also, as we have demonstrated, depending on the architecture on which the code is executed, the types of code optimizations have more or less impact on the performance. Therefore, we are currently generalizing the ideas presented here to incorporate them into a code generation framework that will generate parameterized C code for CPU, the Cell BE, and GPU architectures from a stencil description. The parameters serve as an interface to an autotuner for offline detection of the best parameters (cache and temporal block sizes, loop unrollings, etc.) that will yield the best performance on a specific architecture. Having such a code generation and autotuning framework will not only allow us to obtain good performance (and energy efficiency, a metric that has become even more important recently than pure performance, but has not been discussed in this chapter), but also raise productivity in an increasingly complex landscape of compute architectures.

Acknowledgments

The authors would like to acknowledge the Swiss National Science Foundation, IBM for a Faculty Award, the Sony-Toshiba-IBM Center of Competence at the Georgia Institute of Technology, and the Fraunhofer ITWM in Kaiserslautern, Germany.

Bibliography

[1] T. Brandvik and G. Pullan. Acceleration of a 3D Euler solver using commodity graphics hardware. In *46th AIAA Aerospace Sciences Meeting*. American Institute of Aeronautics and Astronautics, January 2008.

[2] Andreas Christ, Wolfgang Kainz, Eckhart G. Hahn, Katharina Honegger, Marcel Zefferer, Esra Neufeld, Wolfgang Rascher, Rolf Janka, Werner Bautz, Ji Chen, Berthold Kiefer, Peter Schmitt, Hans-Peter Hollenbach, Jianxiang Shen, Michael Oberle, Dominik Szczerba, Anthony Kam, Joshua W. Guag, and Niels Kuster. The Virtual Family development of surface-based anatomical models of two adults and two children for dosimetric simulations. *Physics in Medicine and Biology*, 55(2):N23, 2010.

[3] M. Christen, O. Schenk, P. Messmer, E. Neufeld, and H. Burkhart. Accelerating stencil-based computations by increased temporal locality on modern multi- and many-core architectures. In *High Performance and Hardware Aware Computing: Proceedings of the First International Workshop on New Frontiers in High-performance and Hardware-aware Computing (HipHaC'08), Lake Como, Italy, November 2008 (in conjunction with MICRO-41)*, pages 47–54. Universitätsverlag Karlsruhe, 2008.

[4] M. Christen, O. Schenk, E. Neufeld, P. Messmer, and H. Burkhart. Parallel data-locality aware stencil computations on modern micro-architectures. In *Proceedings of the 2009 IEEE International Symposium on Parallel & Distributed Processing*, pages 1–10. IEEE Computer Society, 2009.

[5] K. Datta, S. Kamil, S. Williams, L. Oliker, J. Shalf, and K. Yelick. Optimization and performance modeling of stencil computations on modern microprocessors. *SIAM Review*, 51(1):129–159, 2009.

[6] K. Datta, S. Williams, V. Volkov, J. Carter, L. Oliker, J. Shalf, and K. Yelick. *Auto-tuning Stencil Computations on Diverse Multicore Architectures*. IN-TECH Publishers, To appear.

[7] M. de Bruijne, D. H. Wielheesen, J. van der Zee, N. Chavannes, and G. C. van Rhoon. Benefits of superficial hyperthermia treatment planning: Five case studies. *International Journal of Hyperthermia*, 23:417–429, Aug 2007.

[8] M.W. Dewhirst, Z. Vujaskovic, E. Jones, and D. Thrall. Re-setting the biological rationale for thermal therapy. *International Journal of Hyperthermia*, 21:779–790, 2005.

[9] M. Frigo and V. Strumpen. Cache oblivious stencil computations. In *ICS '05: Proceedings of the 19th Annual International Conference on Supercomputing*, pages 361–366, New York, NY, USA, ACM, 2005.

[10] Daniel Hackenberg. *Einsatz und Leistungsanalyse der Cell Broadband Engine*. Master's thesis, Fakultät Informatik, Technische Universität Dresden, 2007.

[11] M. Kowarschik, C. Weiß, W. Karl, and U. Rüde. Cache-aware multigrid methods for solving Poisson's equation in two dimensions. *Computing*, 64(4):381–399, 2000.

[12] Christian Lengauer. Loop parallelization in the polytope model. In *CONCUR '93: Proceedings of the 4th International Conference on Concurrency Theory*, pages 398–416, London, UK, Springer-Verlag, 1993.

[13] Z. Li and Y. Song. Automatic tiling of iterative stencil loops. *ACM Trans. Program. Lang. Syst.*, 26(6):975–1028, 2004.

[14] John D. McCalpin. STREAM: Sustainable Memory Bandwidth in High-Performance Computers.

[15] Paulius Micikevicius. 3D finite difference computation on GPUs using CUDA. In *GPGPU-2: Proceedings of 2nd Workshop on General Purpose Processing on Graphics Processing Units*, pages 79–84, New York, NY, USA, ACM, 2009.

[16] E. Neufeld, N. Chavannes, T. Samaras, and N. Kuster. Novel conformal technique to reduce staircasing artifacts at material boundaries for FDTD modeling of the bioheat equation. *Physics in Medicine and Biology*, 52(15):4371, 2007.

[17] Esra Neufeld. *High Resolution Hyperthermia Treatment Planning*. PhD thesis, ETH Zurich, August 2008.

[18] M. M. Paulides, J. F. Bakker, E. Neufeld, J. van der Zee, P. P. Jansen, P. C. Levendag, and G. C. van Rhoon. The HYPERcollar: A novel applicator for hyperthermia in the head and neck. *International Journal of Hyperthermia*, 23:567–576, 2007.

[19] M. M. Paulides, J. F. Bakker, A. P. M. Zwamborn, and G. C. van Rhoon. A head and neck hyperthermia applicator: Theoretical antenna array design. *International Journal of Hyperthermia*, 23(1):59–67, 2007.

[20] H. H. Pennes. Analysis of tissue and arterial blood temperatures in the resting human forearm. *J Appl Physiol*, 1(2):93–122, 1948.

[21] G. Rivera and C. Tseng. Tiling optimizations for 3D scientific computations. In *Supercomputing, ACM/IEEE 2000 Conference*, 2000.

[22] T. Samaras, P. J. Rietveld, and G. van Rhoon. Effectiveness of FDTD in predicting SAR distributions from lucite cone applicator. *IEEE Trans Microw Eng Techn*, 45(11):2059–2063, 2000.

[23] S. Sellappa and S. Chatterjee. Cache-efficient multigrid algorithms. *Lecture Notes in Computer Science*, 2073:107–116, 2001.

[24] Y. Song and Z. Li. A compiler framework for tiling imperfectly-nested loops. In *Proceedings of the 12th International Workshop on Languages and Compilers for Parallel Computing*, pages 185–200. Springer-Verlag, 1999.

[25] SPEAG. SEMCAD X (incl. Manual). *Schmid und Partner Engineering AG (SPEAG), Zurich, Switzerland*, 2006.

[26] G. Sreenivasa, J. Gellermann, B. Rau, J. Nadobny, P. Schlag, P. Deuflhard, R. Felix, and P. Wust. Clinical use of the hyperthermia treatment planning system HyperPlan to predict effectiveness and toxicity. *Int. J. Radiat. Oncol. Biol. Phys.*, 55:407–419, 2003.

[27] J. van der Zee. Heating the patient: A promising approach? *Ann Oncol,* 13(8):1173–1184, 2002.

[28] S. Weinbaum and L. M. Jiji. A new simplified bioheat equation for the effect of blood flow on local average tissue temperature. *Journal of Biomechanical Engineering,* 107(2):131–139, 1985.

[29] S. Williams, J. Shalf, L. Oliker, S. Kamil, P. Husbands, and K. Yelick. Scientific computing kernels on the Cell processor. *Int. J. Parallel Program.,* 35(3):263–298, 2007.

[30] D. Wonnacott. Time skewing for parallel computers. In *Proceedings of the 12th Workshop on Languages and Compilers for Parallel Computing,* pages 477–480. Springer-Verlag, 1999.

[31] P. Wust, B. Hildebrandt, G. Sreenivasa, B. Rau, J. Gellermann, H. Riess, R. Felix, and P. Schlag. Hyperthermia in combined treatment of cancer. *The Lancet Oncology,* 3:487–497, 2002.

Part VII

Bioinformatics

Chapter 13

Enabling Bioinformatics Algorithms on the Cell/B.E. Processor

Vipin Sachdeva

IBM Systems and Technology Group, Indianapolis, IN

Michael Kistler

IBM Austin Research Laboratory, Austin, TX

Tzy-Hwa Kathy Tzeng

IBM Systems and Technology Group, Poughkeepsie, NY

[1]

13.1 Computational Biology and High-Performance Computing

With the discovery of the structure of DNA and the development of new techniques for sequencing the entire genome of organisms, biology is rapidly

---

[1]Portions reprinted with permission, from (Exploring the viability of the Cell Broadband Engine for bioinformatics applications, Sachdeva, V., Kistler, M., Speight, E., and Tzeng, T.-H.K. ISBN: 1-4244-0910-1). ©2007 IEEE.

moving towards a data-intensive, computational science. Biologists search for biomolecular sequence data to compare with other known genomes in order to determine functions and improve understanding of biochemical pathways. Computational biology has been aided by recent advances in both algorithms and technology, such as the ability to sequence short contiguous strings of DNA and from these reconstruct a whole genome [1, 15, 16]. In the area of technology, high-speed micro-array gene and protein chips [12] have been developed for the study of gene expression and function determination. These high-throughput techniques have led to an exponential growth of available genomic data. As a result, the computational power needed by bioinformatics applications is growing exponentially and it is now apparent that this power will not be provided solely by traditional general-purpose processors.

The recent emergence of accelerator technologies like field programmable gate arrays (FPGAs), graphics processing units (GPUs) and specialized processors have made it possible to achieve an order-of-magnitude improvement in execution time for many bioinformatics applications compared to current general-purpose platforms. Although these accelerator technologies have a performance advantage, they are also constrained by the high effort needed in porting the application to these platforms.

In this chapter, we focus on the performance of sequence alignment and homology applications on the Cell/B.E. processor. The Cell/B.E.^{TM} processor, jointly developed by IBM, Sony and Toshiba, is a new member of the IBM PowerPC processor family [6]. This processor was developed for the Sony PlayStation^{TM}3 game console, but its capabilities also make it well suited for various other applications such as visualization, image and signal processing and a range of scientific/technical workloads.

In previous research, we presented *BioPerf* [2], a suite of representative applications assembled from the computational biology community. Using our previous workload experience, we focus on two critical bioinformatics applications—ClustalW (*clustalw*) and FASTA (*ssearch34*). ClustalW is used for multiple sequence alignment, and the FASTA package uses the *ssearch34* Smith-Waterman kernel to perform pairwise alignment of gene sequences. Most Cell/B.E. applications have been developed using the direct memory access (DMA) model, in which explicit DMA commands are used to transfer data to and from synergistic processing elements (SPEs). In the IBM SDK for MultiCore Acceleration Version 3.1 [5], IBM introduced software caches inside SPEs as an alternative to DMA commands for data movement between system memory and the SPEs local store (LS). For ClustalW, we describe implementations we developed for both programming models, including analysis and comparison of the resulting performance. The FASTA application was developed with only the DMA model, and we describe in detail the changes required to utilize the capabilities of the Cell/B.E. processor, along with the performance we were able to achieve.

13.2   The Cell/B.E. Processor

The Cell/B.E. processor is a heterogeneous, multicore chip optimized for compute-intensive workloads and broadband, rich media applications. The Cell/B.E. is composed of one 64-bit power processor element (PPE), eight specialized co-processors called SPEs, a high-speed memory controller and high-bandwidth bus interface, all integrated on-chip. The PPE and SPEs communicate through an internal high-speed element interconnect bus (EIB). The memory interface controller (MIC) provides a peak bandwidth of 25.6 GB/s to main memory. The Cell/B.E. has a clock speed of 3.2 GHz and theoretical peak performance of 204.8 GFLOPS (single precision) and 21 GFLOPS (double precision).

The PPE is the main processor of the Cell/B.E. and is responsible for running the operating system and coordinating the SPEs. Each SPE consists of a synergistic processing unit (SPU) and a memory flow controller (MFC). The SPU is a RISC processor with 128-bit single instruction multiple data (SIMD) registers and a 256 KB LS. The SIMD pipeline can run at four different granularities: 16-way 8b integers, 8-way 16b integers, 4-way 32b integers or single-precision floating-point numbers or 2 64b double-precision floating point numbers. The 256 KB local store is used to hold both the instructions and data of an SPU program. The SPU cannot access main memory directly. The SPU issues DMA commands to the MFC to bring data into the LS or write the results of a computation back to main memory. Thus, the contents of the LS are explicitly managed by software. The SPU can continue program execution while the MFC independently performs these DMA transactions.

13.2.1   Cache Implementation on Cell/B.E.

At the hardware level, SPE programs can only directly access data that resides in the SPE's LS. The SPE must use DMA operations to transfer data from main memory to LS in order to access it, and must issue DMA operations to transfer the results of computation back to main memory. However, a recent feature in the GCC and XLC compilers makes it possible for SPE programs to refer to data in main memory. This feature is called **"Named Address Space support"** or **"ea address space support"** [5]. It allows the programmer to add the type qualifier __ea to variable declarations in an SPE program to indicate that this variable resides in system memory. This allows the PPE and SPE portions of an application to use common variable declarations and share data as easily as two different functions within the PPE portion of the code.

The compilers are able to provide this functionality by generating the DMA operations for the referenced data under the covers. However, issuing a DMA for every access to an __ea variable could severely impact performance, so the compiler implements a software-managed cache in the LS of the SPE. The compiler allocates a region of LS and uses it to hold copies of $\_\_ea$ qualified variables accessed by the SPE program. Several implementations of software data caching for the SPEs have been developed, but we chose to explore the cache support in the compilers because it is nearly transparent to the application programmer.

In our work, we used the GCC compiler and implementation of named address space support [9]. The size of the cache allocated by *spu-gcc* can be set to 8 KB, 16 KB, 32 KB, 64 KB, or 128 KB with the compiler option *-mcache-size=*, where the default size is 64 KB. The compiler does not attempt to maintain coherence of its cached copies with the contents of system memory, so some care must be used in choosing data to access using this mechanism. However, the cache manager does track modifications to cached data at a byte level, and writes these changes back to system memory with an atomic read-modify-write operation, to prevent the loss of concurrent modifications to other bytes on the same cache line. If the programmer carefully partitions the data in system memory such that no two SPEs will write to the same cache line, performance may by improved by avoiding the atomic DMAs in the update phase by specifying the *-mno-atomic-updates* compiler option.

If $\_\_ea$ variables are used for communication between threads, GCC provides the $\_\_cache\_evict(\_\_ea\ void\ *ea)$ function to flush updates from the cache back to system memory for a given $\_\_ea$ variable. Other SPEs that wish to see these updates must also evict the line from their cache before reading the new value. A form of synchronization, such as mailboxes or signotify registers, is typically needed to ensure the first thread completes its updates to memory before the second thread attempts to read the new value. GCC also provides a $\_\_cache\_touch(\_\_ea\ void\ *ea)$ function that can be used to prefetch data into the cache. These and a few other special cache functions are available in the *spu_cache.h* header file.

13.3   Sequence Analysis and Its Applications

Sequence analysis refers to the collection of techniques used to identify similar or dissimilar sequences or subsequences of nucleotides or amino acids. Sequence analysis is one of the most commonly performed tasks in bioinformatics. Within the area of sequence analysis, one of the most well-known and frequently employed techniques is pairwise alignment.

Pairwise alignment is the process of comparing two sequences and involves aligning and inserting gaps in one or both sequences to produce an optimal score. Scores are computed by adding constant or nucleotide (amino-acid) specific match scores while subtracting constant scores for gaps or mismatches. The problem of comparing two entire sequences is called *global alignment*, and comparing portions of two sequences is called *local alignment*. Dynamic programming techniques can be used to compute optimal global and local alignments. Smith and Waterman [13] developed one such dynamic programming algorithm (referred to as "Smith-Waterman") for optimal pairwise global or local sequence alignment. For two sequences of length n and m, the Smith-Waterman algorithm requires $O(nm)$ sequential computation and $O(m)$ space.

Due to the quadratic complexity of the Smith-Waterman algorithm, various attempts have been made to reduce its execution time. One approach has employed MMX and SSE technology common in today's general-purpose microprocessors to achieve significant speedups. Other approaches utilize multiple processors to perform parts of the computation in parallel. Parallel strategies for Smith-Waterman and other dynamic programming algorithms range from fine-grained ones in which processors collaborate in computing the dynamic programming matrix cell-by-cell [8] to coarse-grained ones in which query sequences are distributed among the processors with no communication needs [3].

In our work, we have ported two popular bioinformatics applications, ClustalW and FASTA, to the Cell/B.E. processor. These two applications are representative of the sub-field of sequence analysis in computational biology. ClustalW is a popular tool for multiple sequence alignment, which organizes a collection of input sequences into a hierarchy or tree where more similar sequences are more closely grouped in the tree [14]. Multiple alignment is used to organize data to reflect sequence homology, identify conserved (variable) sites, perform phylogenetic analysis, and other biologically significant results.

The *FASTA* package applies the Smith-Waterman [13] dynamic programming algorithm to compare two input sequences and compute a score representing the alignment between the sequences [10]. This is commonly used in similarity searching where an uncharacterized "query" sequence is scored against vast databases of characterized sequences. A score larger than a threshold value is considered a match. Such a scoring mechanism is useful for capturing a variety of biological information including identifying the coding regions of a gene, identifying similar genes and assessing divergence among other sequences.

To begin our analysis for porting these applications to the Cell/B.E. processor, we used the *gprof* tool to determine the most time-consuming functions for each application. We found that both applications spend more than half of their execution time in a single function: *dropgsw* for FASTA and *forward_pass* for ClustalW. This is a useful fact for an implementation for the Cell/B.E. processor, as it implies that we might obtain a significant speedup for these applications by only porting these functions to run on the SPUs. In addition, both applications perform a large numer of pairwise alignments, which are completely independent and thus can be performed in parallel across the eight SPUs of the Cell/B.E. processor.

Our porting effort was further aided by the availability of Altivec implementations of the key kernels for both our applications. These Altivec implementations make a port to the SPUs much simpler, as many of the Altivec intrinsics map one-to-one to the SPU intrinsics, or can be constructed using a short sequence of intrinsics on the SPU.

13.4   Sequence Analysis on the Cell/B.E. Processor

13.4.1   ClustalW

ClustalW is a progressive multiple sequence alignment application. There are three basic steps to this process. In the first step, all sequences are compared pairwise using the global Smith-Waterman algorithm. A cluster analysis is then performed on each of the scores from the pairwise alignment to generate a hierarchy for alignment call a guide tree. Finally, the multiple alignment is built step by step, adding one sequence at a time, according to the guide tree.

The major time-consuming step of the ClustalW alignment is the all-to-all pairwise comparisons. Depending on the inputs, this step can take 60%-80% of the execution time. The *pairalign* function performs the task of comparing all input sequences against each other, thus performing a total of $\frac{n(n-1)}{2}$ alignments for n sequences.

Our algorithm design focuses on performing the pairalign function on the SPUs, with the rest of the code executing on the power processing units (PPUs). The most time-consuming step of *pairalign* is *forward_pass*, which computes the alignment score for two sequences and the location of the cell in the alignment matrix that produced this score. The open source-release of ClustalW has a scalar version of *forward_pass* which involves multiple branches to compute the value at every matrix cell. Due to the random nature of the inputs, these branches are not easily predicted, and since the SPUs lack dynamic branch prediction, this implementation is not well suited for execution on the SPU. Fortunately, IBM Life Sciences have developed a vectorized implementation of *forward_pass* using Altivec intrinsics which proved to be a much better basis for our port to the SPU. ClustalW, with an Altivec port of *forward_pass* is available from `http://powerdev.osuosl.org/node/43`.

We had to develop SPU implementations for some Altivec intrinsics that are not available on the SPU. The saturated addition intrinsics *vec_adds* and *vec_max* of Altivec are not available for the SPU, and hence have to be constructed using multiple SPU instrinsics. Also, the vectorized code used the *vector status and the control register* for overflow detections, while doing computation with 16-bit (*short*) data types. The overflow detects that the maximum score is greater than the range of the *short* data type, and therefore does a recomputation using integer values. Since this mechanism is not supported inside the SPUs, we changed the code to use 32-bit (*int*) data types. This lowers the efficiency of the vector computation, since now only four values can be packed inside a vector, unlike the eight of the original code, but this was necessary for correct execution without overflow detection.

One of the bottlenecks with the ClustalW code is the alignment score lookup, in which an alignment matrix score is read for finding the cost of match/mismatch among two characters in the sequences. This lookup is a scalar operation, and hence does not perform well on the SPU, since the SPU has only vector registers. For vector execution, four (32 bit) values are loaded into one vector; however, in the code, this step is also preceded by a branch involving multiple conditions, involving both the loop variables for handling of boundary cases. Since the SPUs have only static branch prediction, such a branch, even thought *mostly taken*, was difficult to predict for the SPU. We broke the inner loop of the alignment into several different loops so that the branch evaluation now depends on a single loop variable, and the boundary cases computation can be handled explicitly. This change alone helped us get a more than 2X performance gain.

Besides the innermost kernel execution, we also focused on partitioning of the work among the SPUs. Assuming there are n sequences in the query sequence file, we have a total of $\frac{n(n-1)}{2}$ computations to be performed. To distribute these computations to the SPUs, we packed all the sequences in a single array, with each sequence beginning at a multiple of 16 bytes. This is important, as the MFC can only DMA in/out from 16-byte boundaries. This array, along with an array of the lengths of all the sequences, allows the SPUs to pull in the next sequence without PPU intervention.

The PPU creates a structure to contain the input parameters for the SPU computations, such as matrix type, gap-open and gap-extension penalties, the location of the sequence array in main memory and the length of the sequence array. Then the PPU creates the SPU threads and passes the maximum sequence size through a mailbox message. The SPUs allocate memory only once in the entire computation based on the maximum size, and then wait for the PPU to send a message for them to pull in the parameter structure and begin the computation. Work is assigned to the SPUs using a simple round-robin strategy: each SPU is assigned a number from 0 to 7, and SPU k is responsible for comparing sequence number i against all sequences $i+1$ to n if $i\%numspes = k$. Such a strategy prohibits reuse of the sequence data, but since the communication costs are very low in comparison to the total computation, this strategy seems to work fairly well. For storing of the output values, the SPUs are also passed a pointer to an array of structures, which are 16-byte aligned, in which they can store the output of the *forward_pass* function executed for two sequences.

Next we describe an alternative implementation of ClustalW for the Cell/B.E. using the software managed cache support in the GCC compiler. We use the software managed cache feature to eliminate the need for explicit DMA operations to transfer data between main memory and LS. We focus our discussion on the difference in data transfer operations and performance between these two implementations.

In the DMA implementation, we had to change the data layout in the PPU code to place data structures on 16-byte boundaries and make all data sizes a multiple of 16 bytes, as reqired by the SPE DMA engines. The cache implementation does not require any of these changes, as the __ea qualifier works for arbitrarily aligned variables stored in the PPE address space. Variables in the PPE code that are required for computation by the SPEs are declared with the __ea qualifier in the SPEs code: this includes the original array of pointers, sequence length array, scalar values and the alignment matrix. The only reason for quad-word alignment of the sequences was for DMAs to complete successfully, and was not a requirement for Altivec-enabled code. Some of these variables were local function variables, so they were copied into global variables; these global variables are only declared when ClustalW Cell/B.E. code is being compiled. With these changes, we could get the code to use SPEs for *forward_pass* computation. The vectorized implementation of *forward_pass* computation was reused from the previous DMA implementation.

To improve performance of the cache-only implementation, we experimented with copying the data being accessed on the SPE to a buffer in the LS. Instead of using DMA commands to copy data to SPE LS, we declare the variable as __ea, and then simply copy from the variable, with no alignment restrictions. Copying using caches enhanced programmer productivity to a significant extent, compared with DMA commands, which have 16-byte alignment restrictions. We also experimented with data access and layouts on the PPE side, to see if we get a performance improvement through those changes. In Section 13.4, we detail the performance of the cache implementation with varying input sizes and SPEs and compare it with the DMA implementation described above. We also pinpoint the reasons for the overhead of the cache implementation over the DMA implementation.

13.4.2   FASTA

As mentioned above, our porting strategy for FASTA was to develop an SPU implementation for the most heavily used function, identified by our profiling work as *dropgsw*. The *dropgsw* function performs a pairwise alignment of two sequences using the Smith-Waterman dynamic programming algorithm. The FASTA package already includes an efficient Altivec implementation of the Smith-Waterman algorithm developed by Eric Lindahl. We began with porting this Smith-Waterman Altivec kernel *smith_waterman_altivec_word* to the SPUs by converting the Altivec intrinsics to the corresponding SPU intrinsics. However, two of the Altivec intrinsics used in FASTA are not available on the SPU: *vec_max* and *vec_subs*. For these two cases, we had to construct equivalent functions using short sequences of SPU intrinsics.

To execute the Smith-Waterman kernel on the SPU, the alignment scores are pre-computed on the PPU, and are DMAed to the SPU along with the query and the library sequence. Other parameters such as the alignment matrix and the gap penalties are also included in the context for every SPU.

The Cell/B.E. implementation discussed above is not fully functional as of now: our current implementation requires both sequences to fit entirely in the SPU local store of 256 KB, which limits the sequence size to at most 2048 characters. To do genome-wide or long sequence comparisons, a pipelined approach similar to [7] among the SPUs could be implemented. Each SPU performs the Smith-Waterman alignment for a block, notifies the next SPU through a mailbox message, which then uses the boundary results of the previous SPU for its own block computation. Support of bigger sequences on the Cell/B.E. is a key goal of our future research.

Once a fully functional Smith-Waterman implementation exists on the Cell/B.E., we can employ this kernel in the FASTA package. The FASTA package compares each sequence in a query sequence file with every sequence in a library sequence file, and hence multiple issues for load-balancing could be evaluated. For now, we have a simple round-robin strategy, in which the sequences in the query library are allocated to the SPUs based on the sequence numbers and the SPU number. In many ways, the load-balancing approach will be similar to the one discussed in Section 13.4.1 for ClustalW. For FASTA, we consider extending it for bigger sequences as the more important and difficult problem.

13.5   Results

13.5.1   Experimental Setup

Our experimental setup consists of a QS20 blade, with two Cell/B.E processors running at 3.2 GHz, with a total memory of 1 GB. Sequence similarity applications do not depend on double-precision floating-point performance to a large extent, so using the older QS20 is not a major issue with these applications. For the DMA implementations of ClustalW and FASTA, *ppuxlc* and *spuxlc* versions 8.1 were used. We used *ppu-gcc* and *spu-gcc* release 4.1.1 to compile the PPU and the SPE code respectively, for the ClustalW cache implementation. The optimization level was fixed at *-O3* for both the PPU and the SPE code, for both XLC and GCC compilers.

13.5.2   ClustalW Results and Analysis

Figure 13.1 shows the total time of execution of ClustalW for Cell/B.E. and contemporary architectures for two inputs from the BioPerf suite: 1290.seq is a class-B input with 66 sequences of average length 1082, and 6000.seq is a class-C input with 318 sequences of average length 1043. Results for the Cell/B.E. proceessor are for the best implementation strategy, namely using integer datatypes with no branches. The Opteron and the Woodcrest performance is

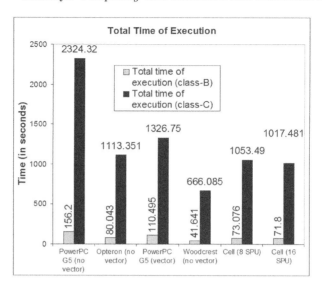

FIGURE 13.1: Comparison of Cell/B.E. performance with other processors for total time of execution.

non-vectorized, as we could not find an open-source SSE-enabled *forward_pass* on these platforms. While the Cell/B.E. outperforms other processors by a significant margin in the time to execute *forward_pass* (not shown in the figure), the total time of execution is slower compared to other processors. This is because the portion of the code executing on the PPU is much slower in comparison to the other superscalar processors. In particular, the PPU uses the scores computed by the *forward_pass* function on the SPUs to generate the guide tree and compute the final alignment. This final alignment computation takes most of the remaining execution time of the application. There are two approaches we could take to address this issue. One is to move additional functions, such as the computaiton of the guide tree and final alignment, to execute on the SPUs. Another approach is to use the Cell/B.E. as an accelerator, in tandem with a modern superscalar processor. This approach has been successfully demonstrated in Los Alamos National Laboratory's Roadrunner system [4], which combines Opteron and Cell/B.E. processors, and could give outstanding results for ClustalW.

In our initial implementation of ClustalW using software managed cache, the SPU used _ea variables to access all data in main memory. Performance measurements for this implementation showed that is was nearly 2X slower than our initial implementation using explicit DMAs. We also explored using different cache sizes and found this to have a negligible effect on performance, even when using just an 8-KB cache for the largest inputs.

To understand which data transfers contribute most to the overhead in the cache implementation, we changed our DMA-only implementation to a mix of DMA and software cache. Figure 13.2 shows the difference in performance as we move from a DMA-only implementation to a mix of DMA and caches; the data were collected using one SPE only, and show the execution time in seconds for the respective implementations. The *DMA + Cache (Sequences)* data show an implementation in which all data are DMAed into SPE, except for the DNA sequences, which are accessed through the software caches. The *DMA + Cache (Alignment Matrix)* data show the performance for an implementation in which all data are DMAed into the SPEs except for the alignment matrix; the alignment matrix is used to compute the scores of aligning every two characters of the query sequence and the library sequence. The *DMA + Cache (Alignment Matrix + Sequences)* data show the the performance for an implementation in which all data is DMAed into the SPEs except for the DNA sequences and the alignment matrix, which are accessed through the software cache. Lastly, the *Cache only* data shows the performance for a cache-only implementation with no DMAs. As can be seen, the difference in performance of the *Cache only* implementation and the *DMA + Cache (Alignment Matrix + Sequences)* implementation is negligible, so we can conclude that the **main overhead in the cache-only implementation is primarily due to the access of the alignment matrix and the DNA sequences**. Cached access to other data such as the scalar values for gap-extension and penalty, as well as storing the results matrix, does not carry any major penalty.

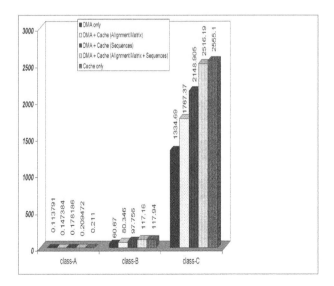

FIGURE 13.2: Performance of DMA implementation with caches for sequences and alignment matrix.

Since the main overhead of the cache implementation is due to alignment matrix and the sequences, we experimented with copying the alignment matrix and the sequences from the software cache into a buffer space in the SPE LS. This was done prior to the SPE computation. This approach leads to a notable performance improvement, and the performances of the DMA and the cache implementations are now very close for both the class-A and the class-B inputs. However, the size of all the sequences for the class-C input is too large to allow it to be copied into the LS, so we must copy each pair of sequences into LS as they are used. As a result, the overhead of the cache implementation for the class-C input is still more than 50% compared to the DMA implementation. This leads to higher overhead of the cache implementation for the largest input size (class-C).

Figure 13.3 shows the difference in performance using multiple SPEs upto a maximum of 16, for both the cache and the DMA implementations for class-B and class-C input of BioPerf. We implemented the same static load-balancing strategy as explained in [11] for both the implementations. As is evident from Figure 13.3, we can see that both implementations scale well with an increasing number of SPEs. **The cache implementation does not lead to increased overhead at a higher number of SPEs**, and shows the same overhead as the runs done with one SPE. These results are for cache sizes of 8 KB. For size-B input, in which we can copy the entire input and the alignment matrix into an LS buffer, the performance of the cache implementation is very close to the DMA implementation. For class-C input, in which only the alignment matrix can be copied entirely into the LS, we still see a performance overhead compared to the DMA implementtaion.

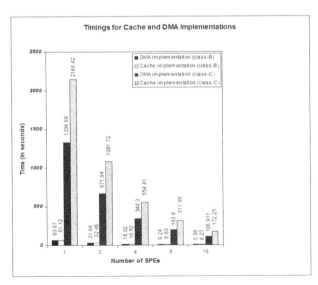

FIGURE 13.3: Timings of ClustalW pairwise alignment step on multiple SPEs with cache and DMA versions.

13.5.3 FASTA Results and Analysis

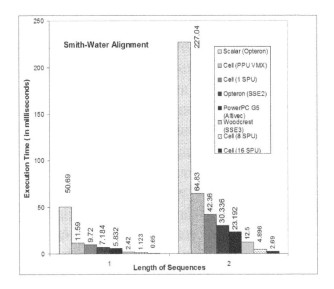

FIGURE 13.4: Performance of Smith-Waterman alignment for different processors.

Figure 13.4 shows the results of the execution of Smith-Waterman on several current general-purpose processors along with our implementation for the Cell/B.E. processor. We executed a pairwise alignment of eight pairs of sequences, using one SPU for each pairwise alignment. The Cell/B.E. processor, despite the absence of instructions explained above, still outperforms every superscalar processor currently in the market. This superior performance is mainly due to the presence of eight SPU cores and the vector execution on the SPUs. We should further state that these results are still preliminary, and further optimization of the kernel performance is still underway. For the Cell/B.E., the codes were compiled with xlc version 8.1 with the compilation flag *-O3*, which gave better or equal performance compared to gcc for both SPU and PPU. For PowerPC G5, we used *-O3 -mcpu=G5 -mtune=G5*, and for Opteron and Woodcrest we used the -O3 flag. Since the PPU also supports Altivec instructions, it is also possible to use the PPU as a processing element, thus enabling nine cores on the Cell/B.E. processor, with even better performance results. Further profiling of our implementation indicates that the computation dominates the total runtime (up to 99.9% considering a bandwidth of 18 GB/s for the SPUs), and hence multi-buffering is not needed for this class of computation.

13.6   Conclusions and Future Work

In this chapter, we have discussed the implementation and results of two popular bioinformatics applications, namely FASTA for the Smith-Waterman kernel and ClustalW. Our preliminary results show that the Cell/B.E. is an attractive avenue for bioinformatics applications. Considering that the total power consumption of the Cell/B.E. is less than half of a contemporary superscalar processor, we consider Cell/B.E. a promising power-efficient platform for future bioinformatics computing.

Our future work will focus on making the applications discussed fully operational, and trying to find other avenues for optimization. We have also begun the work to port HMMER to the Cell/B.E. processor, but we do not yet have results suitable for publication. Our work to date has revealed that one of the critical issues with the HMMER implementation is that the working set size of the key computational kernel, *hmmpfam*, exceeds the space available in the 256-KB LS of the SPU for most HMM and sequence alignments. Thus, solving such inputs will require partitioning of the input among eight SPUs while making sure that the data dependency between the SPUs is still fulfilled correctly. Thus, FASTA and HMMER suffer from similar implementation issues in the use of the multiple SPUs to solve problems which do not fit inside the LS of any one SPU. ClustalW, due to the mostly limited size of its inputs, is fully functional. Besides these critical applications, we would intend to work with other applications in diverse areas such as protein docking, RNA interference, medical imaging and other avenues of computational biology to determine their applicability for the Cell/B.E. processor.

We have also described our implementation of ClustalW running on Cell/B.E. that uses software caches inside the SPEs for data movement between PPE and SPEs. Our initial effort resulted in an overhead of about 2X compared to the DMA only version, but copying the data to the LS prior to using it results in a version which is very close to the DMA performance. Using the software caches enhances the programmer productivity to a significant extent, without a major decrease in a performance. For the largest input size, which does not fit into the SPE LS, we still see an overhead of more than 50% in performance compared to the DMA-only implementation. Our future work will involve investigating the sources of overhead of SPE caches and devising ways to reduce them. Such an implementation will also make it possible to use SPEs for pairwise alignment of very large DNA sequences, without a significant overhead.

Bibliography

[1] E. Anson and E. W. Myers. Algorithms for whole genome shotgun sequencing. In *Proc. 3rd Ann. Int'l Conf. on Computational Molecular Biology (RECOMB99)*, Lyon, France, April 1999. ACM.

[2] D. A. Bader, Y. Li, T. Li, and V. Sachdeva. BioPerf: A benchmark suite to evaluate high-performance computer architecture on bioinformatics applications. In *Proc. IEEE Int'l Symposium on Workload Characterization*, Austin, TX, October 2005.

[3] Z. Galil and K. Park. Parallel dynamic programming. Technical Report CUCS-040-91, Computer Science Department, Columbia Univ., 1991.

[4] IBM. IBM to Build World's First Cell Broadband Engine Based Supercomputer. `http://www-03.ibm.com/press/us/en/pressrelease/20210.wss`, 2006.

[5] IBM. IBM SDK for Multicore Acceleration. `http://www-01.ibm.com/chips/techlib/techlib.nsf/products/IBM_SDK_for_Multicore_Acceleration`, 2008.

[6] J. A. Kahle, M. N. Day, H. P. Hofstee, C. R. Johns, T. R. Maeurer, and D. Shippy. Introduction to the Cell multiprocessor. *IBM Systems Journal*, 49(4/5):589–605, 2005.

[7] Weiguo Liu and Bertil Schmidt. Parallel design pattern for computational biology and scientific computing applications. In *Proc. IEEE Intl. Conf. on Cluster Computing(CLUSTER'03)*, pages 456–459, Hong Kong, December 2003.

[8] W. S. Martins, J. B. Del Cuvillo, F. J. Useche, K. B. Theobald, and G. R. Gao. A multithreaded parallel implementation of a dynamic programming algorithm for sequence comparison. In *Proc. of the Pacific Symposium on Biocomputing*, pages 311–322, HI, January 2001.

[9] Michael Robert Meissner. Adding named address space support to the gcc compiler. In *Proc. 2009 GCC Developers Summit*, Montreal, Canada, June 2009.

[10] W. R. Pearson and D. J. Lipman. Improved tools for biological sequence comparison. *Proceedings of the National Academy of Sciences USA*, 85:2444–2448, 1988.

[11] Vipin Sachdeva, Mike Kistler, Evan Speight, and Tzy-Hwa Kathy Tzeng. Exploring the viability of the Cell Broadband Engine for bioinformatics applications. In *Proc. of the Sixth IEEE Intl. Workshop on High Performance Computational Biology*, 2007.

[12] M. Schena, D. Shalon, R. W. Davis, and P. O. Brown. Quantitative monitoring of gene expression patterns with a complementary DNA microarray. *Science*, 270(5235):467–470, 1995.

[13] T. F. Smith and M. S. Waterman. Identification of common molecular subsequences. *Journal of Molecular Biology*, 147:195–197, 1981.

[14] J. D. Thompson, D. G. Higgins, and T. J. Gibson. CLUSTALW: Improving the senstivity of progressive multiple sequence alignment through sequence weighting, position-specific gap penalties and weight matrix choice. *Nucleic Acids Res.*, 22:4673–4680, 1994.

[15] J. C. Venter and *et al.* The sequence of the human genome. *Science*, 291(5507):1304–1351, 2001.

[16] J. L. Weber and E. W. Myers. Human whole-genome shotgun sequencing. *Genome Research*, 7(5):401–409, 1997.

Chapter 14

Pairwise Computations on the Cell Processor with Applications in Computational Biology

Abhinav Sarje

Iowa State University

Jaroslaw Zola

Iowa State University

Srinivas Aluru

Iowa State University

14.1   Introduction

All-pairs computations occur in numerous and diverse applications across many areas. In many-body simulations and molecular dynamics, computation of pairwise gravitational or electrostatic forces is needed to study system evolution [16]. A similar scenario occurs in computational electromagnetics except that the field is sampled in multiple directions, turning this into a vector computation [15]. In the radiosity method in computer graphics, a scene is divided into patches and the reflection and refraction of light between every pair of patches is of interest to compute the equilibrium light energy distribution [14]. In computational systems biology, correlations between pairs of genes are sought based on their expression levels over a large set of observations [32]. Clustering algorithms typically use a distance metric and use all-pairs distances to guide the clustering process [6]. While the details vary, common to these applications is computation between all pairs of entities from a given set.

There is considerable recent interest in accelerating pairwise computations on Cell processors [5, 31, 33]. In this chapter we consider the general problem of scheduling pairwise computations, given a number of entities, on the Cell processor. The problem can be considered in its native form without loss of generality. In general, one could be carrying out limited pairwise computations in conjunction with a complexity reducing algorithm, or accelerating the per node pairwise computations in a parallel system. Given the ubiquity of applications where such computations occur, it is useful to consider this problem in its abstract form, and develop common algorithmic strategies to extract maximum performance. In this chapter we describe such strategies in the context of applications taken from computational biology.

Cell processor is a heterogeneous multicore processor, consisting of eight main computational cores, synergistic processing elements (SPEs), and one general purpose powerPC processing element (PPE). Each SPE has a limited local store (LS) of 256 KB and all data transfers between the system's main memory and the LS of an SPE need to be carried out through explicit direct memory access (DMA) transfers. Given the constraints of limited LS on the SPEs and the heterogeneity of the Cell processor, harnessing its raw power is a challenging task. In this chapter we describe a scheduling approach to perform the all-pairs computations efficiently in parallel among the SPE cores (Section 14.2). This approach is based on a hierarchical decomposition of the computations into subproblems which can be solved in parallel. Furthermore, scheduling of these computations can be optimized by minimizing the memory transfers needed among the SPEs and the system main memory. We further discuss two specific extensions to this basic strategy: First we describe an extension where a higher level of parallelism can be achieved on a cluster of Cell processors when the number of input entities is large. Then we consider the case where the memory requirement corresponding to even one pair of entities is too large to fit in the LS of an SPE, and describe a strategy of decomposing the individual input entities in addition to the computations. In Section 14.3 we describe application of the pairwise scheduling strategy to an algorithm for reconstruction of gene regulatory networks which employs pairwise mutual information computations. In certain cases when a basic decomposition scheme is inefficient, complex and specialized algorithmic strategies may be needed to perform the computations. As an example, we consider the problem of pairwise genomic alignments in the latter part of the chapter (Section 14.4) and describe parallel algorithms to perform different types of alignments of two biological sequences efficiently on the Cell processor.

14.2 Scheduling Pairwise Computations

We start with a basic problem where given two sets of entities, pairwise computations between all pairs of entities, one taken from each set, is required to be performed. Let us represent an entity as a vector of dimensionality d. The problem can then be stated formally as follows: Given two input matrices M_x and M_y of sizes $n_x \times d$ and $n_y \times d$, where n_x and n_y are the number of vectors in the two matrices respectively, and d is the dimensionality of each vector, we want to compute a matrix D of size $n_x \times n_y$, where entry $D[i, j] = f(M_x[i], M_y[j])$. In general n_x, n_y, and d can be arbitrary, and computing D requires $n_x \cdot n_y$ evaluations of the function f. In the case when $M_x = M_y$ and either f is a symmetric function, i.e. $f(X, Y) = f(Y, X)$, or $f(X, Y)$ is a trivial function of $f(Y, X)$, e.g. $f(X, Y) = -f(Y, X)$, it is sufficient to perform only $\binom{n}{2}$ computations of f in D, where $n = n_x = n_y$. Here we assume that the cost of evaluating f is a function of d.

The above problem is trivially implementable on general purpose processors; however, it becomes challenging on certain parallel architectures like the Cell processor, where the main computational cores have a small memory, and all memory transfers between the main memory and the memory on the cores must be orchestrated explicitly by a programmer. In the following we present an approach for scheduling the all-pairs computations that minimizes the total number of memory transfers performed. Note that the total number of computations is fixed for a particular problem instance, hence, this is the only way to improve the performance of such computations.

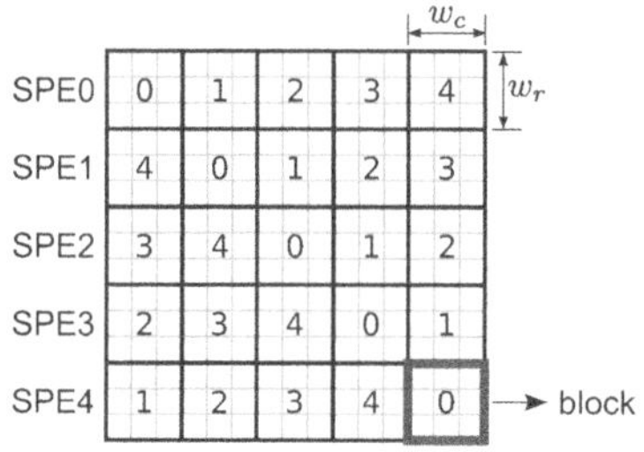

FIGURE 14.1: An example of decomposition of one tile of matrix D for $p_s = 5$ SPEs with $w_r = 3$ and $w_c = 3$. The number of the SPE is indicated beside the row of blocks it is responsible to compute. For each block, the iteration number in which it will be processed is marked.

14.2.1 Tiling

Assume the number of available SPEs is p_s. To compute the matrix D, a decomposition of the computations is necessary due to the limited LS available on the SPEs (256 KB), such that each unit of the decomposition can be computed within the memory limits. In the strategy presented here, we decompose the matrix D hierarchically. First, D is decomposed into sub-matrices which we call *tiles*. Each tile is further divided into $p_s \times p_s$ *blocks*, each block containing w_r rows and w_c columns. Each such block, therefore, corresponds to a unique sub-matrix of D. The size of each tile thus obtained is $(p_s \cdot w_r) \times (p_s \cdot w_c)$. Assume for now that an SPE can store $w = w_r + w_c$ input vectors in its LS $(w_r, w_c \geq 1)$. A tile can hence be processed at once in parallel collectively among all the p_s SPEs. All the tiles are processed one after the other in a sequential manner.

The decomposition and processing scheme of a tile are illustrated in Figure 14.1. Processing of a tile proceeds in p_s iterations. In each iteration, an SPE is assigned a block of the tile, and its task is to evaluate the function f for each position in the corresponding sub-matrix of D. To do so, it requires the corresponding w_r row vectors from the input matrix M_x, and w_c column vectors from the input matrix M_y. An SPE computes a row of such blocks in the tile, hence the same input row vectors are needed by it throughout all the p_s iterations. After transferring these w_r row vectors from the main memory at the beginning of the first iteration, the SPE retains them in its LS for all the subsequent iterations. The input column vectors, on the other hand, differ for each block assigned to the SPE. Therefore, after initial downloading from the main memory, they are shifted between the SPEs at the end of every iteration in a cyclic fashion.

This tiling scheme raises two questions: (1) what should be the size of a block, and (2) in what order should the resulting tiles be computed? Explicit DMA transfers are required to move data from/to the main memory to/from an SPE, and between two SPEs. The maximum valid size of a single transfer

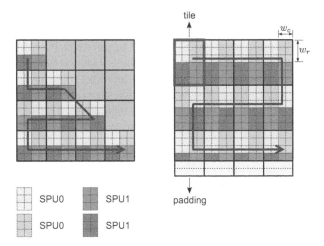

FIGURE 14.2: Tile decomposition of the output matrix D with $p_s = 2$ and $w_c \leq w_r$ when D is symmetric (left) and in the general case (right). The snake-like order in which the tiles are processed is marked for the row-wise traversal.

is 16 KB for the Cell processor. Denote this transfer block size by B. The goal in determining block size and tile ordering is to minimize the overall number of DMA transfers.

14.2.2 Tile Ordering

A tile is processed in parallel among the SPEs once the respective column and row vectors are loaded from the main memory to the LS on the SPEs, and the output matrix D is fully computed by processing one tile after another. Note that a tile represents a set of row and column input vectors that can be collectively stored on all the SPEs. To begin computation of a tile, these vectors need to be transferred from the main memory to the SPEs, if they are not already present on the SPEs. Once loaded, computations within a tile require DMA transfers only among the SPEs, which are significantly faster compared to those from/to the main memory. Consequently, the only way to reduce the number of DMA transfers as the processing moves from tile to tile is to reuse vectors from the previous tile as much as possible. Observe that two adjacent tiles may share the same row vectors, or the same column vectors, but not both. Hence, a tile ordering that reuses $\max(w_r, w_c)$ vectors as much as possible minimizes the number of DMA transfers. A snake-like traversal order of the tiles achieves this, where a row-wise traversal is used if $w_c \leq w_r$, and a column-wise traversal is used otherwise. Figure 14.2 depicts this tile traversal order for optimizing the number of DMA transfers.

14.2.3 Tile Size

To access the main memory, SPEs need to use explicit DMA transfers. Although such transfers can be seen as extremely fast one-sided communication operations, in most cases they need to be followed by a mailbox message or signal notification. Each such synchronization generates some overhead which partially can be hidden by multi-buffering techniques. During the processing of a tile of the matrix D, DMA transfers arise for the following three steps:

1. Fetching of input vectors from the main memory to the LS of an SPE—this communication is initiated by the SPEs based on the effective addresses of the input vectors obtained from the PPE and is completely one-sided. This is performed at the beginning of the processing of a tile.

2. Transferring the output data from the LS of an SPE to the main memory—here the SPEs write to the effective address of the output destination and then notify PPE about it through a mailbox message to initiate further processing by the PPE. This is performed for each block in a tie.

3. Shifting of column vectors between LS of SPEs—each SPE writes its current column vectors to the LS of its neighboring SPE and uses signal notifications to synchronize the entire process. This is also performed for each block (except the first) in a tile.

Since a single DMA transfer is limited to size B, transfers of larger size are handled by DMA lists which break down the transfer into multiple DMA requests. Because the element interconnect bus (EIB) connecting the various components on the Cell processor is a high-bandwidth bidirectional bus, it is advantageous to transfer maximal sized data through one request, over a number of small sized transfers (even though EIB supports variable length packets). Given the above, w_r and w_c can be derived to minimize the total number of DMA transfers. Without loss of generality, assume row-wise traversal is used.

Let $\mathcal{M}$ denote the total available free memory, in bytes, in the LS of an SPE. In this scheme we use double buffering for the output data when processing a block on the SPEs. In addition to this extra memory for output, to implement the shifting of column vectors between the SPEs, an additional buffer is needed to store w_c column vectors. Therefore, we can calculate the maximum number of vectors that can fit into the LS of an SPE using the following equation for memory usage:

$$\mathcal{M} = (2 \cdot w_c + w_r) \cdot d \cdot c_i + 2 \cdot w_c \cdot w_r \cdot c_o, \tag{14.1}$$

where an input element and an output element are encoded with c_i and c_o bytes, respectively. For a given w_r and w_c, the number of tiles obtained by partitioning matrix D is $n_r \times n_c$, where the number of tile rows $n_r = \left\lceil \frac{n}{p_s \cdot w_r} \right\rceil$

and the number of tile columns $n_c = \left\lceil \frac{n}{p_s \cdot w_c} \right\rceil$. In the following we calculate the number of DMA transfers for the three steps mentioned above.

14.2.3.1 Fetching Input Vectors

To load the input vectors for a tile from the main memory, the number of DMA transfers corresponding to an SPE, as a function of d, can be expressed for rows as

$$L_{rt}(d) = \left\lceil \frac{c_i \cdot w_r \cdot d}{B} \right\rceil, \tag{14.2}$$

and similarly for columns as

$$L_{ct}(d) = \left\lceil \frac{c_i \cdot w_c \cdot d}{B} \right\rceil. \tag{14.3}$$

Therefore the total number of row vector transfers for an SPE is

$$L_r(d) = n_r \cdot L_{rt}(d), \tag{14.4}$$

and the number of column vector transfers, following the row-wise snake-like traversal, is:

$$L_c(d) = (n_c \cdot n_r - n_r + 1) \cdot L_{ct}(d). \tag{14.5}$$

14.2.3.2 Shifting Column Vectors

Within a tile, column vectors are shifted between the SPEs. In one shifting operation, w_c column vectors are transferred, and a total of $p_s - 1$ such operations are performed by an SPE, resulting in

$$L_{st}(d) = \left\lceil \frac{c_i \cdot w_c \cdot d}{B} \right\rceil \cdot (p_s - 1) \tag{14.6}$$

number of DMA transfers for a tile. This gives a total of

$$L_s(d) = n_r \cdot n_c \cdot L_{st}(d) \tag{14.7}$$

transfers for column vector shifting.

14.2.3.3 Transferring Output Data

Finally, on completion of the computation of a block of a tile of D, an SPE must perform DMA transfers to store the computed output to the main memory. This requires

$$L_{ot}(d) = \left\lceil \frac{c_o \cdot w_r \cdot w_c}{B} \right\rceil \cdot p_s \tag{14.8}$$

transfers for one tile, resulting in a total of

$$L_o(d) = n_r \cdot n_c \cdot L_{ot}(d) \tag{14.9}$$

output transfers.

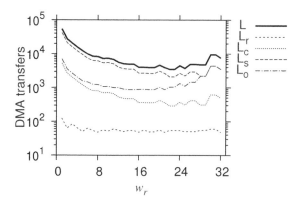

FIGURE 14.3: The total number of DMA transfers as a function of w_r. Y-axis is in log-scale.

14.2.3.4   Minimizing Number of DMA Transfers

The overall total number of DMA transfers to compute the output D, for an SPE, is obtained by summing up Equations 14.4, 14.5, 14.7, and 14.9:

$$L(d) = L_r(d) + L_c(d) + L_s(d) + L_o(d). \qquad (14.10)$$

In Figure 14.3 the total number of DMA transfers $L(d)$ is plotted as a function of parameter w_r, for the case when $\mathcal{M} = 200$ KB, $n = 1,000$, $d = 1,500$, and $p_s = 8$. It can be clearly observed that by properly choosing w_r and w_c, the number of DMA transfers can be reduced by more than one order of magnitude. Thus, optimizing the choice of w_r and w_c significantly contributes to the performance of this tiling strategy.

It should be noted that there is a small difference between the general case and the case when output matrix D is symmetric. Because in the latter case we need to generate only the lower (or upper) triangular part of D, the number of DMA transfers for all components except of row vector transfers will change. Moreover, some of the tiles will fall on the main diagonal, and consequently, either the computational load for the corresponding SPEs will be uneven, or a part of the computations will be done unnecessarily. If the extra work is avoided, the time taken to compute a tile would be the same since some SPEs will be computing while the others will be idle, and to optimize this would require reassignment of the blocks to be computed among the SPEs to guarantee a maximum possible load balance. Alternatively, following the same basic scheduling scheme within a tile described above, the amount of extra computations performed can be minimized by setting $w_r = w_c$ to obtain square blocks. This latter option is shown as the symmetric case in Figure 14.2.

In the following, we discuss extensions to this basic decomposition scheme for two specific cases: (1) when the number of input vectors n is large, and

PPE0	0	1	2	3		
PPE1		0	1	2	3	
PPE2			0	1	2	3
PPE3	3			0	1	2
PPE4	2	3			0	1
PPE5	1	2	3			0

FIGURE 14.4: An example of partitioning of matrix D for six processors (six PPEs). For each partition the iteration number in which it will be processed is marked.

(2) when dimensionality d of the vectors is large. These cases can be dealt with further decomposition of the output matrix D and the input vectors as shown below.

14.2.4 Extending Tiling across Multiple Cell Processors

With large n when the problem either does not fit in the system memory of one Cell processor or is too compute intensive, use of multiple processors to compute the problem in parallel among them is necessitated. In the following we describe an extension to the basic tiling scheme to parallelize the strategy across a cluster of multiple Cell processors. This extension involves additional decomposition of D at the highest level in the hierarchy. Let p denote the number of Cell processors available. Then, the output matrix D is first partitioned into $p \times p$ blocks of sub-matrices $D_{i,j}$ ($0 \le i, j < p$), of size $\frac{n}{p} \times \frac{n}{p}$ each (call these blocks *partitions*). The algorithm to compute D then proceeds in $\lceil \frac{p+1}{2} \rceil$ iterations. In each iteration, a PPE is assigned a partition. Its task is to compute the pairwise information for each position in the assigned partition. To do so, it requires the input vectors of all rows or columns in the assigned partition. When D is symmetric, for the partitions on the main diagonal the same vectors represent both rows and columns. For other partitions, the row vectors and column vectors are distinct. Assume in the following that D is symmetric without loss of generality. Call a set of $\frac{p \cdot (p+1)}{2}$ partitions containing only one of $D_{i,j}$ or $D_{j,i}$ for each pair (i,j) to be the complete set of unique partitions. Then the assignment of the partitions to PPE processors is done as follows: In iteration i, PPE with rank j is assigned the partition $D_{j,(j+i) \mod p}$. This scheme is illustrated in Figure 14.4. It is straightforward to see that this scheme computes all unique partitions.

Note that the same row vectors are needed by a PPE throughout all iterations. A scheme similar to the one for processing blocks in a tile, as presented in Section 14.2.1, is used at this level. To begin with, the n input vectors are assigned to the p PPEs in a partition decomposition. These serve as vectors for both rows and columns during iteration 0. Each processor retains the same set of row vectors in subsequent iterations, while the column vectors are shifted upwards at the end of each iteration, i.e. PPE with rank j sends its column vectors to PPE with rank $(j - 1 + p) \mod p$. This scheme assumes the message-passing model of parallel computing.

The assignment of the partitions to the processors creates the same workload with the following exceptions: In iteration 0, the partitions assigned fall on the main diagonal of D, for which only the lower (or upper) triangular part is needed. As all PPEs deal with diagonal partitions in the same iteration, it simply means that this iteration will take roughly half the compute time as others. The other exception may occur during the last iteration. To see this, consider that the PPEs collectively compute p partitions in each iteration. The total number of unique partitions is $\frac{p \cdot (p+1)}{2}$. The following two cases are possible:

1. p is odd. In this case, the number of iterations is $\lceil \frac{p+1}{2} \rceil = \frac{p+1}{2}$. The total number of partitions computed is $\frac{p \cdot (p+1)}{2}$, which is the same as the total number of unique partitions. Since the algorithm guarantees that all unique partitions are computed, each unique partition is computed only once.

2. p is even. In this case, the number of iterations is $\lceil \frac{p+1}{2} \rceil = \frac{p}{2} + 1$, causing the total number of partitions computed to be $p \cdot \left(\frac{p}{2} + 1 \right)$, which is $\frac{p}{2}$ more than the number of unique partitions. It is easy to argue that this occurs because in the last iteration, half the PPEs are assigned partitions that are transpose counterparts of the partitions assigned to the other half (marked with darker shading in Figure 14.4).

When p is even, the computational cost can be optimized by recognizing this exception during the last iteration, and having each PPE compute only half of the partition assigned to it so that the PPE which has the transpose counterpart computes the other half. This will save half an iteration, significant only if p is small. For large p, one could ignore this cost and run the last iteration similar to others.

14.2.5   Extending Tiling to Large Number of Dimensions

Thus far we have assumed that at least two vectors can fit in the LS of an SPE. In some cases this assumption does not hold due to high dimensionality of the input data, when not even a single input vector may fit in the LS of an SPE, necessitating further decomposition of the input data. In this section we describe an extension of the tiling decomposition to the dimension of the input vectors. This strategy is valid only in certain cases as we will explain later. Assume that at least one element corresponding to a dimension of the input vectors can fit in the LS an SPE. The input vectors are decomposed

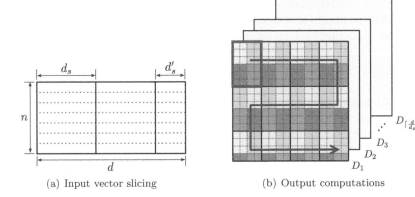

(a) Input vector slicing   (b) Output computations

FIGURE 14.5: Decomposition of input vectors (left) of dimension d into slices, each of dimensions d_s (last slice consists of $d'_s \leq d_s$ dimensions). Computations on each slice give partial results as matrices D_i, $1 \leq i \leq \left\lceil \frac{d}{d_s} \right\rceil$. Aggregation of each D_i may be required to obtain final output matrix D.

into *slices*, each with d_s dimensions ($d_s < d$) so that at least two such vector slices can fit in the LS of an SPE. This gives a total of $\left\lceil \frac{d}{d_s} \right\rceil$ slices of the input. The number of dimensions in the first $\left\lfloor \frac{d}{d_s} \right\rfloor$ slices obtained is d_s, and the dimension of the last slice is $d'_s = \left(d - d_s \cdot \left\lfloor \frac{d}{d_s} \right\rfloor \right)$. This scheme is depicted in Figure 14.5.

Each slice is processed independently following the tiling procedure described previously resulting in partial results for the output matrix D. Observe that the input vector slices from one slice cannot be reused in processing any other slice, obviating any possibility of reducing the number of DMAs when switching from one slice to the next. The total number of DMA transfers during the computations when incorporating slicing is obtained using Equation 14.10 as

$$L'(d_s) = \left\lfloor \frac{d}{d_s} \right\rfloor \cdot L(d_s) + L(d'_s). \tag{14.11}$$

This raises the question of finding d_s.

One of the restrictions the Cell architecture puts on a programmer is that the DMA transfer sizes need to be multiples of 128 bytes, and aligned to the cache line size which is 128 bytes as well. This requirement forms a natural constraint on the choice of d_s, so a set of possible values for d_s is then defined to be

$$\Delta = \left\{ d_x : d_x \in \left[\frac{128}{c_i}, d_{max} \right] \wedge (d_x \cdot c_i)\%128 = 0 \right\}, \tag{14.12}$$

where d_{max} is the maximal size of a vector slice that can fit in the LS of an

SPE, which can be computed from Equation 14.1. Finally, to find a particular value of d_s to obtain a minimum number of DMA transfers, we minimize the function $L'(d_x)$ by exploring the possible values of d_s from Δ:

$$d_s = \underset{d_x \in \Delta}{\mathrm{argmin}}\left(L'(d_x)\right). \qquad (14.13)$$

Once computations on all the slices are completed, reduction operations may be required to accumulate the partial results obtained from each slice (as depicted in Figure 14.5) to compute the final output matrix D.

Such a decomposition with input vector slicing is valid only if the function f to be computed consists of associative operations that can be carried out independently on each of the dimensions of the input vectors, making it decomposable to compute partial results from slices of the input vectors. In some cases, the decomposition of f is trivial, e.g. Minkowski distance (L_p-norm distance) and Spearman's rank correlation [7] computations, while in some others, the decomposition may require complex algorithmic strategies, possibly with need of auxiliary storage, e.g. Pearson correlations and mutual information (MI) computations using B-spline estimator [9] or kernel estimators [21]. In these cases, the pairwise computation for each slice of a pair of vectors is independent of other slices from the same input vectors. In some cases, this may not be true. For instance, the computation of each slice may need to employ the result obtained from previous slices. Later, in Section 14.4, we will demonstrate such a case with the pairwise sequence alignment problem, where the computations are not independent and performing them in parallel involves a complex algorithmic strategy.

The tiling strategy described above with or without the two specific extensions discussed is applicable to any application dealing with all-pairs computations on the Cell processor. We demonstrate this in the next section through an application taken from computational systems biology—reconstruction of gene regulatory networks, employing all-pairs computations.

14.3   Reconstructing Gene Regulatory Networks

Gene regulatory networks are an attempt to develop a systems level model of the complex interactions among various products of genes, such as proteins and RNAs, which coordinate to execute cellular processes in an organism. To construct such models by using observations of gene expressions as a biological process unfolds, one relies on experimental data from high-throughput technologies such as microarrays [1], or short-read sequencing [28], which measure a snapshot of all gene expression levels under a particular condition or in a time series. Construction of gene regulatory networks is an intensely studied problem in systems biology and many methods have been developed for it. In this section we discuss utilization of the tiling approach described earlier to obtain an efficient implementation of an algorithm to reconstruct gene regulatory networks based on pairwise MI computations [32, 33] on the Cell processor.

14.3.1 Computing Pairwise Mutual Information on the Cell

MI is arguably the best measure of correlation between random variables [8]. The method for gene network reconstruction given in [32] employs, in the first stage, all-pairs MI computations to construct a distance matrix. This stage is the dominant compute-intensive part of the algorithm making it an apt candidate to accelerate on the Cell processor using the tiling strategy for pairwise computations.

Formally, MI between random variables X and Y is defined using entropy as:

$$I(X;Y) = H(X) + H(Y) - H(X,Y). \tag{14.14}$$

This definition can be directly used only if the marginal and joint probability distributions of X and Y are known. This is hardly ever the case; therefore, in practice MI is estimated based on random variables observation vectors. Several different MI estimators have been proposed (e.g. [20]); however, the B-spline approach [9] has been shown to be computationally efficient while providing very good accuracy of the estimation. In this approach, to obtain MI directly from Equation 14.14, probability distributions are approximated by classifying observations of random variables into q categories, where each observation can be assigned simultaneously to k categories with different weights. To obtain the weights, B-spline functions of order k are used. For an observation, a B-spline function B_k^q returns a vector of size q with k positive weights that indicate which categories the observation should be assigned to. Given two vectors a and b, each with d observations of a random variable, joint probability of each pair of categories is obtained using the following equation:

$$P = \frac{1}{d} \sum_{i=0}^{d-1} (B_k^q(a_i) \times B_k^q(b_i)). \tag{14.15}$$

The resulting matrix P is further plugged into entropy calculations in Equation 14.14 to obtain the corresponding estimate of MI. Hence, in this case, the function f consists of non-trivial computations, which, nevertheless, can be effectively accelerated on the Cell processor. To optimize the MI computations on the Cell processor, a number of algebraic transformations can be performed, and then the modified kernel can be implemented using SPE intrinsic vector instructions. Because typically k equals 3 or 4, the loop that occurs in the vector product in Equation 14.15 can be unrolled for these specific values of k, while keeping a generic implementation for less frequently used values, giving a further performance improvement on the Cell processor.

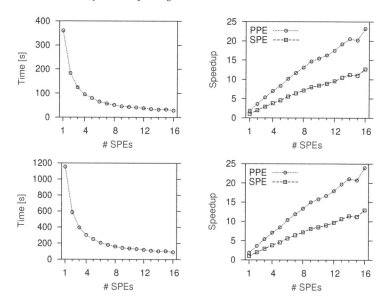

FIGURE 14.6: Execution time (left) and relative speedups (right) with respect to single PPE core, and a single SPE core as a function of number of SPEs for the data set with 911 observations (top) and 2,996 observations (bottom).

14.3.2   Performance of Pairwise MI Computations on One Cell Blade

In this section we present some basic results from experiments to validate the tiling approach on the Cell platform applied to gene regulatory network construction. We present the scaling on a QS20 Cell blade with a varying number of SPEs. The implementation used for this purpose was done using C++ with Cell SDK 3.1. All computations performed for this application were in double precision.

To demonstrate how the method scales with respect to the number of SPE cores on a single QS20 blade, we used two data sets containing 911 and 2,996 microarray experiments with 512 genes each. For all these experiments we used the following default parameters: $b = 10$, $k = 4$ and $q = 10$. The implementation was executed on one PPE core using from one up to 16 SPE cores. These results are summarized in Figure 14.6.

As shown, the method maintains a linear scalability up to 16 SPEs irrespective of the number of dimensions d. Moreover, the implementation is able to obtain about 80% efficiency when all 16 SPEs are used validating the tiling scheme.

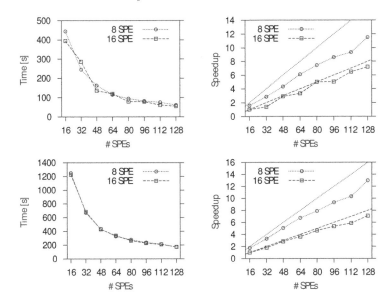

FIGURE 14.7: Execution time (left) and relative speedups (right) to compute matrix D as a function of number of SPE cores used on the Cell cluster, where each PPE uses eight SPEs (Cell 8 SPE), and 16 SPEs (Cell 16 SPE), for the data set with 911 observations (top) and 2,996 observations (bottom).

14.3.3   Performance of MI Computations on Multiple Cell Blades

Here we show how this extension to the tiling scheme scales on a Cell-based cluster with a varying number of blades. The gene regulatory network construction implementation on the Cell processor was extended with message passing interface (MPI) to incorporate this higher level of parallelism across multiple processors. These experiments were conducted on a cluster of eight QS20 dual-Cell blades, connected through a Gbit-Ethernet interconnect.

For this experiment, we used two data sets consisting of 2,048 genes with 911 and 2,996 microarray experiments, respectively. In Figure 14.7 we present the performance results of the executions on the Cell cluster as a function of the number of SPE cores used. We consider two granularities for a single Cell blade—one PPE using all 16 SPEs, and two PPEs using eight SPEs each. We see a linear scalability in these cases also. The granularity where one PPE uses 16 SPEs performs better because in this case the smaller number of MPI processes (eight, one on each blade) result in a lesser amount of communication over the very high latency interconnect of the cluster.

14.4   Pairwise Genomic Alignments

In this section we describe a specialized pairwise computation—pairwise genomic alignments. In the previous sections we presented the tiling scheme to schedule a large number of pairwise computations, and then discussed a slicing strategy for the special case when the input vectors are of high dimensionality and the function f to be computed is decomposable into independent parts for the vector slices. Pairwise genomic alignment computations are not decomposable into independent parts to compute in parallel. Therefore, here we present a specialized parallel algorithm for computing alignments for two genomic sequences on the Cell processor.

Computing alignments between genomic sequences, as a means to elucidate different kinds of relationships between them, is a fundamental tool arising in numerous contexts and applications in computational biology. A number of algorithms for sequence alignments have been developed in the past few decades, the most common being the ones for pairwise global alignments (aligning sequences in their entirety [23]) and local alignments (aligning sequences that each contain a substring which are highly similar [27].) Some applications require more complex types of alignments. One such example is when aligning an mRNA sequence transcribed from a *eukaryotic* gene with the corresponding genomic sequence to infer the gene structure [12]. A gene consists of alternating regions called *exons* and *introns*, while the transcribed mRNA corresponds to a concatenated sequence of the exons. This requires identifying a partition of the mRNA sequence into consecutive substrings (the exons) which align to the same number of ordered, non-overlapping, non-consecutive substrings of the gene, a problem known as *spliced alignment*. Another important problem is that of *syntenic alignment* [18], for aligning two sequences that contain conserved substrings that occur in the same order (such as genes with conserved exons from different organisms, or long syntenic regions between genomes of two organisms).

14.4.1   Computing Alignments

Dynamic programming is the most commonly used method for computing pairwise alignments [12,18,23,27], and takes time proportional to the product of the lengths of the two input sequences[1]. Various parallel algorithms have also been developed for these methods, and some of these deal with parallelization across multiple processing units [4,10,11]. In the rest of the chapter we present algorithms for computing pairwise alignments in parallel on the

---

[1]The original Smith-Waterman algorithm [27] has cubic complexity, but it is widely known that this can be implemented in quadratic time, as is shown in [13]. Also, [12] presents an algorithm for spliced alignment with cubic complexity, but this problem can be treated as a special case of syntenic alignment and solved in quadratic time [26].

Cell processor. We first describe the basic algorithms using dynamic programming for three kinds of alignments—global/local, spliced, and syntenic. We then present a parallel algorithm for computing the alignment scores along with actual alignments in parallel across the SPEs with linear memory usage.

14.4.1.1 Global/Local Alignment

In the following we first provide a brief description of the sequential dynamic programming algorithm for computing global alignments. The same scheme can be adapted to apply to the specialized spliced and syntenic alignments, which we will describe later. Consider two sequences $X = x_1 x_2 \ldots x_{d_1}$ and $Y = y_1 y_2 \ldots y_{d_2}$ over an alphabet Σ, and let '$-$' denote the gap character. Note that we are considering two sequences which can be of different lengths, d_1 and d_2, as opposed to the vectors of same lengths considered in the previous parts of the chapter. A global alignment of the two sequences is a $2 \times \mathcal{N}$ matrix, where $\mathcal{N} \geq max(d_1, d_2)$, such that each row represents one of the sequences with gaps inserted in certain positions and no column contains gaps in both sequences. The alignment is scored as follows: a score function, $s : \Sigma \times \Sigma \to \mathbb{R}$, prescribes the score for any column in the alignment that does not contain a gap. We assume the function s returns the score for any pair of characters from Σ in constant time. *Affine gap penalty* functions are commonly used to determine the scores of columns involving gaps, so that a sequence of gaps is assigned less penalty than treating them as individual gaps—this is because biologically a mutation affecting a short segment of a genomic sequence is more likely than several individual base mutations. Such a function is defined as follows: for a maximal consecutive sequence of x gaps, a penalty of $h + g \cdot x$ is applied. Thus, the first gap in a maximal sequence is charged $h + g$, while the rest of the gaps are charged g each. When $h = 0$, the scoring function is called a *linear gap penalty* function, and when $g = 0$, it is called a *constant gap penalty* function. The score of the alignment is the sum of scores over all the columns. We will assume the use of affine gap penalty function in the following.

The global alignment problem with affine gap penalty function can be solved using three $(d_1 + 1) \times (d_2 + 1)$ sized dynamic programming tables, denoted $\mathcal{C}$, $\mathcal{D}$ (for deletion), and $\mathcal{I}$ (for insertion). An element $[i, j]$ in a table stores the optimal score of an alignment between $x_1 x_2 \ldots x_i$ and $y_1 y_2 \ldots y_j$ with the following restrictions on the last column of the alignment: x_i is matched with y_j in $\mathcal{C}$, a gap is matched with y_j in $\mathcal{D}$, and x_i is matched with a gap in $\mathcal{I}$. The tables can be computed using the following recursive equations, which can be applied row by row, column by column, or anti-diagonal by anti-diagonal (also called minor diagonal):

$$\mathcal{C}[i, j] = s(x_i, y_j) + \max \begin{cases} \mathcal{C}[i-1, j-1] \\ \mathcal{D}[i-1, j-1] \\ \mathcal{I}[i-1, j-1] \end{cases} \qquad (14.16)$$

$$\mathcal{D}[i,j] = \max \begin{cases} \mathcal{C}[i,j-1] - (h+g) \\ \mathcal{D}[i,j-1] - g \\ \mathcal{I}[i,j-1] - (h+g) \end{cases} \qquad (14.17)$$

$$\mathcal{I}[i,j] = \max \begin{cases} \mathcal{C}[i-1,j] - (h+g) \\ \mathcal{D}[i-1,j] - (h+g) \\ \mathcal{I}[i-1,j] - g. \end{cases} \qquad (14.18)$$

The first row and column of each table are initialized to $-\infty$ except in the cases: $\mathcal{C}[0,0] = 0$; $\mathcal{D}[0,j] = h + g \cdot j$; and $\mathcal{I}[i,0] = h + g \cdot i$ ($1 \leq i \leq d_1$; $1 \leq j \leq d_2$). The maximum of the scores among $\mathcal{C}[d_1,d_2]$, $\mathcal{D}[d_1,d_2]$, and $\mathcal{I}[d_1,d_2]$ gives the optimal global alignment score. By keeping track of a pointer from each entry to one of the entries that resulted in the maximum while computing its score, an optimal alignment can be constructed by retracing the pointers from the optimal score at bottom right to the top left corner of the tables. This procedure is known as *traceback*.

14.4.1.2   Spliced Alignment

During the synthesis of a protein, mRNA is formed by transcription from the corresponding gene, followed by removal of the introns and splicing together of the exons. In order to identify genes on a genomic sequence, or to infer the gene structure, one can align the processed products, such as mRNA, EST and cDNA, to the original DNA sequence. To solve this spliced alignment problem, a solution similar to the one for global alignments is described here. While Gelfand *et al.*'s [12] algorithm has a $O(d_1^2 \cdot d_2 + d_1 \cdot d_2^2)$ run-time complexity, an $O(d_1 \cdot d_2)$ algorithm can be easily derived as a special case of Huang and Chao's $O(d_1 \cdot d_2)$ time syntenic alignment algorithm [18] by disallowing unaligned regions in one of the sequences. This algorithm uses the three tables, as for global alignment, along with a fourth table $\mathcal{H}$ which represents those regions of the DNA sequence which are excluded from the aligned regions, i.e. they correspond to introns or other unaligned regions. A large penalty ε is used in table $\mathcal{H}$ to prevent short spurious substrings in the larger sequence from aligning with the other sequence. Intuitively, a sequence of contiguous gaps with a penalty greater than the threshold ε is replaced by a path in the table $\mathcal{H}$, representing this region to be unaligned. The recursive equations to compute the four tables are as follows:

$$C[i,j] = s(x_i, y_j) + \max \begin{cases} C[i-1, j-1] \\ D[i-1, j-1] \\ I[i-1, j-1] \\ H[i-1, j-1] \end{cases} \quad (14.19)$$

$$D[i,j] = \max \begin{cases} C[i, j-1] - (h+g) \\ D[i, j-1] - g \\ I[i, j-1] - (h+g) \\ H[i, j-1] - (h+g) \end{cases} \quad (14.20)$$

$$I[i,j] = \max \begin{cases} C[i-1, j] - (h+g) \\ D[i-1, j] - (h+g) \\ I[i-1, j] - g \\ H[i-1, j] - (h+g) \end{cases} \quad (14.21)$$

$$H[i,j] = \max \begin{cases} C[i-1, j] - \varepsilon \\ D[i-1, j] - \varepsilon \\ H[i-1, j]. \end{cases} \quad (14.22)$$

14.4.1.3 Syntenic Alignment

Syntenic alignment, used to compare sequences with intermittent similarities, is a generalization of spliced alignment allowing unaligned regions in both the sequences. This is used to discover an ordered list of similar regions separated by dissimilar regions which do not form part of the final alignments. This technique is applicable to comparison of two genes with conserved exons, such as counterpart genes from different organisms. A dynamic programming algorithm for this has been developed by Huang and Chao [18]. Similar to spliced alignment, a large penalty ε is used to prevent alignment of short substrings. This dynamic programming algorithm also has four tables, but with an extension in the table H that both sequences can have substrings excluded from aligning. Table definitions for C, D, and I remain the same as Equations 14.19–14.21 for spliced alignment. The definition of table H is modified as follows:

$$H[i,j] = \max \begin{cases} C[i-1, j] - \varepsilon \\ D[i-1, j] - \varepsilon \\ C[i, j-1] - \varepsilon \\ I[i, j-1] - \varepsilon \\ H[i-1, j] \\ H[i, j-1]. \end{cases} \quad (14.23)$$

As seen from this equation, the table $\mathcal{H}$ can derive scores either from an entry in previous row, or a previous column. This directionality information is important to retrieve the alignment and needs to be stored explicitly. Another way to view this extra information is to split the table $\mathcal{H}$ into two, $\mathcal{H}_h$ and $\mathcal{H}_v$, where they have the restrictions of alignment paths going only horizontally or only vertically, respectively.

14.4.2   A Parallel Alignment Algorithm for the Cell BE

In the literature many attempts have been made to implement sequence alignments on the Cell processor [2, 19, 24, 26, 29, 30]. Most of these methods for sequence alignments on the Cell are restricted to computing the alignment score using the basic Smith-Waterman algorithm [27] for local alignments. Although computation of alignment scores is useful in statistical analyses to assess the alignment quality, or to find a small subset of sequences which have a high similarity with the query sequence, these implementations do not compute the actual alignment, which is necessary to gain biological insight into the genomic sequences begin aligned. Moreover, they work for smaller sequence sizes since only a small memory is available on each SPE to store the whole sequences and the dynamic programming tables to be computed. Therefore, here we focus on parallel algorithms to compute a single optimal pairwise alignment on the Cell processor. To perform the alignments in parallel, the input sequences and the dynamic programming table computations need to be decomposed among the various SPEs, which also need to synchronize the computations among themselves.

In the following we describe these decomposition and communication strategies, and focus on a parallel approach incorporating a linear space strategy to increase the size of problems that can be solved using the collective SPE memory and also infer an optimal alignment. Hirschberg [17] presented a divide-and-conquer algorithm to obtain an optimal alignment while using linear space, and this strategy is incorporated in the parallel algorithm discussed below. This scheme should be sufficient for most global/local/spliced alignment problems as the sequences are unlikely to exceed several thousand bases. This alignment method for the Cell processor is based on the parallel algorithm by Aluru *et al.* [4], which we review below.

14.4.2.1   Parallel Alignment using Prefix Computations

The parallel algorithm for computing global alignments in linear space given in [4] consists of two phases: (1) *problem decomposition phase*, and (2) *subproblem alignment phase*. In the first phase, the alignment problem is divided into p_s non-overlapping subproblems, where p_s is the number of processing elements. Once the problem is decomposed, each processing element performs a linear space alignment algorithm, computing an optimal alignment for the corresponding subproblem. The result from each processing element is then simply concatenated to obtain an optimal alignment of the actual problem. We describe the problem decomposition phase below.

Initially, the sequence X is provided to all the processors and Y is equally divided among them—each processor receives a distinct block of $\frac{d_2}{p_s}$ consecutive columns to compute. Define p_s *special columns*, where C_l is the column number $(l+1) \times \frac{d_2}{p_s}$ $(0 \le l \le p_s - 1)$, of a table to be the last columns of the blocks allocated to each processing element, except for the last one. The intersections of an optimal alignment path with these special columns define the segment of the first sequence to be used within a particular processing element independently of other blocks, thereby splitting the problem into p_s subproblems.

To compute the intersections of an optimal path with the special columns, the information on the special columns is explicitly stored. In addition to the score values, for each entry of a table a *pointer* is also computed. This represents the table and row number of the entry in the closest special column to the left that lies on an optimal path from $C[0,0]$ to the entry. The pointer information is also explicitly stored for the special columns. Conceptually, these pointers give the ability to perform a traceback through special columns, without considering other columns. The entries of the dynamic programming tables are computed row by row using the parallel prefix operation as described in [4], in linear space—storing only the last computed row, and the special columns, thereby using $O(d_1 + \frac{d_2}{p_s})$ space.

On completion, a traceback procedure using the stored pointers along the special columns is used to split the problem into p_s subproblems in $O(p_s)$ time. The problem decomposition phase is visualized in Figure 14.8. Once the problem is divided among the processors, in the second phase each processing element performs an alignment on its corresponding segments of sequences while adopting Hirschberg's technique [17,22] to use linear space. We describe this later in context of adapting to the Cell processor.

14.4.2.2 Wavefont Communication Scheme

To obtain an efficient implementation on the Cell processor, we describe a hybrid parallel algorithm that is a combination of the special columns based parallel alignment algorithm described above and Edmiston's [10] wavefront communication pattern which we describe in the following. The parallel communication strategy, commonly known as the *wavefront communication scheme*, was first proposed in the context of sequence alignment by Edmiston *et al.* It has since been employed by many parallel algorithms including sequence alignment on the Cell [2, 26, 30].

Let us again denote the number of available SPEs by p_s. In this method, each table is divided into a $\beta \times p_s$ matrix of blocks where β is the number of blocks in one column. Therefore, each block contains at most $\lceil \frac{d_2}{p_s} \rceil$ columns and r rows, where r is a pre-chosen block size $(\beta = \frac{d_2}{r})$. Let $\mathcal{B}_{i,j}$ denote a block, where $0 \le i < \beta$ and $0 \le j < p_s$. Each SPE is assigned a unique

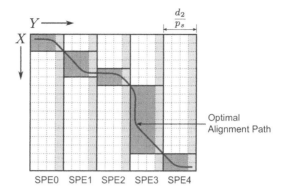

FIGURE 14.8: An example of the block division in parallel-prefix based special columns technique—The second sequence is divided into vertical blocks, which are assigned to different processors. Special columns constitute the shaded rightmost column of each vertical block. The intersections of an optimal alignment path with the special columns are used for problem division. The shaded rectangles around the optimal alignment path represent the subdivisions of the problem for each processor.

column of blocks to compute: SPE j computes the blocks $\mathcal{B}_{i,j}$. The blocks are simultaneously computed one anti-diagonal at a time. All blocks on an anti-diagonal can be computed simultaneously as they depend only on blocks on the previous two anti-diagonals—computation of a block $\mathcal{B}_{i,j}$ on the anti-diagonal t, where $t = i + j$, only depends on blocks $\mathcal{B}_{i-1,j}$ and $\mathcal{B}_{i,j-1}$ from the anti-diagonal $t-1$, and block $\mathcal{B}_{i-1,j-1}$ from the anti-diagonal $t-2$. Because of the block assignment to the SPEs, each of them only needs to receive the last column of a block (plus an additional element) from the SPE. An illustration of this wavefront pattern is shown in Figure 14.9. Each SPE receives all of the first sequence (length d_1) and a distinct $\frac{d_2}{p_s}$ length substring of the second sequence. The total number of rounds of block computations in this scheme is equal to the number of block anti-diagonals, which is equal to $p_s + \beta - 1$, although each SPE computes exactly β blocks. The SPEs need to synchronize with each other after transferring their corresponding rightmost column to the next SPE. For an efficient implementation, this can be achieved using the signal notification registers on the SPEs. The dynamic programming tables can thus be computed on each of the SPEs.

14.4.2.3   A Hybrid Parallel Algorithm

The hybrid parallel alignment algorithm on the Cell processor presented here is a combination of the special columns based parallel alignment algorithm with Edmiston's wavefront communication pattern described above. In the wavefront alignment scheme, each processing element works on a block of the tables independently, communicating the last column to the next pro-

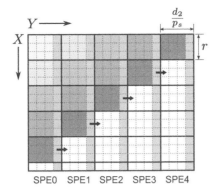

FIGURE 14.9: An example of the block division in the wavefront technique—SPE j is assigned a column of blocks $\mathcal{B}_{i,j}$, $0 \leq i < \beta$, as shown by the labels below each column. Block computations follow diagonal wavefront pattern, where for anti-diagonal t, blocks $\mathcal{B}_{i,j}$ such that $i+j = t$ are computed simultaneously in parallel (shown by blocks in the same shade of gray). SPE j $(0 \leq j < p_s)$ sends the rightmost computed column in its assigned block to SPE $j + 1$ for the next iteration.

cessing element when done and then starts computation on its next block; the parallel prefix approach requires the processing elements to communicate a single element when computing each row. If implemented as such on the Cell processor, these short but frequent communications for each row increase channel stalls in the SPEs, which are reduced to one bulk communication per block of size r in the wavefront scheme. Each communication leads to a synchronization event among the SPEs. To make the most use of parallelism on the Cell processor, such events should be minimized. Moreover, the block size can be optimized for DMA transfer in the wavefront communication scheme, which makes it a better choice for the Cell. Furthermore, adopting the space-saving method is particularly important for the Cell processor due to the small LS on each SPE.

The parallel decomposition phase of the special columns based algorithm [4] is modified to incorporate the wavefront communication scheme and store only the special column for each SPE block. This also enables use of double buffering in moving input column sequence, and overlapping of DMA transfers with block computations. Each SPE transfers portions of the second sequence allotted to it by the PPE from the main memory to its LS. For each computation block, it transfers blocks of the first sequence and performs the table computations in linear space, while storing all of the last column. Once done, it transfers the recently computed block of last column data to the next SPE and continues computation on the next block. Once the special columns containing the pointers to the previous special columns are computed, the segments of the first sequence are found which are to be aligned to the segments

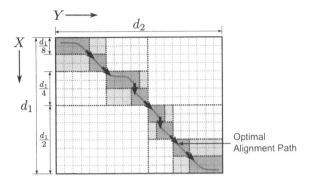

FIGURE 14.10: An example of the recursive space saving scheme—In Hirschberg's technique, the problem is recursively divided into subproblems around an optimal alignment path, while using linear space. The small arrows show the direction in which an optimal alignment path crosses the divisions.

of the second sequence on the corresponding SPEs, thereby decomposing the problem into p_s independent subproblems. This is followed by the sequential alignment phase for each subproblem executing in parallel among the SPEs. This is based on Hirschberg's space-saving technique which we describe next.

14.4.2.4   Hirschberg's Technique for Linear Space

Once the alignment problem is decomposed into subproblems among all the SPEs, each SPE simultaneously computes optimal alignments for its local subproblem making use of Hirschberg's space-saving technique [17,22], which reduces the space usage from $O(d_1 \cdot d_2)$ to $O(d_1 + d_2)$ while enabling retrieval of an optimal alignment. This method is a divide-and-conquer technique where the problem is recursively divided into subproblems, the results of which are combined to obtain an optimal alignment of the original problem [3]. In this scheme, one of the input sequences is divided into two halves, and tables are computed for each half aligned with the other input sequence. This is done in the normal top down and left-to-right fashion for the upper half and in a reverse bottom up and right-to-left manner (aligning the reverses of the input sequences) for the lower half. For these computations, it is sufficient to store a linear array for the last computed row. Once the middle two rows are obtained from the corresponding two halves, they are combined to obtain the optimal alignment score, dividing the second sequence at the appropriate place where the optimal alignment path crosses these middle rows. Care needs to be taken to handle the gap continuations across the division, and the possibility of multiple optimal alignment paths. The problem is subsequently divided into two subproblems, and this is repeated recursively for each subproblem. An illustration of the recursion using this scheme is given in Figure 14.10.

On completion, each SPE contains an optimal alignment of its subproblem, and writes it to the main memory through DMA transfers. A concatenation of these alignments gives an overall optimal alignment.

14.4.2.5 Algorithms for Specialized Alignments

For a parallel algorithm for spliced and syntenic alignments on the Cell processor, the same techniques as described above can be followed. In these cases, four tables need to be computed, and the same wavefront scheme is followed for the computations in the problem decomposition phase. A parallel algorithm for solving the syntenic alignment problem is described in [11], which is similar to the parallel-prefix based alignment algorithm described earlier, and hence can be adapted to the Cell using the same hybrid algorithm given above.

14.4.2.6 Memory Usage

The LS space usage for table computations (apart from space needed for input sequences and output alignment) on a single SPE during the problem decomposition phase is $\left(d_1 + \frac{d_2}{p_s} \right) \cdot \lambda \cdot c_o$ bytes, where λ is the number of dynamic programming tables used (three in the case of global/local alignment and four for spliced and syntenic alignments), and c_o is the number of bytes needed to represent a single element of a single table (alignment score and pointer). The computation space usage during second phase is lower: $\left(\frac{d_2}{p_s} \cdot \lambda \cdot c_o' \right)$ bytes, where c_o' is number of bytes required to store a single table entry in this phase (alignment score).

14.4.3 Performance of the Hybrid Alignment Algorithms

Here we present some basic performance results for the hybrid parallel algorithm for genomic alignments on the Cell. More detailed results can be found in [25, 26]. The implementation used for these results was developed on the IBM Cell SDK 3.0, and run on a QS20 Cell Blade. For these tests, the block size r was chosen to be 128 to optimize the DMA transfers.

The run-times and the corresponding speedups of global alignment for varying number of SPEs are shown in Figure 14.11. Synthetic input sequences of lengths 2,048×2,048 were used for this purpose. The speedups shown are obtained by comparing the parallel Cell implementation with the parallel implementation running on a single SPE on the Cell processor, and an optimal sequential implementation of Hirschberg's space-saving technique based global alignment algorithm for a single SPE on the Cell processor (to completely eliminate the parallel decomposition phase). On one SPE, the parallel implementation obviously performs worse than the serial implementation, since it includes the additional problem decomposition phase which computes the whole table to merely return the entire problem as the subproblem to solve

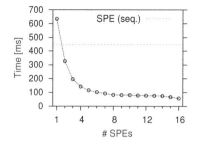 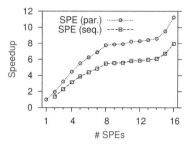

FIGURE 14.11: The execution times (left) and the corresponding speedups (right) of global alignment implementation for an input of size 2,048×2,048. Speedups are obtained against the parallel implementation running on single SPE, and a sequential implementation on one SPE. There is a small gain in the range of eight to 12 SPEs. This is because when using more than eight SPEs, communication goes off-chip on the blade. This with the low computation time creates an off-chip communication to computation tradeoff.

sequentially. This is used to study the scaling of the algorithm, and we obtain a relative speedup of 11.25 and absolute speedup of almost 8 on 16 SPEs.

It can be seen in the run-time/speedup graph (Figure 14.11) that the run-times only show a marginal improvement as the number of SPEs is increased from eight to 12, as opposed to the near linear scaling exhibited below eight and beyond 12. The latency for data transfer from one Cell processor to the other Cell processor on the blade (*off-chip* communication) is much higher than any data transfer between components on a single processor (*on-chip* communication), and these communication times are significant in this case compared to the computational running time of the implementation. On using more than eight SPEs, both the processors on the Cell blade are used and data need to be transferred from one processor to the other. Due to the higher off-chip communication latency, the run-time using nine SPEs is similar (or even worse in case of other alignment problems discussed in later sections) to the run-time using eight SPEs. A trade-off is created with the off-chip communication time and computation time on the two processors. When amount of computation exceeds the communication time, the run-time further starts to decrease, thereby increasing the speedups as seen in Figure 14.11 for more than 12 SPEs.

We present the execution times and speedups of spliced and syntenic alignment implementation on real biological data in Figure 14.12. For spliced alignment, the phytoene synthase gene from *Lycopersicum* (tomato) is aligned with the mRNA corresponding to this gene's transcription, and for syntenic alignment, we perform alignment of a copy of the phytoene synthase gene from *Lycopersicum* (tomato) and *Zea mays* (maize). Again we see similar results to those obtained for global alignment.

To assess absolute performance of the Cell implementation of the alignment

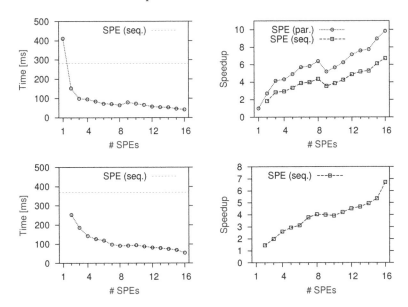

FIGURE 14.12: The execution times (left) and speedups (right) of spliced alignment (top) for phytoene synthase gene from *Lycopersicum* with its mRNA sequence (1,792×872), and syntenic alignment (bottom) for the phytoene synthase gene from *Lycopersicum* (tomato) and *Zea mays* (maize) (1,792×1,580) on a varying number of SPEs. For spliced alignment, both absolute and relative speedups are shown, while only absolute speedup is shown for syntenic alignment.

algorithms, the metric of number of cells in the dynamic programming tables updated per second (CUPS) is used and we show the results for the three alignment implementations in Figure 14.13, with a varying number of SPEs.

14.5 Ending Notes

In this chapter we described efficient and scalable strategies to orchestrate all-pairs computations on the Cell architecture, based on decomposition of the computations and the input entities. We considered the general case to schedule the computations on a Cell processor and extended the strategies to incorporate the cases when the number of input entities is large, and when the sizes of individual entities is too large to fit within the memory limitations of the SPEs. We demonstrated its application in context of an application to reconstruct gene regulatory networks. Further we presented specialized paral-

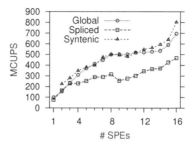

FIGURE 14.13: Cell updates per second for the global, spliced and syntenic alignments on input sizes of 2,048×2,048, 1,792×872, and 1,792×1,580, respectively, are shown for a varying number of SPEs and are given in MCUPS (10^6 CUPS). CUPS for one SPE are shown for the parallel implementation running on a single SPE.

lel algorithms for computing global/local, spliced, and syntenic alignments of two genomic sequences on the Cell processor.

The performance results of the algorithms presented show that the Cell processor is a good platform to accelerate various kinds of applications dealing with pairwise computations. The all-pairs computations strategies can be applied to many applications from a wide range of areas which require such computations to be performed. Depending on the type of the application, the various algorithms presented in this chapter can be adapted to suit the needs and develop an efficient implementation on the Cell platform.

Acknowledgment

The Cell cluster located at the STI Center in the Georgia Institute of Technology, Atlanta, USA, was used for all the experiments presented in this chapter.

Bibliography

[1] The chipping forecast II. Special Supplement. *Nature Genetics*, 2002.

[2] A. M. Aji, W. Feng, F. Blagojevic, et al. Cell-SWat: modeling and scheduling wavefront computations on the Cell Broadband Engine. In

Proceedings of the Conference on Computing Frontiers (CF '08), pages 13–22, 2008.

[3] S. Aluru. *Handbook of Computational Molecular Biology (Chapman & Hall/CRC Computer and Information Science Series)*. Chapman & Hall/CRC, 2005.

[4] S. Aluru, N. Futamura, and K. Mehrotra. Parallel biological sequence comparison using prefix computations. *Journal of Parallel and Distributed Computing*, 63:264–272, 2003.

[5] N. Arora, A. Shringarpure, and R. Vuduc. Direct N-body kernels for multicore platforms. In *Proceedings of the International Conference on Parallel Processing (ICPP '09)*, pages 379–387, 2009.

[6] P. Berkhin. *Grouping Multidimensional Data*, Chapter: A survey of clustering data mining techniques, pages 25–71. Springer, 2006.

[7] J. C. Caruso and N. Cliff. Empirical size, coverage, and power of confidence intervals for Spearman's rho. *Educational and Psychological Measurement*, 57:637–654, 1997.

[8] T. M. Cover and J. A. Thomas. *Elements of Information Theory*. Wiley, 2nd edition, 2006.

[9] C. O. Daub, R. Steuer, J. Selbig, et al. Estimating mutual information using B-spline functions—an improved similarity measure for analysing gene expression data. *BMC Bioinformatics*, 5:118, 2004.

[10] E. W. Edmiston, N. G. Core, J. H. Saltz, et al. Parallel processing of biological sequence comparison algorithms. *International Journal of Parallel Programming*, 17(3):259–275, 1988.

[11] N. Futamura, S. Aluru, and X. Huang. Parallel syntenic alignments. *Parallel Processing Letters*, 63(3):264–272, 2003.

[12] M. S. Gelfand, A. A. Mironov, and P. A. Pevzner. Gene recognition via spliced sequence alignment. *Proc. National Academy of Sciences USA*, 93(17):9061–9066, 1996.

[13] O. Gotoh. An improved algorithm for matching biological sequences. *Journal of Molecular Biology*, 162(3):705–708, 1982.

[14] P. Hanrahan, D. Salzman, and L. Aupperle. A rapid hierarchical radiosity algorithm. In *Proceedings of the Conference on Computer Graphics and Interactive Techniques (SIGGRAPH '91)*, pages 197–206, 1991.

[15] B. Hariharan, S. Aluru, and B. Shanker. A scalable parallel fast multipole method for analysis of scattering from perfect electrically conducting surfaces. In *Proceedings of the ACM/IEEE Conference on Supercomputing (SC '02)*, pages 1–17, 2002.

[16] B. Hendrickson and S. Plimpton. Parallel many-body simulations without all-to-all communication. *Journal of Parallel and Distributed Computing*, 27:15–25, 1995.

[17] D. S. Hirschberg. A linear space algorithm for computing maximal common subsequences. *Communications of the ACM*, 18(6):341–343, 1975.

[18] X. Huang and K. Chao. A generalized global alignment algorithm. *Bioinformatics*, 19:228–233, 2003.

[19] S. Isaza, F. Sánchez, G. Gaydadjiev, et al. Preliminary analysis of the Cell BE processor limitations for sequence alignment applications. In *Proceedings of the International Workshop on Embedded Computer Systems (SAMOS '08)*, pages 53–64, 2008.

[20] S. Khan, S. Bandyopadhyay, A. Ganguly, S. Saigal, et al. Relative performance of mutual information estimation methods for quantifying the dependence among short and noisy data. *Physical Review E*, 76(2 Pt. 2):026209, 2007.

[21] Y. Moon, B. Rajagopalan, and U. Lall. Estimation of mutual information using kernel density estimators. *Physical Review E*, 52(3):2318–2321, 1995.

[22] E. W. Myers and W. Miller. Optimal alignments in linear space. *Computer Applications in Biosciences*, 4(1):11–17, 1988.

[23] S. B. Needleman and C. D. Wunsch. A general method applicable to the search for similarities in amino acid sequence of two proteins. *Journal of Molecular Biology*, 48:443–453, 1970.

[24] V. Sachdeva, M. Kistler, E. Speight, and T. K. Tzeng. Exploring the viability of the Cell Broadband Engine for bioinformatics applications. *Parallel Computing*, 34(11):616–626, 2008.

[25] A. Sarje and S. Aluru. Parallel biological sequence alignments on the Cell Broadband Engine. In *Proceedings of the International Symposium on Parallel and Distributed Processing (IPDPS '08)*, pages 1–11, 2008.

[26] A. Sarje and S. Aluru. Parallel genomic alignments on the Cell Broadband Engine. *IEEE Transactions on Parallel and Distributed Systems*, 20(11):1600–1610, 2009.

[27] T. Smith and M. Waterman. Identification of common molecular subsequences. *Journal of Molecular Biology*, 147:195–197, 1981.

[28] T. T. Torres, M. Metta, B. Ottenwalder, et al. Gene expression profiling by massively parallel sequencing. *Genome Research*, 18(1):172–177, 2008.

[29] H. Vandierendonck, S. Rul, M. Questier, and K. Bosschere. Experiences with parallelizing a bio-informatics program on the Cell BE. In *Proceedings of the High Performance Embedded Architectures and Compilers (HiPEAC'08)*, volume 4917, pages 161–175, 2008.

[30] A. Wirawan, K. C. Keong, and B. Schmidt. Parallel DNA sequence alignment on the Cell Broadband Engine. In *Workshop on Parallel Computational Biology*, volume 4967 of *LNCS*, pages 1249–1256. Springer, 2008.

[31] A. Wirawan, B. Schmidt, and C. K. Kwoh. Pairwise distance matrix computation for multiple sequence alignment on the Cell Broadband Engine. In *Proceedings of the International Conference on Computational Science (ICCS '09)*, pages 954–963, 2009.

[32] J. Zola, M. Aluru, and S. Aluru. Parallel information theory based construction of gene regulatory networks. In *Proceedings of the International Conference on High Performance Computing (HiPC '08)*, volume 5375 of *LNCS*, pages 336–349, 2008.

[33] J. Zola, A. Sarje, and S. Aluru. Constructing gene regulatory networks on clusters of Cell processors. In *Proceedings of the International Conference on Parallel Processing (ICPP '09)*, pages 108–115, 2009.

Part VIII

Molecular Modeling

Chapter 15

Drug Design on the Cell BE

Cecilia González-Álvarez

Barcelona Supercomputing Center-CNS

Harald Servat

Barcelona Supercomputing Center-CNS
Universitat Politècnica de Catalunya

Daniel Cabrera-Benítez

Barcelona Supercomputing Center-CNS

Xavier Aguilar

PDC Center for High Performance Computing

Carles Pons

Barcelona Supercomputing Center-CNS
Spanish National Institute of Bioinformatics

Juan Fernández-Recio

Barcelona Supercomputing Center-CNS

Daniel Jiménez-González

Barcelona Supercomputing Center-CNS
Universitat Politècnica de Catalunya

15.1   Introduction

Historically, discovery of new therapeutic drugs has relied on trial- and error-approaches, natural products, or even serendipitous findings (the discovery of penicillin being one the most famous examples). The field rapidly expanded in the past century with the development of modern chemistry and the industrialization of the drug discovery process. However, the process of developing a new drug is costly, very long, and most often unsuccessful. Indeed, the pace of discovery of new therapeutic drugs has been consistently decreasing for the last years, and most of the current drugs are acting on a small number of biological targets [45]. This is particularly contradictory in a time in which technical advances have helped characterize complete genomes, including human ones, and have made available a humongous amount of information potentially useful for drug discovery. In this context, existing computational methods for target identification and characterization, ligand binding prediction (docking), and virtual screening aim to expand the current set of available targets and boost the discovery of new active compounds. Nevertheless, the application of these methods usually comes at a high computational cost. For instance, average computational times for virtual screening with state-of-the-art DOCK have been reported to be around one ligand per minute on a Xeon 3.0 GHz processor [47]. Therefore, a standard small-compound library, like the ZINC database [32] with over 13 million purchasable compounds (as of October 2009), would need a minimum of 90 days on 100 Xeon 3.0 GHz processors to be fully screened. With more computational power, provided by parallel architectures like cell broadband engine (Cell BE), more conformations per compound could be considered, thus expanding the conformational space of the chemical compounds. Furthermore, the size of the libraries could be enlarged, aiming to better cover the chemical space.

For those reasons, in previous years several biomedical applications have been ported and adapted to Cell BE. For instance, in [29] an implementation for the Cell BE of a fully atomistic molecular dynamics program, a simulation methodology widely used for drug design, was presented. This implementation was based on the distribution of the calculations of the Lennard-Jones, electrostatic forces and potentials among the synergistic processing elements (SPEs). The performance obtained was 20 times faster than in an Opteron PC. Other molecular dynamics codes like GROMACS have been ported to Cell BE [44]. In that work, the authors showed how they adapted the water interaction kernel 112 of GROMACS to the Cell BE using vectorization, parallelization across SPEs, the quick access to the LS of an SPE- and high-speed element interconnect bus (EIB) rings of the Cell BE. Although they achieved

11x speedup inside the kernel code, the application as a whole, consisting of several kernels and several thousand lines of code, did not show a large speedup change. SimBioSys [4] has ported eHits, a flexible ligand docking software they developed, to the Cell BE architecture, achieving 36x speedup on the application compared to an implementation on a conventional processor (not mentioned) with the same frequency.

Other different architectures as well as the Cell BE have been used in drug design. In [46], the authors adapted EUDOC, a molecular docking program, to the IBM Blue Gene/L. The code for EUDOC was optimized by increasing the use of single instruction multiple data (SIMD) instructions and was parallelized with message passing interface (MPI). They achieved 34x speedup compared to a computing cluster using 396 Xeon 2.2 GHz processors.

The work done in [49] described a molecular docking code, PIPER, ported to a graphic processing unit (GPU). The Fast Fourier Transform (hereafter FFT), correlation, scoring and filtering parts of the code were adapted to run inside the GPU, achieving an end-to-end speedup of at least 17.7x compared to a single core and 6.1x compared to four cores of a 2 Ghz quad-core Intel Xeon. Finally, in [36] a parallel implementation of AutoDock using MPI was presented. It achieved a speedup around 89x versus its sequential version using 96 processors of a 256-processor IBM Power6-based cluster.

The objective of this chapter is to overview the main drug design applications and evaluate the portability of those applications to the Cell BE architecture. In order to achieve this goal, a porting analysis of six applications that covers the different approaches to the drug design problem is shown, and two practical case studies are analyzed in detail.

The rest of the chapter is organized as follows: Section 15.2 an overview of the bioinformatics and drug design fields. The Cell BE porting analysis is shown in Section 15.3. The experimental setup is described in Section 15.4. Then, FTDock and Moldy case studies are presented in Sections 15.5 and 15.6, respectively. Finally, the conclusions are covered in Section 15.7.

15.2 Bioinformatics and Drug Design

The application of computers to biological sciences (a multi-disciplinary field generally called "Bioinformatics") has always kept as one of its ultimate goals the design of new therapeutic drugs based on the analysis of the huge amount of sequence, structural, and functional data that are currently available. The drug design field has benefited from the development of new algorithms and computer tools to model and characterize biological processes at molecular level. The most important computer techniques used for drug design can be divided into ligand-based and receptor-based approaches. On the one hand, ligand-based approaches try to find potential molecules that could be chemically and structurally similar to a given active compound. While this is quite effective in cases where an active drug is known, its usefulness is more limited for cases where no active compound is available. On the other hand, receptor-based approaches aim to find molecules that can bind into the active site of a target biomolecule (the receptor, typically a protein or enzyme), for which structural and functional information is available (ideally at atomic resolution). These computational approaches are boosting the structure-based drug design field and have already contributed to the discovery and optimization new lead compounds in combination with experimental techniques (X-ray crystallography, nuclear magnetic resonance (NMR), site-directed mutagenesis) [26]. The most important receptor-based approaches from the point of view of computational requirements are: i) virtual ligand screening, which can reduce the number of possible compounds to be studied experimentally; ii) docking, which can provide a rationale for the binding mode of known active compounds [50]; and iii) molecular mechanics (MM), which can refine existing models and give accurate energetic descriptions. From a technical point of view, all of these approaches present two fundamental aspects, which affect efficacy, implementation, and porting to different platforms such as Cell BE: a search engine that samples all possible conformations, and a scoring function capable of identifying the correct conformations among all the sampled ones. Regarding the scoring functions, the most widely used approaches are: i) knowledge-based scoring functions, which are derived from statistical analysis of structural data and have the advantage of a low computational cost; ii) empirical scoring functions, which estimate the energy of the system from parameters obtained from experimental values, at higher computational cost; and iii) forcefield-based scoring functions, with more sophisticated physics-based methods to describe the energetic of the interactions [27] that are highly CPU-demanding.

All these scoring functions can be implemented in combination with very different sampling strategies, which can strongly affect the efficacy and portability to other platforms. In most methods, sampling and scoring are intimately integrated and thus it is difficult to separate them in independent processes. However, for analyzing the factors that can affect porting to Cell BE, we can divide the search algorithms in two broad categories: systematic (full exploration of the space in a finite grid) and stochastic sampling (where models are explored with a given probability). We will provide in the following sections a porting analysis and examples of implementation to Cell BE for each one of these categories. We will give now an overview of the most popular receptor-based modeling methods in drug design, focusing on the specific sampling algorithms. Table 15.1 summarizes a set of bioinformatics applications that implement those methods.

TABLE 15.1: Some of the most popular computational methods/approaches used in drug design.

Method	Type of Search	Program	Free License	Source Code
Protein-ligand	Incremental	FlexX	For 8 weeks	No
		DOCK 6	Yes	Yes
	Random	ICM-Docking	Reduced version	No
		MCDOCK	-	-
	Genetic Algorithm	AutoDock	Yes	Yes
		GOLD	No	No
Protein-protein	Systematic	FTDock	Yes	Yes
		ZDOCK	Yes	No
		Hex	Yes	No
	Random	ICM-Docking	Reduced version	No
		Haddock	Yes	Yes
		RosettaDock	Yes	Yes
Molecular mechanics	Molecular Dynamics	Moldy	Yes	Yes
		AMBER	No	Yes
		GROMACS	Yes	Yes
		NAMD	Yes	Yes
		X-PLOR	Yes	Yes
		Tinker	Yes	Yes
		ECEPPAK	Yes	Yes
	Normal Mode Analysis	elNémo	Yes	Yes

15.2.1 Protein-Ligand Docking

Predicting the binding mode between a small molecule (ligand or candidate) and a protein (receptor or target) by computational docking is currently one of the main bioinformatics tasks during the drug design process. For instance, in virtual ligand screening, millions of ligands present in databases are evaluated, aiming to identify those that are most likely to bind to the target of interest, which are known as hits or drug candidates. Most of the currently available structure-based methods for virtual ligand screening make use of the high-throughput application of protein-ligand docking techniques. For this, it is essential to have fast and accurate docking methods that may include some degree of flexibility in both receptor and ligand. Regarding the sampling approach, a popular technique is the *fragment-based/incremental search*, in which the ligand grows incrementally in the active site of the receptor, thus facilitating the treatment of flexibility in a stepwise manner and avoiding a combinatorial explosion. Two of the most popular methods are FlexX [7] and DOCK [38]. Many other methods use *stochastic sampling*, such as Monte-Carlo (MC) search or genetic algorithms. In MC-based methods, random changes are applied to the molecules that will be accepted or rejected according to a probability-based criterion (e.g. Metropolis) taking into account the improvement with respect to the previous candidate structure. Examples of methods are ICM-Docking [12] and MCDOCK [39]. Genetic algorithms start with an initial population of molecules, defined by a set of variables (known as genes) that represent the state of the system. The population recombines and evolves as genetic operators are applied until a fitness level (or a certain amount of steps) has been reached. This strategy is used in AutoDock [43] and GOLD [51]. The latter relies on conformational and non-bonded contact information from the Cambridge Structural Database [1].

15.2.2   Protein-Protein Docking

Protein-protein docking aims to predict the structure of a complex given the structures of the individual proteins that are known to interact. Due to the growing gap between the number of known protein interactions and the structural information at atomic detail, computer methods are increasingly needed in order to propose highly accuracy models that can help the drug design process. Indeed, one of the big challenges in drug design is the identification of small molecules capable of inhibiting protein-protein interactions [52], and for that, protein-protein docking will be essential. Most of the currently available protein-protein docking methods use either a *systematic sampling* or a *stochastic search*.

The methods using *systematic sampling* typically treat the molecules as rigid-bodies, limiting the search to the rotational and translational space, in an exhaustive way. Flexibility is ignored, at least in a first stage. FFT algorithms to speedup calculations are commonly used. The docking program FT protein docking (FTDock) [8] is perhaps the most popular FFT-based method. Another docking method is ZDOCK [17], which computes the grid correlations based on shape complementarity, desolvation energy, and electrostatics, with excellent prediction success rates. The program Hex [10] uses Spherical Polar Fourier correlations to accelerate the calculations. The use of knowledge-based scoring functions as in RPScore [42] or ZDOCK 3.0 [41], or empirical potentials as in pyDock [24], can greatly improve the success rates of the predictions.

A second group of methods uses *stochastic search*, typically based on MC methods. The advantage is that accurate energy-based scoring functions can be easily integrated into the search system, although at higher computational cost. One of the most efficient stochastic programs is ICM-DISCO [30], which achieved top success rates in the CAPRI blind test [2,40]. The program Haddock [11] is an efficient docking method based on a set of CNS [22] scripts, and can also deal with nucleic acids and other biomolecules. RosettaDock [15] is a successful docking program also based on the above-described approaches.

15.2.3 Molecular Mechanics

Molecular mechanics (MM) relies on a classical description of bonded and non-bonded forces between the atoms, based on Newton's equations of motion. Due to its high computational cost it is not usually the first choice in a typical drug design project, but it is used for a more detailed refinement of existing models obtained by other techniques (homology-modeling, docking, or virtual screening) to extract dynamic properties of the system, or to obtain accurate energetic descriptions. The most popular MM approaches are molecular dynamics (MD), MC and normal mode analysis (NMA). Currently, MD provides the most accurate description of the conformational behavior and the thermodynamic and kinetic properties of biomolecules. Moldy [35] is a free open source molecular MD program from the CCP5 project [3]. Alternative MD packages are AMBER [23], CHARMM [21], GROMACS [9,25], and NAMD [13], which scales well for parallel supercomputing on high-end parallel platforms. Other popular MM programs are CNS/X-PLOR [18], used for macromolecular structure determination by X-ray crystallography and NMR; Tinker [16]; and ECEPPAK [14], which uses internal coordinate mechanics and MC search. Different approaches aim to simplify the dynamics of the system by essential dynamics or NMA, as elNémo [5]. It is also possible to combine MM with quantum mechanics methods, but at a very high computational cost.

15.3 Cell BE Porting Analysis

This section shows a porting analysis for the following drug design applications: DOCK6, AutoDock, AMBER, GROMACS, NAMD lite[1], and Tinker applications. This analysis has been done based on profiling results of those applications on Intel Core2 Duo's computers, using `gcc` with optimization flags `-O3 -march=core2` and a representative input data to evaluate the applications. Profiling analysis has been done using `gprof`, `oprofile`, and `Valgrind`.

In particular, the porting analysis consists of:

1. Determining which are the most time-consuming functions and their percentage of execution time.

2. Analyzing if the data structures needed by each function can fit on the local scale (LS) of one SPE (256 Kbytes for code and data).

---

[1] Reduced and sequential version of the NAMD application.

TABLE 15.2: Cell BE porting analysis for some of the most popular computational methods/approaches used in drug design.

Program	Most Time-Consuming Functions	Exec. Time	Workload Fits in LS
DOCK 6	`compute_score`	57%	Depends on the molecules
	`vector_to_dockmol`	25%	
AutoDock	`eintcal`	38%	No
	`trilinterp`	24%	No
AMBER	`short_ene`	60%	Depends on the molecules
GROMACS	`nb_kernel312`	61%	No
NAMD lite	`force`	45%	Depends on the molecules
	`force_compute_nbpairs_gridcell`	37%	
	`cell_interactions_even`	16%	
Tinker	`hessian`	72%	No
	`gradient`	20%	No

3. If it is possible to apply double buffering, independently of the fact that data into the LS.

4. Chances of vectorizing the code for the pwer processing element (PPE) and/or the SPEs.

Table 15.2 summarizes the porting analysis of those applications. Therefore, considering the most time-consuming functions and their percentage of execution time, the overall ideal speedups are: 5.5x for DOCK 6, 2.6x for AutoDock, 2.5x for AMBER, 2.6x for GROMACS, 50x for NAMD lite, and 12.5x for Tinker. Note that a larger overall speedup will be achieved if more parts of the applications can be optimized and/or offloaded to an SPE.

The applications can be parallelized in different grain levels depending on the size of the data structures (and code!) and on synchronization issues.

Function offloading (FO) can be applied if the chosen workloads fit into the LS. In the case of the analyzed applications, the workloads do not fit or depend on the molecule size. Therefore, if the workload does not fit into the LS, the programmer will have to divide the data structures in blocks. If the blocking technique is not even possible, changes on the data structures should be considered in order to offload a function. An example of workload that has to be modified is the `e_vdW_Hb` structure of the `eintcal` function of the AutoDock application. Otherwise, the `eintcal` function will have to be run on the PPE. Finally, the double-buffering technique may be used to overlap communications with computation [20].

Furthermore, vectorization can be applied to several parts of the code, to be executed either on the PPE or on the SPE. However, due to random memory accesses, vectorization may not be efficiently applied without data reordering. This particular issue should be evaluated in each case.

15.4 Experimental Setup

The case study experiments have been run on a dual-processor Cell BE based blade. It contains two SMT-enabled Cell BE processors at 3.2 GHz with 1GB DD2.0 XDR RAM (512 MB per processor). The system runs Linux Fedora Core 6, kernel 2.6.20 (NUMA enabled) with 64 KB LS page mapping.

All codes have been developed in C and use the SPE IBM SDK. Codes running on the SPE components are compiled using spu-gcc 4.1.1 with -O3 optimization option. Codes running on the PPE are compiled using gcc 4.1.1 20061011 (Red Hat 4.1.1-30) and -O3 -maltivec optimization options. The OpenMP implementation used is the one supported by that version of GCC. The MPI implementation library used is OpenMPI 1.2. Performance has been analyzed using gprof profiler, gettimeofday, and time-based decrementers.

The FTDock application has been analyzed with one of the enzyme/inhibitor tests of the benchmark used in [31]. A 128^3 grid has been used in that analysis and the execution has been stopped after 1000 iterations of the main loop.

The Moldy application has been analyzed using the water system example from the Moldy package, extending the number of molecules to 10000.

15.5 Case Study: Docking with FTDock

15.5.1 Algorithm Description

FTDock uses the shape recognition algorithm of [34], measuring shape complementarity by Fourier correlation [31]. The algorithm uses FFTs to scan all the possible spatial translations of two rotating rigid molecules.

Given two molecules A and B, A being the largest one, the first step is the discretization of A in an N^3 grid. Then, molecule B is rotated in all the possible orientations to execute a translational scan of this molecule relative to molecule A. To perform the scan, for each rotation molecule B is also discretized and the surface complementarity is evaluated with the correlation of the molecules.

For each rotation, the correlation cost of the translational scan of molecule B is $O(N^6)$, though it can be solved in $O(N^3 \log N^3)$ using FFTs as follows:

$$
\begin{aligned}
F_A &= FFT(f_A) \\
F_B &= FFT(f_B) \\
F_C &= (F_A^*)(F_B) \\
f_C &= iFFT(F_C)
\end{aligned}
$$

FFT and *iFFT* denote forward and inverse 3D FFT, respectively. f_A and f_B functions define the values of the grids for discretized molecules A and B, respectively. (F_A^*) denotes complex conjugate and $(F_A^*)(F_B)$ denotes complex multiplication. For each rotation, FTDock scans f_C to find the three best correlation scores using the scoring filter function. The number of partial results depends on the number of possible orientations scanned, that is, $360 \times 360 \times 180/\alpha$, where α is the angular deviation. α is 15 as in [31].

15.5.2 Profiling and Implementation

The profiling information obtained running FTDock on the PPE of a Cell BE has shown that the most time-consuming functions are: the 3D FFT/iFFTs (54% of total execution time), the complex multiplication $((F_A^*)(F_B))$ (12%), the scoring filter function (21%), and the discretization function (12%). That exposes an ideal overall speedup of 100x on the FTDock application. Furthermore, code analysis indicates that rotations can be executed in parallel because they are data independent.

3D FFT/iFFTs have been implemented on top of the 1-stride 1D FFT routine of the IBM SDK on the SPE, and distributing all the grid planes in parallel among the SPEs (details of the implementation in [48]). In Chapter 7 a 1D FFT implementation of large data sets is explained but is not within the scope of the FTDock application. Therefore, for each grid orientation, and for each plane on one orientation, 1D FFT is applied to each row of the plane, and then, the plane is transposed. That implementation uses a SIMD version of the Eklundh Matrix transposition algorithm [28] for the plane transpositions. There are other parallel 3D FFT implementations for the Cell BE: the FFTW Cell BE version [6], the efficient parallel implementation presented in [19], the prototype one offered by the IBM SDK 3.1, and also the 3D FFT implementation for large grid sizes in Chapter 8. However, in the Cell BE FTDock implementation, the 3D FFT/iFFTs are specialized. This specialized version computes the 3D FFT of molecule B, the complex multiplication, and the 3D iFFT of the complex multiplication result operations are joined so that: (i) exploitation of temporal data locality on the LS of a SPE is improved, and (ii) redundant plane transpositions of the 3D FFT followed by the corresponding 3D iFFT can be saved.

The scoring filter function has been offloaded so that each SPE looks for the best three local scores and transfers them to main memory to do a reduction on the PPE. Finally, the PPE sorts all those local scores using Quicksort. That

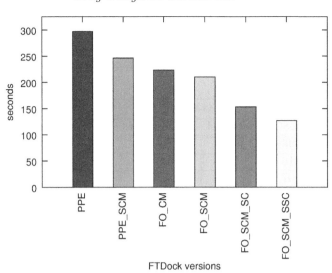

FIGURE 15.1: FTDock execution times. Labels are explained in the text.

sorting process does not have a significant cost because the total number of scores to sort is small. The scoring filter function uses a Straight insertion sorting algorithm [37] that implements a three-nested loop. This loop largely penalizes the function performance due to the lack of a branch predictor on the SPEs. To overcome this issue, the sorting algorithm has been vectorized, removing the three-nested loop.

Finally, regarding coarser grains of parallelism, the discretization function has been parallelized on the PPEs using OpenMP. Besides, the main rotation loop is executed in parallel using the MPI programming model.

15.5.3 Performance Evaluation

Figure 15.1 shows the execution time of different versions of the FTDock application, running on the PPE and one SPE. The baseline FTDock (PPE label in the figure) has scalar versions of complex multiplication and scoring filter on the PPE. The PPE_SCM version implements a SIMD version of the complex multiplication on the PPE. FO_CM and FO_SCM are the FTDock versions with the scalar and SIMD complex multiplication, respectively, offloaded to the SPE, and the scoring filter function on the PPE. Finally, FO_SCM_SC and FO_SCM_SSC are the FO_CM and FO_SCM versions with the scalar and SIMD scoring filter function also offloaded to the SPE.

In general, FO and vectorization of the complex multiplication and scoring filter significantly improve the performance. In that case, FTDock running on the PPE (first column in the figure) achieves nearly 2x of speedup. Figure 15.1

TABLE 15.3: Execution times in seconds for different versions of FTDock. Speedup (sp) compared to the vectorial PPE version is shown. Each label identifies a version of the FTDock which are explained in the text.

SPEs	Execution Time (seconds)				
#	PPE_SCM	FO_CM	FO_SCM	FO_SCM_SC	FO_SCM_SSC (sp)
1	246	223	210	153	127 (1.9)
2	216	167	151	82	70 (3.1)
4	207	139	131	47	41 (5.0)
8	196	131	131	30	27 (7.3)

shows that memory hierarchy is a bottleneck when the PPE accesses the data. That can be seen comparing PPE and FO_CM bars and FO_SCM and FO_SCM_SC, where the FO significantly reduces the execution time. The reason is that the load/store pending queue of cache miss accesses (MSHR) limits the performance of the PPE version, and that causes the PPE pipeline to stall. However, direct memory access (DMA) transfers do not have to deal with that issue. Second, vectorization is important in both PPE and SPE, achieving around 20% of improvement for the code analyzed.

Table 15.3 shows the execution time of those versions for one, two, four, and eight SPEs. The last column of the table also shows the speedup achieved when offloading and vectorization are done compared to the SIMD PPE version.

Finally, the FTDock parallelization using MPI and OpenMP has been analyzed. In the case of the MPI parallelization, each MPI process executes a rotation of the main loop of FTDock, and then a reduction of the results is performed. Therefore, up to two MPI processes, one per PPE of the two Cell BEs of the blade, can be run in parallel, and up to eight SPEs per MPI process can be used. Indeed, the performance of two MPI processes and $n = 8$ SPEs per MPI is better than using one MPI process and $2n = 16$ SPEs of the two Cell BEs. In general, that happens for any $2n$ SPEs placed on the two Cell BEs. The main reason is that DMA transfers between the two Cell BEs (LS or main memory) are done on the 3D FFT/iFFTs, increasing the contention of the EIB and the main memory [33]. Furthermore, FTDock reaches a speedup between 1.1x and 1.2x parallelizing the discretization function with OpenMP. In particular, the stand-alone function cannot achieve more than 1.5x. This is due to the large number of branches that may not be predicted by the PPE branch-prediction hardware. Nevertheless, a second thread can make forward progress and increase PPE pipeline usage and system throughput.

15.6   Case Study: Molecular Dynamics with Moldy

15.6.1   Algorithm Description

Computation in Moldy is divided into two main parts. The first one computes short-range interactions between close molecules, while the second one deals with long-range interactions between molecules and the system. Both parts calculate Newton-Euler equations of motion for dynamics.

For big systems (i.e. 5000-10000 molecules), the critical part of the MD application is the evaluation of the reciprocal-space part of the Ewald sum for the long-range forces. That evaluation is performed with the computation of Equation 15.1, that is mainly a summation over the reciprocal K lattice vectors. N_{sites} stands for the number of charged sites of the system. More information about the equation and its terms is in [35].

$$\frac{1}{\epsilon_0 V} \sum_{k>0} \frac{1}{k^2} e^{-\frac{k^2}{4\alpha^2}} \left\{ \left| \sum_{i=1}^{N_{sites}} q_i \cos(k \cdot r_i) \right|^2 + \left| \sum_{i=1}^{N_{sites}} q_i \sin(k \cdot r_i) \right|^2 \right\} \quad (15.1)$$

This equation is evaluated within the function `ewald()` with $O(kN_{sites})$ cost. A pseudo-code of that function is shown in Figure 15.2.

```
1    function ewald
2    Precompute list of K-vectors
3    Precompute sin and cos arrays of k · rᵢ
4    forall k := 1 : K do
5       Calculation of prefactors for energy, force
6       forall i := 1 : N_sites do
7          Construct qᵢ cos(k · rᵢ) and qᵢ sin(k · rᵢ) for current k
8       endfor
9       Evaluation of potential energy term
10      forall i := 1 : N_sites do
11         Evaluation of site forces
12      endfor
13   endfor
```

FIGURE 15.2: Ewald function algorithm.

15.6.2   Profiling and Implementation

The Ewald summation consumes more than 90% of execution time for big systems. The profiling analysis shows that `qsincos` computation (lines 6-8 of Figure 15.2) and `forces` computation (lines 10-12) are the most time-consuming parts of Ewald.

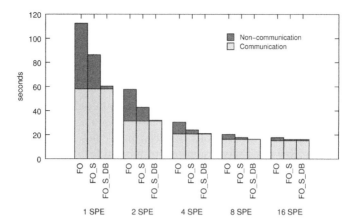

FIGURE 15.3: Moldy execution times. PPE execution time is not shown. Labels are explained in the text.

The K iterations of the main loop of Ewald computation (lines 4-13 of Figure 15.2) can be evenly distributed among all the SPEs, and a reduction is performed by the PPE to correctly update the results.

LS cannot hold the complete sinus and cosinus arrays needed on the computations of each iteration of the `ewald()` main loop. Thus, for each SPE and each new iteration, each SPE has to load those arrays using several DMA transfers since they do not fit in the LS. This function generates two arrays of results that also have to be temporally transferred to main memory. Then, once the `qsincos` calculation (lines 6-8 of Figure 15.2) is performed for the N_{sites}, the computation of potential energy and site forces vectors can continue. Those arrays are transferred to main memory as well. All DMA transfers are done using the double-buffering technique. Finally, the most time-consuming parts of the `ewald()` function (i-loops on the Figure 15.2) have been vectorized.

15.6.3   Performance Evaluation

Figure 15.3 shows the results of three different versions of the Moldy parallelization on the Cell BE for one, two, four, eight and 16 SPEs, using single floating-point precision. Those versions include: FO, FO and vectorization using the SIMD code of the SPE (FO_S), and FO, vectorization, and double buffering (FO_S_DB). For each bar, the communication and non-communication parts of the execution time are shown. The non-communication part is the computation not hidden by the transfers to main memory. The PPE execution time is *circa* 250 seconds and is not shown in the figure.

Results show that just FO on one SPE of the most time-consuming parts of the applications improves significantly the performance of the application on the Cell BE (FO compared to PPE). The main reason is that SPE DMA transfers exploit the memory and EIB bandwidth better than memory accesses from the PPE [48]. Vectorization is also a significant source of improvement if computation and communication are not overlapped (FO compared to FO_S). Furthermore, it is observed that double buffering almost hides the computation time.

Finally, the communication time dominates the total time of all the executions of Figure 15.3. This is due to: i) the LS of an SPE is not big enough for the working set of MD applications and the same data have to be accessed again and again, and ii) there is not enough computation exploiting temporal data locality. That means, in terms of scalability, that performance results are fine until four SPEs. From that point, the EIB is saturated, the communication time gets stuck, and almost all the computation time is overlapped.

15.7 Conclusions

Computational cost is a major bottleneck in the development of new and more accurate drug design computing methods. However, parallel architectures such as Cell BE may help expand the conformational space of the chemical compounds. In this chapter we have outlined the main applications on drug design, analyzed the porting of six of them, and shown two practical case studies, FTDock and Moldy, which are a docking and a molecular dynamics application, respectively.

There are four steps on porting drug design applications to Cell BE to exploit this architecture. The first step is the analysis of the application workload to determine which kind of partitioning can be applied to fit the data on the LS of the SPEs. The second step is the offloading of the most time-consuming functions. Offloading is likely to have an immediate benefit for the application because all memory accesses are done using DMA operations. Furthermore, the programmer should reuse data as much as possible among different functions offloaded to the same SPE. The third step is to apply the double-buffering technique on the DMA accesses to the main memory in order to overlap computation with communication on the SPE. The fourth and final step is the vectorization of the code on the PPE and/or the SPEs. The performance impact of those steps on the application has been shown in this chapter. Particularly, offloading has the most significant impact on application performance gain, followed by double buffering and vectorizing, depending on the application characteristics.

Finally, the Cell BE porting of both case studies has been compared to other architectures. This comparison reflects the advantages of using Cell BE in the drug design field. Regarding FTDock, a 3x speedup is achieved compared to a parallel version running on a POWER5 multicore with two 1.5 GHz POWER5 chips with 16 GB of RAM. Besides, Moldy on the Cell BE consumes less power and takes the same time as an MPI parallelization on four Itanium Montecito processors of SGI Altix 4700 (eight threads).

Acknowledgments

This work has been performed on the Cell BE based blades in the Barcelona Supercomputing Center, obtained through the IBM SUR Program. The work has been supported by the European Commission (SARC project contract no. 27648), the Spanish Ministry of Education (contract no. TIN2007-60625, CSD2007-00050, and BIO2008-02882), and HiPEAC (no. IST-004408).

Bibliography

[1] The Cambridge Structural Database (CSD). Project WebSite: http://www.ccdc.cam.ac.uk/products/csd (accessed November 12, 2009).

[2] CAPRI: Critical Assessment of PRediction of Interaction. Project Web-Site: http://capri.ebi.ac.uk (accessed November 12, 2009).

[3] Collaborative Computational Project 5—The Computer Simulation of Condensed Phases. Project WebSite: http://www.ccp5.ac.uk (accessed November 12, 2009).

[4] eHITS. Project WebSite: http://www.simbiosys.com (accessed November 12, 2009).

[5] elNémo: The Elastic Network Model. Project WebSite: http://www.igs.cnrs-mrs.fr/elnemo (accessed November 12, 2009).

[6] Fastest Fourier Transform in the West. Project WebSite: http://www.fftw.org (accessed November 12, 2009).

[7] FlexX. Project WebSite: http://www.biosolveit.de/flexx (accessed November 12, 2009).

[8] Fourier Transform Rigid-body Docking. Project WebSite: http://bmm.cancerresearchuk.org/docking/ftdock.html (accessed November 12, 2009).

[9] GROMACS. Project WebSite: http://www.gromacs.org (accessed November 12, 2009).

[10] Hex Protein Docking. Project WebSite: http://www.loria.fr/~ritched/hex (accessed November 12, 2009).

[11] High Ambiguity Driven Biomolecular DOCKing. Project WebSite: http://www.nmr.chem.uu.nl/haddock (accessed November 12, 2009).

[12] ICM-docking. Project WebSite: http://www.molsoft.com/docking.html (accessed November 12, 2009).

[13] NAMD Scalable Molecular Dynamics. Project WebSite: http://www.ks.uiuc.edu/Research/namd (accessed November 12, 2009).

[14] NIH ECEPPAK. Project WebSite: http://cbsu.tc.cornell.edu/software/eceppak (accessed November 12, 2009).

[15] RosettaDock. Project WebSite: http://www.rosettacommons.org (accessed November 12, 2009).

[16] Tinker—Software Tools for Molecular Design. Project WebSite: http://dasher.wustl.edu/tinker (accessed November 12, 2009).

[17] ZDOCK: Protein Docking. Project WebSite: http://zlab.bu.edu/zdock (accessed November 12, 2009).

[18] A.T. Brunger, P.D. Adams, G.M. Clore, W.L. Delano, P. Gros, R.W. Grosse-Kunstleve, J.-S. Jiang, J. Kuszewski, M. Nilges, N.S. Pannu, R.J. Read, L.M. Rice, T. Simonson, and G.L. Warren. X-PLOR—CNS Crystallography & NMR System. Project WebSite: http://cns-online.org/v1.21 (accessed November 12, 2009).

[19] D.A. Bader and V. Agarwal. FFTC: Fastest Fourier Transform for the IBM Cell Broadband Engine. *14th IEEE International Conference on High Performance Computing (HiPC), Springer-Verlag LNCS*, 4873:172–184, 2007.

[20] Daniel A. Brokenshire. Maximizing the power of the cell broadband engine processor: 25 tips to optimal application performance. http://www.ibm.com/developerworks/power/library/pa-celltips1, Jun 2006 (accessed November 12, 2009).

[21] B.R. Brooks, C.L. Brooks, A.D. Mackerell, L. Nilsson, R.J. Petrella, B. Roux, Y. Won, G. Archontis, C. Bartels, S. Boresch, A. Caflisch, L. Caves, Q. Cui, A.R. Dinner, M. Feig, S. Fischer, J. Gao, M. Hodoscek, W. Im, K. Kuczera, T. Lazaridis, J. Ma, V. Ovchinnikov, E. Paci, R. W. Pastor, C.B. Post, J.Z. Pu, M. Schaefer, B. Tidor, R.M. Venable, H.L. Woodcock, X. Wu, W. Yang, D.M. York, and M. Karplus.

CHARMM: the biomolecular simulation program. *Project WebSite: http://yuri.harvard.edu*, 30(10):1545–1614, Jul 2009. (accessed November 12, 2009).

[22] A.T. Brünger, P. D. Adams, G.M. Clore, W.L. DeLano, P. Gros, R.W. Grosse-Kunstleve, J.S. Jiang, J. Kuszewski, M. Nilges, N.S. Pannu, R.J. Read, L.M. Rice, T. Simonson, and G.L. Warren. Crystallography & NMR system: A new software suite for macromolecular structure determination. *Acta Crystallogr D Biol Crystallogr*, 54(Pt 5):905–921, Sep 1998.

[23] D.A. Case, T.A. Darden, T.E. Cheatham, C.L. Simmerling, J. Wang, R.E. Duke, R. Luo, M. Crowley, R.C. Walker, W. Zhang, K.M. Merz, B. Wang, S. Hayik, A. Roitberg, G. Seabra, I. Kolossvry, K.F. Wong, F. Paesani, J. Vanicek, X. Wu, S.R. Brozell, T. Steinbrecher, H. Gohlke, L. Yang, C. Tan, J. Mongan, V. Hornak, G. Cui, D.H. Mathews, M.G. Seetin, C. Sagui, V. Babin, and P.A. Kollman. AMBER 10. Project WebSite: http://www.gromacs.org, 2008 (accessed November 12, 2009).

[24] Tammy Man-Kuang Cheng, Tom L. Blundell, and Juan Fernández-Recio. pyDock: Electrostatics and desolvation for effective scoring of rigid-body protein-protein docking. *Proteins*, 68(2):503–515, Aug 2007.

[25] Markus Christen, Philippe H. Hünenberger, Dirk Bakowies, Riccardo Baron, Roland Bürgi, Daan P. Geerke, Tim N. Heinz, Mika A. Kastenholz, Vincent Kräutler, Chris Oostenbrink, Christine Peter, Daniel Trzesniak, and Wilfred F. van Gunsteren. The GROMOS software for biomolecular simulation: GROMOS05. *J Comput Chem*, 26(16):1719–1751, Dec 2005.

[26] David E. Clark. What has virtual screening ever done for drug discovery? *Expert Opinion on Drug Discovery*, 3:841–851, 2008.

[27] M. Cohen-Gonsaud, V. Catherinot, G. Labesse, et al. From molecular modeling to drug design. In *Practical Bioinformatics*, pages 35–71. Springer-Verlag, 2004.

[28] J.O. Eklundh. A fast computer method for matrix transposing. *IEEE Transactions on Computers*, 21:801–803, 1972.

[29] G. De Fabritis. Performance of the cell processor for biomolecular simulations. *Comp Phys Commun 176, 660*, 176, 2007.

[30] Juan Fernández-Recio, Maxim Totrov, and Ruben Abagyan. Soft protein-protein docking in internal coordinates. *Protein Sci*, 11(2):280–291, Feb 2002.

[31] Henry A. Gabb et al. Modelling protein docking using shape complementary, electrostatics and biochemical information. *J Mol Biol*, 272, 1997.

[32] John J. Irwin and Brian K. Shoichet. ZINC–A free database of commercially available compounds for virtual screening. *J Chem Inf Model,* 45(1):177–182, 2005.

[33] Daniel Jiménez-González, Xavier Martorell, and Alex Ramírez. Performance analysis of cell broadband engine for high memory bandwidth applications. *ISPASS,* 2007.

[34] Ephraim Katchalski-Katzir, Isaac Shariv, Miriam Eisenstein, Asher A. Friesem, Claude Aflalo, and Ilya A. Vakser. Molecular surface recognition: Determination of geometric fit between proteins and their ligands by correlation techniques. *Proc Natl Acad Sci USA Biohphysics,* 89:2195–2199, 1992.

[35] Keith Refson. *Moldy User's Manual.* 24 May 2001.

[36] Prashant Khodade, R. Prabhu, Nagasuma Chandra, Soumyendu Raha, and R. Govindarajan. Parallel implementation of *AutoDock. J. Appl Crystallography,* 40(3):598–599, Jun 2007.

[37] D. Knuth. The art of computer programming: Sorting and searching. *Addison-Wesley Publishing Company,* 3, 1973.

[38] P. Therese Lang, Scott R. Brozell, Sudipto Mukherjee, Eric F. Pettersen, Elaine C. Meng, Veena Thomas, Robert C. Rizzo, David A. Case, Thomas L. James, and Irwin D. Kuntz. DOCK 6: Combining techniques to model RNA-small molecule complexes. Project Web-Site: http://dock.compbio.ucsf.edu/DOCK_6/index.htm, 15(6):1219–1230, Jun 2009 (accessed November 12, 2009).

[39] M. Liu and S. Wang. MCDOCK: A Monte Carlo simulation approach to the molecular docking problem. *J Comput Aided Mol Des,* 13(5):435–451, Sep 1999.

[40] Raúl Méndez, Raphaël Leplae, Leonardo De Maria, and Shoshana J. Wodak. Assessment of blind predictions of protein-protein interactions: Current status of docking methods. *Proteins,* 52(1):51–67, Jul 2003.

[41] Julian Mintseris, Brian Pierce, Kevin Wiehe, Robert Anderson, Rong Chen, and Zhiping Weng. Integrating statistical pair potentials into protein complex prediction. *Proteins,* 69(3):511–520, Nov 2007.

[42] G. Moont, H.A. Gabb, and M.J. Sternberg. Use of pair potentials across protein interfaces in screening predicted docked complexes. *Proteins,* 35(3):364–373, May 1999.

[43] Garrett M. Morris, Ruth Huey, William Lindstrom, Michel F. Sanner, Richard K. Belew, David S. Goodsell, and Arthur J. Olson. AutoDock4

and AutoDockTools4: Automated docking with selective receptor flexibility. Project WebSite: http://autodock.scripps.edu, 30(16):2785–2791, Dec 2009 (accessed November 12, 2009).

[44] Stephen Olivier, Jan Prins, Jeff Derby, and Ken Vu. Porting the GROMACS molecular dynamics code to the cell processor. In *IPDPS*, pages 1–8. IEEE, 2007.

[45] John P. Overington, Bissan Al-Lazikani, and Andrew L. Hopkins. How many drug targets are there? *Nat Rev Drug Discov*, 5(12):993–996, Dec 2006.

[46] Y.-P. Pang, T.J. Mullins, B.A. Swartz, J.S. McAllister, B.E. Smith, C.J. Archer, R.G. Musselman, A.E. Peters, B.P. Wallenfelt, and K.W. Pinnow. EUDOC on the IBM Blue Gene/L system: Accelerating the transfer of drug discoveries from laboratory to patient. *IBM J Res Dev*, 52(1/2):69–81, 2008.

[47] Nicolas Sauton, David Lagorce, Bruno O. Villoutreix, and Maria A. Miteva. MS-DOCK: Accurate multiple conformation generator and rigid docking protocol for multi-step virtual ligand screening. *BMC Bioinformatics*, 9:184, 2008.

[48] Harald Servat, Cecilia González-Álvarez, Xavier Aguilar, Daniel Cabrera-Benítez, and Daniel Jiménez-González. Drug design issues on the cell be. In Per Stenström, Michel Dubois, Manolis Katevenis, Rajiv Gupta, and Theo Ungerer, editors, *HiPEAC*, volume 4917 of *Lecture Notes in Computer Science*, pages 176–190. Springer, 2008.

[49] B. Sukhwani and M. Herbordt. GPU acceleration of a production molecular docking code. In *Proceedings of 2nd Workshop on General Purpose Processing on Graphics Processing Units*, pages 19–27. ACM Press, 2009.

[50] Peter L. Toogood. Inhibition of protein-protein association by small molecules: Approaches and progress. *J Med Chem*, 45(8):1543–1558, Apr 2002.

[51] Marcel L. Verdonk, Gianni Chessari, Jason C. Cole, Michael J. Hartshorn, Christopher W. Murray, J. Willem, M. Nissink, Richard D. Taylor, and Robin Taylor. Modeling water molecules in protein-ligand docking using GOLD. Project WebSite: http://www.ccdc.cam.ac.uk/products/life_sciences/gold, 48(20):6504–6515, Oct 2005 (accessed November 12, 2009).

[52] James A. Wells and Christopher L. McClendon. Reaching for high-hanging fruit in drug discovery at protein-protein interfaces. *Nature*, 450(7172):1001–1009, Dec 2007.

Chapter 16

GPU Algorithms for Molecular Modeling

John E. Stone

Beckman Institute, University of Illinois at Urbana-Champaign

David J. Hardy

Beckman Institute, University of Illinois at Urbana-Champaign

Barry Isralewitz

Beckman Institute, University of Illinois at Urbana-Champaign

Klaus Schulten

Department of Physics, University of Illinois at Urbana-Champaign

16.1   Introduction

Over the past decade, graphics processing units (GPUs) have become fully programmable parallel computing devices capable of accelerating a wide variety of data-parallel algorithms used in computational science and engineering. The tremendous computational capabilities of GPUs can be employed to accelerate molecular modeling applications, enabling molecular dynamics simulations and their analyses to be run much faster than before and allowing the use of scientific techniques that are impractical on conventional hardware platforms. In this chapter, we describe the key attributes of GPU hardware architecture and the algorithm design techniques required to successfully exploit the computational capabilities of GPUs for scientific applications. We illustrate these techniques for molecular modeling applications, describing the process involved in adapting or redesigning existing algorithms for high performance on GPUs. Many of these techniques can be applied beyond the molecular modeling field and should be of interest to anyone contemplating adaptation of existing algorithms to GPUs.

16.2   Computational Challenges of Molecular Modeling

In every living cell, the processes of life are carried out by millions of biomolecules. To study the function of biomolecules, scientists create mechanical models, with a molecule represented as a collection of atoms with given coordinates, masses, charges, and bond connectivity. In molecular dynamics simulations [18], the bonded and non-bonded interactions between different types of atoms are defined, then Newton's equations of motion are integrated to determine the motion of the atoms. The simulations capture key details of structural changes and interactions, which can lead to understanding of biological processes as varied as muscle contraction, protein synthesis, harvesting of sunlight, and viral infection.

Over the past three decades, the scale of molecular dynamics simulations has increased with available computational power from sizes of hundreds of atoms and timescales of picoseconds, up to millions of atoms and timescales of microseconds. Longer timescales provide the sampling needed to reliably represent biologically relevant motions; larger model structures capture key machines in living cells, for example, the ribosome. A molecular modeling study can be divided into three steps: *system setup, simulation,* and *analysis.*

System Setup. To prepare a model system for simulation, a modeler starts with experimentally derived coordinate data of one or more large biomolecules and adds the components of a natural environment, namely, water, ions, lipids, and other small molecules. This step involves checking that all components are arranged consistently with the known properties of the system. Electrostatics calculations need to be performed to place ions energetically most favorably near charged parts of a system. There may be multiple rounds of iteration toward system setup. GPU-accelerated calculation in this step yields today results in seconds or minutes versus hours or days previously, while only requiring a small amount of compute hardware for on-demand calculations. Methods that can run on available GPU hardware are likelier to be run on-demand, and so provide shorter turnaround time than batch jobs. The authors have used GPUs to accelerate ion placement in the preparation of several large simulations [5, 24].

Simulation. The bulk of calculation in a modeling study arises in the molecular dynamics simulation itself, the generation of the molecular dynamics trajectory. Large production simulations are often run in parallel on queued high performance computing (HPC) cluster systems. A simulation must generally be completed within weeks to months—practical wall-clock time lengths for research projects. The actual calculation can require the use of hundreds or thousands of cluster nodes, resources not available to all scientists. GPU-accelerated simulations can reduce the cost of the required compute hardware, giving more scientists access to simulation studies, in particular increasing access for experimentalist [10, 22].

Analysis of Simulation Results. As a scientist examines a long simulation trajectory, she may realize that a range of analysis jobs need to be performed in addition to the analysis planned when designing the simulation. The various analyses that are ultimately required must process a massive amount of trajectory data generated by the simulation. Analyses are usually performed on a subset of trajectory data, a regular temporal sampling over the trajectory, but this sampling must still be detailed enough to capture the relevant events and behaviors of the simulated system. Similar GPU acceleration advantages hold here as for the system-setup step, but with a calculation that may be thousands of times longer. The speedup from a GPU-accelerated analysis cluster can make it possible for a researcher to greatly improve sampling in the analysis of long simulation trajectories, and at lower cost than a central processing unit (CPU)-based cluster. Many jobs that arise when examining a trajectory are ideally sized for high-speed analysis on GPU-accelerated workstations, or on a small number of GPU-accelerated cluster nodes, eliminating wait time for access to large HPC clusters.

16.3 GPU Overview

Over the past decade GPU hardware and software have evolved to support a high degree of programmability. This has led to the use of GPUs as accelerators for computationally demanding science and engineering applications involving significant amounts of data parallelism [11,13,16]. Many algorithms involved in molecular modeling and computational biology applications can be adapted to GPU acceleration, achieving significant performance increases relative to contemporary multicore CPUs [6,19,22,23]. Below we describe details of GPU hardware and software organization that impact the design of molecular modeling algorithms.

16.3.1   GPU Hardware Organization

Throughout their development, GPUs have been designed primarily for accelerating rasterization and shading of three-dimensional geometry. The rasterization process contains a significant amount of inherent data parallelism, and GPU hardware architects have historically exploited this in hardware designs, using large arrays of arithmetic units to accelerate graphics computations. Recently, the demand for increased realism and fidelity in graphics-intensive applications, such as games and scientific and engineering software, has led GPU architectures toward designs composed of hundreds of fully programmable processing units [17]. The individual processing units contained in modern GPUs now support all standard data types and arithmetic operations, including 32-bit and 64-bit IEEE floating point [11]. State-of-the-art GPUs can achieve peak single-precision floating-point arithmetic performance of 1.5 trillion floating-point operations per second (TFLOPS), with double-precision floating-point rates reaching approximately half that speed [15]. GPUs also contain large high-bandwidth memory systems, designed to service the tremendous bandwidth requirements of graphics workloads. These memory systems achieve bandwidths of over 100 GB/sec in recent devices, and are well suited to the needs of computational workloads such as dense linear algebra that are frequently memory bandwidth bound.

Unlike CPUs, which are generally designed for latency-sensitive workloads with coarse- or medium-grained parallelism, GPUs are designed for throughput-oriented workloads that are extremely data parallel and relatively latency insensitive. GPUs achieve high throughput through the use of hardware multithreading and concurrent execution of tens of thousands of threads. Heavy reliance on multithreading for hiding latency enables GPUs to employ modest-sized on-chip caches, leaving much of the die area for arithmetic units. GPUs further reduce transistor count and die area by driving groups of processing units from shared instruction decoders in a single-instruction multiple-data (SIMD) model, executing the same instruction in lock-step, but on different input data. The physical group of threads or work items that execute together in lock-step in hardware is known as a "warp" in NVIDIA CUDA terminology or a "wavefront" in AMD terminology. Many warps or wavefronts are multiplexed onto the same hardware to hide memory and execution latency, and to fully utilize different types of execution hardware. GPUs contain multiple memory systems, each designed to serve different access patterns. GPUs contain a large high-latency, high-bandwidth "global" memory, which can be read and written by any thread. Special "texture" filtering hardware provides cached, read-only access to GPU global memory, improving performance for random memory access patterns and for multidimensional interpolation and filtering. Recent devices, such as the NVIDIA "Fermi" GPU architecture, have added L1 and L2 caches to improve performance for read-write accesses to global memory, particularly for atomic update operations. GPUs also include thread-private registers, as well as two other types of small, low-latency on-chip memory systems, one for inter-thread communication and reduction operations (called "shared memory" in CUDA or "local memory" in OpenCL) and another for read-only data (called "constant memory"). GPUs are designed such that each group of SIMD processing units is serviced by an independent instance of these fast on-chip memories, greatly amplifying their aggregate memory bandwidth.

16.3.2   GPU Programming Model

Although GPUs are powerful computers in their own right, they must be managed by application software running on the host CPU. Application programs use GPUs as co-processors for the highly data-parallel parts of their workloads, invoking GPU computations on-demand. The GPUs and CPUs cooperate to perform memory transfers and execute parallel computations. The CPU orchestrates the work by preparing the parallel decomposition, performing data marshalling, and reorganizing any data structure memory layout required by the functions (known as "kernels") that run on the GPU. The most popular GPU programming tools available today are NVIDIA's CUDA [14] and OpenCL [12], an industry-standard many-core data-parallel programming interface. We give only a brief introduction to GPU programming here; more detailed discussions of CUDA and OpenCL are available in the literature [13, 21].

The key facilities provided by GPU programming toolkits include routines for enumerating and managing available GPUs, performing GPU memory allocations, transferring data between the host machine and GPUs, launching GPU kernels, querying execution progress, and checking for errors. Unlike multicore CPUs or MPI-based parallel algorithms, GPUs require extremely fine-grained parallelism, orders of magnitude finer than is usually required with other parallel programming platforms. Since GPUs are throughput-oriented devices that depend on hardware multithreading to hide memory access and instruction execution latencies, GPU algorithms must instantiate on the order of 10,000 to 30,000 independent work items in order to keep all of the processing units occupied and to provide adequate latency hiding.

Once a fine-grained parallel decomposition has been created, the next step is to organize it into so-called work groups of independent work items that can be efficiently executed together in a number of warps or wavefronts, on a set of SIMD processing units. Figure 16.1 shows an example of how multidimensional computational domains can be decomposed into grids of data-

GPU Parallel Decomposition

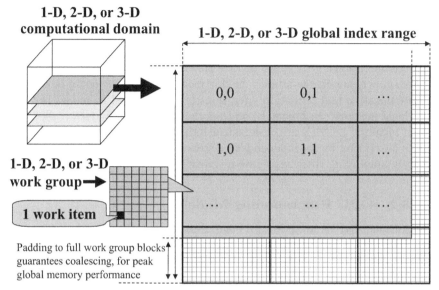

FIGURE 16.1: In both CUDA and OpenCL, data-parallel problems are decomposed into a large number of independent work items. Blocks of work items execute together on the same group of SIMD processing units. In CUDA, independent work items are termed "threads," work groups are called "thread blocks," and a global index range is called a "grid."

parallel work items. The parallel work is described by a d-dimensional global index space (where d is one, two, or three) that is assigned to the constituent work items or threads that execute on the GPU. The total number of work items in each dimension of the global index space must be an exact multiple of each associated work group dimension. The assignment of indices to individual work groups and work items is done by the GPU. Since they have a well-defined relationship, it is easy for each GPU thread to determine what data it operates on and what computations to perform using these indices. The index spaces used by CUDA and OpenCL are similar in concept to node ranks in MPI and other parallel programming interfaces.

Since GPUs have tremendous arithmetic capabilities, memory bandwidth often ends up being the primary performance limiter. Some algorithms, such as certain types of reaction-diffusion problems [1], may be inherently memory bandwidth bound, limiting the best GPU speedup to around a factor of ten relative to a CPU algorithm (resulting directly from the ratio of their respective memory bandwidths). Successful GPU algorithm designs optimize memory performance through the use of data structures, memory layouts, and traversal patterns that are efficient for the underlying hardware. GPU global memory systems are attached by wide data paths that typically require memory accesses to be organized so they are at least as wide as the memory channel (thereby achieving peak transfer efficiency) and are aligned to particular address boundaries (thereby reducing the number of transfers needed to service a request). The requirements for so-called coalesced memory accesses lead to data structures organized in fixed sized blocks that are a multiple of the minimum required coalesced memory transfer size. As a result of the need to maintain memory address alignment, it is often beneficial to pad the edges of data structures and parallel decompositions so they are an even multiple of the memory coalescing size, even if that means generating threads and computing work items that are ultimately discarded. The illustration in Figure 16.1 shows an example of how a two-dimensional grid of work items is padded to be a multiple of the coalescing size.

16.4   GPU Particle-Grid Algorithms

Many molecular modeling algorithms calculate physical quantities on three-dimensional grids corresponding to a volume of space surrounding a molecular structure of interest. Examples include electrostatic potential maps, molecular orbitals, and time-averaged spatial occupancy maps of gases from molecular dynamics simulations. GPUs are well suited to these types of computations, as many such algorithms can be made to be data parallel and highly arithmetic intensive through careful data structure design and optimization. Many related algorithms accumulate values at lattice points summed from hundreds to thousands of contributions, spanning one or more frames in a time series or molecular dynamics trajectory.

16.4.1   Electrostatic Potential

A map of the electrostatic potential for a molecular system is generally represented on a three-dimensional lattice. The lattice can be specified by an origin, three basis vectors providing the extent of the lattice in each dimension, and the number of lattice subdivisions along each basis vector. The electrostatic potential V_k for lattice point k located at position $\vec{r}_k$ is given by

$$V_k^{\text{direct}} = \sum_i C \frac{q_i}{|\vec{r}_i - \vec{r}_k|} , \qquad (16.1)$$

with the sum taken over the atoms modeled as point charges, where an atom i having partial charge q_i is located at position $\vec{r}_i$, and C is a constant. For a system of N atoms and a lattice of M points, the computational complexity of direct summation is $O(MN)$.

The quadratic complexity of evaluating expression (16.1) becomes intractable for larger systems and longer trajectories. The summation can be computed with linear computational complexity $O(M + N)$ by introducing a switching function S to make the potential zero beyond a fixed cutoff distance R_0,

$$V_k^{\text{cutoff}} = \sum_i C \frac{q_i}{|\vec{r}_i - \vec{r}_k|} S(|\vec{r}_i - \vec{r}_k|), \qquad (16.2)$$

with $0 \leq S(r) \leq 1$ and $S(r) = 0$ whenever $r \geq R_0$. Each atom now contributes to the electrostatic potential of just the sphere of lattice points enclosed by the R_0 cutoff radius. Also, since the density of a molecular system is bounded, there is an upper bound on the number of atoms within the R_0-sphere centered at a given lattice point. The choice of a specific switching function depends on the application. For example, when using Eq. (16.2) to compute the short-range part of an $O(N)$ multilevel summation approximation to Eq. (16.1), S is chosen to be a piecewise defined polynomial that is smooth at R_0 [6].

A sequential algorithm for direct summation requires nested loops over the lattice points and atoms. The algorithmic formulation has a scatter-based memory access pattern, dominated by memory writes, if the atoms are traversed in the outer loop, and has a gather-based memory access pattern, dominated by memory reads, if the lattice points are traversed in the outer loop. Although either approach can be made reasonably efficient for execution on a single CPU core, the gather-based approach is expected to give better performance, because it requires fewer overall memory accesses, NM atom reads and M lattice point writes, compared with the scatter-based approach requiring N atom reads and NM lattice point reads and writes, since each lattice point is iteratively accumulating its value. The coordinates of the lattice points can be computed on-the-fly as the algorithm progresses, avoiding the need to store individual lattice coordinates except while they are actually in use. In a sequential algorithm for cutoff summation, the scatter-based approach is more efficient, since the regularity of the lattice makes it efficient to determine the exact sphere of points that an atom affects.

16.4.2 Direct Summation on GPUs

A scatter-based direct summation algorithm cannot be efficiently implemented in parallel on GPUs due to the need for concurrent updates to lattice points. Although recent GPUs support many types of atomic update operations, they have much lower performance than conflict-free updates using coalesced global memory accesses. The gather-based algorithm is ideal for GPU computation due to its inherent data parallelism, arithmetic intensity, and simple data structures. More sophisticated algorithms can be constructed using design principles derived from techniques used in this example. Since direct summation contains no branching or distance cutoff tests in its inner loop, it may also be used as a rough measure of the peak floating-point arithmetic performance available to other closely related particle-grid algorithms.

The choice of parallel decomposition strategy for GPU particle-grid algorithms is determined primarily by the need to achieve high arithmetic intensity. Doing so requires making optimal use of memory bandwidth for the atomic data and for the lattice values. Directly relating the work item index space to the indices of the lattice simplifies computation, particularly on the GPU, where the scheduling hardware generates work item indices automatically. Similarly, the use of a lattice array memory layout directly related to the work item index space simplifies the task of achieving coalesced reads and writes to global GPU memory and helps avoid bank conflicts when accessing the on-chip shared memory. The illustration in Figure 16.1 shows how a three-dimensional lattice can be decomposed into independent work items and how the global pool of work items or threads are grouped together into blocks that execute on the same SIMD execution unit, with access to the same on-chip shared memory and constant memory.

On-the-fly calculation of such quantities as lattice coordinates make better use of the GPU resources, since the registers and fast on-chip memory spaces are limited while arithmetic is comparatively inexpensive. Whenever neighboring lattice points share a significant amount of input data, multiple lattice points can be computed by a single work item, using an optimization technique known as *unroll-and-jam* [3]. This optimization allows common components of the distances between an atom and a group of neighboring lattice points to be reused, increasing the ratio of arithmetic operations to memory operations at the cost of a small increase in register use [6, 20, 22]. Manual unrolling of these lattice point evaluation loops can provide an additional optimization, increasing arithmetic intensity while reducing loop branching, indexing, and control overhead.

Besides the memory accesses associated with lattice points, a good algorithm must also optimize the memory organization of atomic coordinates and any additional data, such as atomic element type, charge, radii, or force constants. The needs of specific molecular modeling algorithms differ depending on what group of atoms contribute to the accumulated quantity at a given lattice point. For algorithms that accumulate contributions from all atoms in each lattice point (e.g., direct Coulomb summation or computation of molecular orbitals), atom data are often stored in a flattened array with elements aligned and padded to guarantee coalesced global memory accesses. Arrays containing additional coefficients, such as quantum chemistry wavefunctions or basis sets, must also be organized to facilitate coalesced memory accesses [23]. In cases where coefficient arrays are traversed in a manner that is decoupled from the lattice or atom index, the use of on-chip constant memory or shared memory is often helpful.

The performances of efficient GPU implementations of direct summation algorithms have previously been shown to be as high as a factor of 44 times faster than a single core of a contemporary generation CPU for simple potential functions [16, 22]. Potential functions that involve exponential terms for dielectric screening have much higher arithmetic intensity, enabling GPU direct summation speedups of over a factor of 100 in some cases [21].

16.4.3 Cutoff Summation on GPUs

The cutoff summation algorithm must, like direct summation, be reformulated as a data-parallel gather algorithm to make it suitable for GPU computing. The computation involves performing a spatial hashing of the atoms into a three-dimensional array of bins. The summation of Eq. (16.2) may then be calculated by looping through the lattice points, and, for each lattice point, looping over the atoms within the surrounding neighborhood of nearby bins and evaluating an interaction whenever the pairwise distance to the atom is within the cutoff.

The CPU performs the spatial hashing of the atoms and copies the array of bins to the GPU. A work group is assigned a cubic region of lattice points to minimize the size of the surrounding neighborhood of bins that interact with the lattice points and to help ameliorate the effects of branch divergence from the cutoff distance test. The neighborhood for a work group can be determined from a lookup table stored in the constant memory cache. The GPU kernel uses a work group to collectively load a bin from the neighborhood into the on-chip shared memory and then loop through the cached atom data to evaluate their interactions. Computing multiple lattice points with a single work item reuses the shared memory data and increases the arithmetic intensity.

Generally, if data can be presorted to guarantee access with uniform- or unit-stride access, it is relatively straightforward to design a GPU kernel so that a work group collectively loads segments of arrays from global memory on demand, enabling data to be shared among all cooperating work items. This can be an effective strategy for using on-chip shared memory as a form of program-managed cache, and has previously been shown to achieve performance within 27% of a hardware cache, in the case of molecular orbital computation [23]. For GPU algorithms that include just the contributions from atoms within a specified cutoff distance, atomic coordinates and related data may be stored in spatially hashed data structures. The complete set of candidate atoms that must be considered for the summation of a given lattice point can be accessed as a function of the work group index and associated lookup tables or indexing arithmetic [6, 20].

The performances of GPU cutoff summation algorithms have previously been shown to be as high as a factor of 32 times faster than a single core of a contemporary CPU [6]. Performance varies with the cutoff distance, the lattice spacing, the average atom bin fill, and the branch divergence within warps or wavefronts due to the cutoff distance test.

16.4.4   Floating-Point Precision Effects

The summation of a large set of numbers, in this case the individual electrostatic potential contributions, can result in loss of precision due to the effects of floating-point roundoff and cancellation. Early GPUs offered only single-precision arithmetic, requiring the use of techniques such as Kahan summation, alignment of exponents and fusion significands of multiple floating-point words, or sorting of values prior to summation to improve arithmetic precision [2,7]. Although recent GPUs fully support double-precision floating-point arithmetic, these precision-enhancing techniques are worth noting since they may also be used to improve accuracy even for double-precision operations and to make the resulting sum less dependent on the order of summation. The use of mixed precision might offer the best tradeoff between accuracy and performance, with double precision reserved for those parts of the computation most affected by roundoff error, such as the accumulation of the potentials, while single precision is used for everything else.

16.5   GPU N-Body Algorithms

Molecular dynamics calculates forces between the atoms at every simulation time-step. Calculation of the non-bonded forces that occur between all pairs of the N atoms (the N-body problem) dominates the overall computational cost per time-step. The naive approach of directly evaluating $\frac{1}{2}N(N-1)$ pairwise interactions becomes intractable for large systems, which motivates the introduction of a cutoff distance and switching functions for the pair potentials, making the computational complexity linear in N. Although the algorithms for these N-body calculations share similarities with those of the previous section, there is less regularity in evaluating interactions between pairs of atoms, a smaller ratio of arithmetic operations to memory operations, and more exceptional cases, making GPU N-body algorithms more challenging to design and optimize than those for particle-grid calculation.

16.5.1   N-Body Forces

The force field used to model biomolecules combines terms from covalently bonded atoms and terms from pairwise non-bonded interactions. The non-bonded forces are calculated as the negative gradients of the electrostatic potential,

$$U^{\text{elec}}(\vec{r}_1, \ldots, \vec{r}_N) = \frac{1}{2} \sum_{i=1}^{N} \sum_{j \notin \chi(i)} C \frac{q_i q_j}{|\vec{r}_j - \vec{r}_i|}, \tag{16.3}$$

and the Lennard-Jones potential,

$$U^{\mathrm{LJ}}(\vec{r}_1, \dots, \vec{r}_N) = \frac{1}{2} \sum_{i=1}^{N} \sum_{j \notin \chi(i)} 4\epsilon_{ij} \left(\left(\frac{\sigma_{ij}}{|\vec{r}_j - \vec{r}_i|} \right)^{12} - \left(\frac{\sigma_{ij}}{|\vec{r}_j - \vec{r}_i|} \right)^{6} \right), \quad (16.4)$$

where atom i has position $\vec{r}_i$ and charge q_i, the constant C is fixed, the constants ϵ_{ij} and σ_{ij} are determined by the atom types of i and j, and the set $\chi(i)$ are the atom indices j that are excluded from interacting with atom i, typically including atom i and atoms that are covalently bonded to atom i.

The individual interaction terms decay quickly as the pairwise distance increases, so the computational work for the N-body problem is typically reduced from quadratic to linear complexity by introducing a cutoff distance and switching functions for electrostatic and Lennard-Jones potentials, similar to Eq. (16.2). The long-range electrostatic forces can be determined through complementary fast methods, as discussed in Section 16.5.3.

The basic algorithm for computing the non-bonded forces from (smoothly) truncated interaction potentials first performs a spatial hashing of atoms into bins and then accumulates forces on each atom due to the pairwise interactions within each bin and between the atoms in neighboring bins. The computational cost of evaluation is $O(N)$, with a constant that depends on the cutoff distance and the bin size. Scalable parallelism generally requires a spatial decomposition of the atoms into bins, which then permits non-bonded calculation with communication requirements limited to nearest neighbor processors. The atom positions are updated at each time-step from the numerical integration of Newton's equations of motion.

Many algorithmic enhancements are available to improve performance of the basic algorithm. The covalent bond topology, particularly the bonds to hydrogen, can be used to reduce the number of pairwise distance checks against the cutoff. Pairlists can be generated to explicitly list the (i, j) interactions available within the cutoff, omitting the excluded pairs. An extended interaction distance can be employed to reduce the frequency for updating the assignment of atoms to bins and regenerating the pairlists, which in turn reduces parallel communication.

16.5.2   N-Body Forces on GPUs

Although the density of systems of biomolecules with explicit solvent is fairly uniform, the calculation is less regular than a particle-grid calculation and depends on how full each bin is. When using GPUs to calculate the non-bonded forces, it is natural to assign each work item to calculate the force on a single atom. The hashing of atoms into bins is generally best done on the CPU, with the bins then copied to the GPU memory. A work group might be assigned to perform the calculations for one bin up to some clustering of bins, as needed to fully occupy the work group. The algorithm would proceed with a work group looping through the surrounding neighborhood of bins,

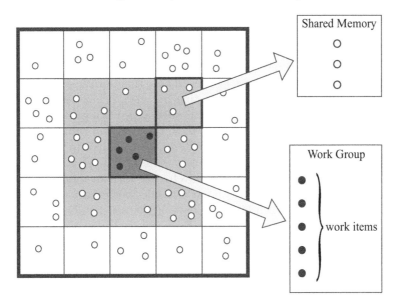

FIGURE 16.2: The atoms are spatially hashed into bins, which are assigned to work groups. The work group for a bin (dark-gray shading) loops over the neighboring bins (light-gray shading), loading each into shared memory. Each work item loops over the atoms in shared memory (white-filled circles), accumulating forces to its atom (black-filled circles).

cooperatively loading each bin into shared memory for processing, as depicted in Figure 16.2. Each work item loops over the atoms stored in shared memory, accumulating the force for any pairwise interactions that are within the cutoff distance. The memory bandwidth is effectively amplified by having each work item reuse the atom data in shared memory.

The atom bin input data must include position and charge for each atom. The constants for Lennard-Jones interactions are based on atom type and determined through combining rules (involving arithmetic and/or geometric averaging of the individual atom parameters) to obtain parameters ϵ_{ij} and σ_{ij} in Eq. (16.4) for the (i, j) interaction. Although there are few enough atom types to fit a square array of $(\epsilon_{ij}, \sigma_{ij})$ pairs into the constant memory cache, accessing the table will result in simultaneous reads to different table entries, which will degrade performance. A better approach might be to perform the extra calculation on-the-fly to obtain the combined ϵ_{ij} and σ_{ij} parameters. The individual atom parameters could either be stored in the bins or read from constant memory without conflict.

A sequential code typically makes use of Newton's Third Law by evaluating each interaction only once and updating the forces on both atoms. Doing this cannot be readily adapted to GPUs without incurring additional cost to safeguard against race conditions, either by synchronizing memory access to avoid write conflicts to the same memory location or maintaining additional buffer storage for later summation. It is typically more efficient for the GPU to evaluate each (i, j) interaction twice.

Branch divergence is a bigger problem here than for the particle-grid case, due not just to pairwise distance testing but also to bins that are not completely full. If multiple partially full bins are mapped to a work group, then the straightforward mapping of bin slots to work items will result in warps with a mixture of active and inactive work items. Performance might be improved by moving the available work to the lowest numbered work items, which could then possibly leave entire warps (or wavefronts) without useful work, thereby permitting them to finish immediately. Another optimization technique is to sort the atoms within each bin, perhaps using either a Hilbert curve or a recursive bisection partitioning, to make the pairwise distances calculated across a warp (or wavefront) more likely to collectively all pass or all fail the cutoff test.

Care must be taken to correctly handle the excluded non-bonded interactions. The particle-grid calculation of electrostatic potentials poses no such difficulties, and it is fine for the GPU to disregard any near-singularities resulting from an atom being too close to a grid point and have the CPU eliminate any out-of-range values. However, attempting a similar procedure for force calculation would destroy the precision of the force computation with cancellation errors caused by adding and later subtracting values of very large magnitude for excluded pairs of atoms in close proximity. The GPU must either explicitly test for exclusions or use an inner cutoff distance, within which the force magnitude is clamped to its value at the inner cutoff and the CPU is then used to correct the force for all nearby interactions and exclusions. It is important to clamp the force within the inner cutoff distance rather than simply truncating the force to zero. Differences in the rounding behavior and operation precision between the CPU and GPU make it possible for a pairwise distance approximately equal to the inner cutoff distance (within some rounding tolerance) to test differently on the two different architectures. In this case, the truncation to zero approach could cause a large discontinuous jump in the force.

Single precision is generally found to be sufficient for forces as long as the atom velocities are not too fast. The previously mentioned techniques can be applied to improve precision, such as summing forces to double-precision buffers. Single precision can be used for the atom input data if positions are stored using their relative displacement from the bin center.

16.5.3 Long-Range Electrostatic Forces

Accurate modeling of biomolecules requires calculation of the long-range contribution from the electrostatics potential. The most popular methods for computing the long-range electrostatics are particle-mesh Ewald (PME) [4] and particle-particle particle-mesh (P3M) [8]. Both of these methods make use of a local interpolation of the long-range electrostatic potential and its derivatives from a grid, where the grid of potentials has been calculated by a three-dimensional fast Fourier transform (FFT) with an overall computational cost of $O(N \log N)$. The grid calculations for PME and P3M are amenable to GPUs, and GPU libraries exist for calculating FFTs. Both of these methods still need calculation of a short-range part, as has already been discussed. However, the short-range interactions require the evaluation of the error function $\text{erf}(r)$, which is costly to calculate without special hardware support from the GPU.

Rather than calculating $\text{erf}(r)$ directly, an alternative approach is to evaluate the interactions through table lookup from texture cache memory. Advantages to using a table lookup, besides avoiding expensive function evaluations, include hardware-accelerated interpolation from texture memory, the elimination of special cases like the switching function for the Lennard-Jones potential, and the automatic clamping of the force to appropriate values at the outer and inner cutoffs. Disadvantages include some reduction in the precision of forces and the fact that the floating-point performance will continue to outpace the performance of the texture cache memory system on future hardware.

16.6 Adapting Software for GPU Acceleration

The challenges involved in effectively using GPUs often go beyond the design of efficient kernels. Modifications to the structure of existing application code are often required to amortize GPU management overhead, and to accumulate enough work to fully occupy all of the processing units of a GPU when a kernel is launched. Other challenges can involve scheduling of work across multiple GPUs on the same host or on different nodes of HPC clusters, and even mundane tasks such as GPU resource management and error handling. Below we describe some of these issues and the solutions we have devised as a result of adapting our applications NAMD [18] and VMD [9] for GPU acceleration.

16.6.1 Case Study: NAMD Parallel Molecular Dynamics

We develop and support a parallel molecular dynamics simulation package, NAMD, as a service to the computational biology research community. NAMD is designed to run on a wide variety of machines, including laptop and desktop computers, clusters of commodity servers, and even high-end supercomputers containing hundreds of thousands of processor cores. To achieve high scalability, NAMD uses a fine-grained parallel decomposition, allowing sophisticated mechanisms in the underlying Charm++ runtime system to dynamically balance load and optimize performance. The event-driven execution model provided by Charm++ (and used by NAMD) performs work in response to incoming messages from other processors.

The NAMD parallel decomposition is relatively fine grained, resulting in relatively small amounts of work being executed in response to a single incoming message. While this strategy reduces latency and provides many other benefits, it is at odds with the needs of contemporary GPU hardware and software. Current GPUs require tens of thousands of independent work items or threads in order to successfully hide memory and instruction execution latencies, and must therefore be presented with a large amount of data-parallel work when a kernel is launched. The amount of work triggered (or otherwise available) as a result of any individual incoming message is too small to yield efficient execution on the GPU, so incoming messages must be aggregated until there is enough work to fully occupy the GPU. Once enough work has accumulated, data are copied to the GPU, and a GPU kernel is launched.

It is undesirable to allow the GPU to idle while waiting for work to arrive; a refinement of the above approach partitions work into two categories: the portion resulting from communications with peers and the portion that depends only on local data. At the beginning of a simulation time-step, computations involving only local data can begin immediately. By launching local computations on the GPU immediately, the CPU can continue aggregation of incoming work from peers, and the GPU is kept fruitfully occupied. One drawback to this technique is that it will add latency to computations resulting from messages that arrive after the local work has already been launched, potentially delaying peers waiting on results they need. In the ideal case, local work would be scheduled immediately, but at a low priority, to be immediately superceded by work arriving from remote peers, thereby optimizing GPU utilization and also minimizing latency [19].

Next-generation GPUs will allow multiple kernels to execute concurrently [15], enabling work to be submitted to the GPU in much smaller batches, and making it possible to simultaneously process local and remote workloads. Despite the lack of work prioritization capabilities in current GPU hardware and software interfaces, it should be possible to take advantage of this capability to reduce the latency of work for peers while maintaining full GPU utilization. One possible approach for accomplishing this would submit local work in small batches just large enough to keep the GPU fully occupied. Work contained in messages arriving from remote peers would be immediately submitted to the GPU, thereby minimizing latency. GPU work completion events or periodic polling would trigger additional local work to be submitted on the GPU to keep it productively occupied during periods when no remote messages arrive. If future GPU hardware and software add support for work prioritization, then this scheme could be implemented simply by assigning highest priority to remote work, thereby minimizing latency [19].

Another potential complication that arises with GPU computing is how best to decompose work and manage execution with varing ratios of CPU cores and GPUs. Today, most applications that take advantage of GPU acceleration use a relatively simple mapping of host CPU threads to GPU cores. Existing GPU programming toolkits such as CUDA and OpenCL associate a GPU device with a context that is specific to a particular host thread. This creates pressure on application developers to organize the host-side application code so that only one thread accesses a particular GPU device. Parallel message passing systems such as MPI are completely unaware of the existence of GPUs, and on clusters with an unequal number of CPU cores and GPUs, multiple processes can end up competing for access to the same GPU. Since current-generation GPUs can only execute a single kernel from a single context at a time, and kernels run uninterrupted until completion, sharing of GPUs between multiple processes can create performance anomalies due to mutual exclusivity of access and the unpredictable order in which requests from competing processes are serviced. An application can determine what processes (e.g., MPI ranks) are running on the same physical node, and thereby avoid performance problems through self-managed software scheduling of GPU sharing. This is the approach taken by NAMD.

16.6.2   Case Study: VMD Molecular Graphics and Analysis

In addition to NAMD, we develop VMD, a matching tool for visualization, structure building and simulation preparation, and analysis of the results from molecular dynamic simulations. Unlike NAMD, which is most often used on parallel computers, VMD is typically used on laptop and desktop computers. Due to this prevailing usage model, the visualization and analysis techniques incorporated into VMD have historically been focused on methods that perform well on a moderate number of CPU cores. With the advent of GPU computing, VMD can now provide interactive visualization and analysis features that would have been impractical previously, using the same GPU hardware that is already leveraged for accelerating interactive graphical rendering of large molecular complexes. VMD uses GPUs to accelerate the interactive calculation and animated display of molecular orbitals from quantum chemistry simulations, with a single GPU achieving performance levels one hundred times that of a typical CPU [23].

Despite the tremendous computational power of GPUs, the requirement to perform computations at rates appropriate for interactive visualization tasks often exceeds the capabilities of a single device, necessitating the use of multiple GPUs in parallel. In highly controlled environments such as HPC clusters, hardware configurations are often carefully selected with each node containing one or more identical GPUs and each node consistent with its peers to help optimize performance and stability. Conversely, the laptop and desktop computers most used for applications such as VMD often contain an accumulation of multiple generations of GPU hardware of varying capability. In the case of laptops, where power consumption is of great concern, it is becoming common for multiple GPUs to be incorporated, with one relatively high-efficiency low-functionality GPU, and another performance-oriented high-capability GPU that consumes far more power. In such a system, peak application performance is attained by using all of the available CPU cores, and both GPUs in concert. Due to the potential for significant differences in performance between diverse GPUs, the use of dynamic load balancing and task scheduling mechanisms is critical to effective use of such GPUs within a single computation.

Another problem that can occur as the result of differing capabilities or capacities of GPUs on highly integrated platforms is resource exhaustion errors caused by competition or interference from other applications running on the same system. Unexpected errors deep within a multi-GPU code are difficult to handle, since errors must be handled by the thread detecting the problem, and actions to correct the error may require cooperation with other threads. In VMD, we use an error handling mechanism that is built into the dynamic work distribution system, enabling errors to be propagated from the thread encountering problems to its peers. This mechanism allows orderly cleanup in the case of fatal errors, and allows automatic redistribution of failed work in cases where the errors relate to resource exhaustion affecting only one GPU device.

16.7   Concluding Remarks

We have presented some of the most computationally expensive algorithms used in molecular modeling and explained how these algorithms may be reformulated as arithmetic-intensive, data-parallel algorithms capable of achieving high performance on GPUs. We have also discussed some of the software engineering issues that arise when adding support for GPU acceleration in existing codes, as done with NAMD for parallel molecular dynamics [18] and VMD for molecular visualization and analysis [9]. The algorithm design techniques and complementary software engineering issues are applicable to many other fields.

In the coming years, we expect the GPU hardware architecture to continue to evolve rapidly and become increasingly sophisticated, while gaining even greater acceptance and use in HPC. New capabilities are anticipated that will reduce the "performance cliffs" encountered by suboptimal access of the GPU memory systems, which will ease the difficulty of adapting algorithms to make effective use of GPU acceleration. Nevertheless, the algorithm design techniques presented here are likely to remain relevant for achieving optimal performance.

Acknowledgments

This work was supported by the National Institutes of Health under grant P41-RR05969. The authors also wish to acknowledge additional support provided by the NVIDIA CUDA center of excellence at the University of Illinois at Urbana-Champaign. Performance experiments were made possible by a generous hardware donation by NVIDIA.

Bibliography

[1] Anton Arkhipov, Jana Hüve, Martin Kahms, Reiner Peters, and Klaus Schulten. Continuous fluorescence microphotolysis and correlation spectroscopy using 4Pi microscopy. *Biophys. J.*, 93:4006–4017, 2007.

[2] David H. Bailey. High-precision floating-point arithmetic in scientific computation. *Comput. Sci. Eng.*, 7(3):54–61, 2005.

[3] David Callahan, John Cocke, and Ken Kennedy. Estimating interlock and improving balance for pipelined architectures. *J. Paral. Distr. Comp.*, 5(4):334 – 358, 1988.

[4] U. Essmann, L. Perera, M. L. Berkowitz, T. Darden, H. Lee, and L. G. Pedersen. A smooth particle mesh Ewald method. *J. Chem. Phys.*, 103:8577–8593, 1995.

[5] James Gumbart, Leonardo G. Trabuco, Eduard Schreiner, Elizabeth Villa, and Klaus Schulten. Regulation of the protein-conducting channel by a bound ribosome. *Structure*, 17:1453–1464, 2009.

[6] David J. Hardy, John E. Stone, and Klaus Schulten. Multilevel summation of electrostatic potentials using graphics processing units. *J. Paral. Comp.*, 35:164–177, 2009.

[7] Yun He and Chris H. Q. Ding. Using accurate arithmetics to improve numerical reproducibility and stability in parallel applications. In *ICS '00: Proceedings of the 14th International Conference on Supercomputing*, pages 225–234, New York, NY, USA, 2000. ACM Press.

[8] R. W. Hockney and J. W. Eastwood. *Computer Simulation Using Particles*. McGraw-Hill, New York, 1981.

[9] William Humphrey, Andrew Dalke, and Klaus Schulten. VMD—Visual molecular dynamics. *J. Mol. Graphics*, 14:33–38, 1996.

[10] Volodymyr V. Kindratenko, Jeremy J. Enos, Guochun Shi, Michael T. Showerman, Galen W. Arnold, John E. Stone, James C. Phillips, and Wen-mei Hwu. GPU clusters for high-performance computing. In *Cluster Computing and Workshops, 2009. CLUSTER '09. IEEE International Conference on*, pages 1–8, New Orleans, LA, USA, 2009.

[11] Erik Lindholm, John Nickolls, Stuart Oberman, and John Montrym. NVIDIA Tesla: A unified graphics and computing architecture. *IEEE Micro*, 28(2):39–55, 2008.

[12] Aaftab Munshi. OpenCL Specification Version 1.0, December 2008. http://www.khronos.org/registry/cl/.

[13] John Nickolls, Ian Buck, Michael Garland, and Kevin Skadron. Scalable parallel programming with CUDA. *ACM Queue*, 6(2):40–53, 2008.

[14] NVIDIA. *CUDA Compute Unified Device Architecture Programming Guide*. NVIDIA, Santa Clara, CA, USA, 2007.

[15] NVIDIA's next generation CUDA compute architecture: Fermi. White Paper, NVIDIA, 2009. Available online (Version 1.1, 22 pages).

[16] John D. Owens, Mike Houston, David Luebke, Simon Green, John E. Stone, and James C. Phillips. GPU computing. *Proc. IEEE*, 96:879–899, 2008.

[17] John D. Owens, David Luebke, Naga Govindaraju, Mark Harris, Jens Kruger, Aaron E. Lefohn, and Timothy J. Purcell. A survey of general-purpose computation on graphics hardware. *Comput. Graph. Forum*, 26(1):80–113, 2007.

[18] James C. Phillips, Rosemary Braun, Wei Wang, James Gumbart, Emad Tajkhorshid, Elizabeth Villa, Christophe Chipot, Robert D. Skeel, Laxmikant Kale, and Klaus Schulten. Scalable molecular dynamics with NAMD. *J. Comp. Chem.*, 26:1781–1802, 2005.

[19] James C. Phillips, John E. Stone, and Klaus Schulten. Adapting a message-driven parallel application to GPU-accelerated clusters. In *SC '08: Proceedings of the 2008 ACM/IEEE Conference on Supercomputing*, Piscataway, NJ, USA, 2008. IEEE Press.

[20] Christopher I. Rodrigues, David J. Hardy, John E. Stone, Klaus Schulten, and Wen-mei W. Hwu. GPU acceleration of cutoff pair potentials for molecular modeling applications. In *CF'08: Proceedings of the 2008 Conference on Computing Frontiers*, pages 273–282, New York, NY, USA, 2008. ACM.

[21] John E. Stone, David Gohara, and Guochun Shi. OpenCL: A parallel programming standard for heterogeneous computing systems. *Computing in Science and Engineering*, 12:66–73, 2010.

[22] John E. Stone, James C. Phillips, Peter L. Freddolino, David J. Hardy, Leonardo G. Trabuco, and Klaus Schulten. Accelerating molecular modeling applications with graphics processors. *J. Comp. Chem.*, 28:2618–2640, 2007.

[23] John E. Stone, Jan Saam, David J. Hardy, Kirby L. Vandivort, Wen-mei W. Hwu, and Klaus Schulten. High performance computation and interactive display of molecular orbitals on GPUs and multi-core CPUs. In *Proceedings of the 2nd Workshop on General-Purpose Processing on Graphics Processing Units, ACM International Conference Proceeding Series*, volume 383, pages 9–18, 2009.

[24] Leonardo G. Trabuco, Christopher B Harrison, Eduard Schreiner, and Klaus Schulten. Recognition of the regulatory nascent chain TnaC by the ribosome. *Structure*, 18:627–637, 2010.

Part IX

Complementary Topics

Chapter 17

Dataflow Frameworks for Emerging Heterogeneous Architectures and Their Application to Biomedicine

Umit V. Catalyurek

The Ohio State University

Renato Ferreira

Universidade Federal de Minas Gerais, Brazil

Timothy D. R. Hartley

The Ohio State University

George Teodoro

Universidade Federal de Minas Gerais, Brazil

Rafael Sachetto

Universidade Federal de Minas Gerais, Brazil

17.1 Motivation

Thanks to the increasing transistor density found in integrated circuits due to Moore's Law, graphics processing units (GPUs), the Cell Broadband Engine (CBE), and traditional microprocessors with increasingly high levels of core-level parallelism are now state-of-the-art commodity components. Their computational throughput, flexibility, and power efficiency is such that designing systems with these components is the best way to reach the scientific and biomedical community's ever increasing need for processing cycles. Indeed, Los Alamos National Laboratory's Cell-accelerated Roadrunner cluster was the first supercomputer to break the petaflops barrier, and held the title of the fastest supercomputer in the world until just recently, losing it to Oak Ridge National Laboratory's Jaguar cluster.

In order to leverage the high-performance potential of these new technologies, applications must be built with multi-level hardware parallelism in mind, including inter-node, thread-level, and instruction-level parallelism. Here, we demonstrate that the dataflow programming model, together with efficient dataflow runtime systems, such as Anthill [34] and DataCutter [3], is well-suited to address computational platform heterogeneity, while allowing efficient use of all levels of hardware parallelism. Coarse-grain dataflow runtime systems enable effective orchestration of multiple accelerators in conjunction with multicore CPUs, while fine-grain runtime systems enable effective use of individually programmable, heterogeneous compute cores of processors like the CBE. In both cases, the dataflow model allows the abstraction of the heterogeneous resources, enabling the developer to encapsulate architecture-specific details.

The dataflow programming model allows efficient parallelization for a broad class of applications, as it explicitly provides for several dimensions of application parallelism. In the dataflow model, applications consist of a set of processing stages that communicate through directed data channels. For runtime systems which can manage distributed computational resources, explicitly defining task boundaries in this manner explicitly enables application *task, pipelined,* and *data* parallelism. The goal of dataflow runtime systems is to match this application parallelism to hardware parallelism in an efficient manner. Irregular applications are challenging to develop for heterogeneous and distributed systems using traditional programming and runtime frameworks, as they do not explicitly address computational resource heterogeneity and application irregularity.

This chapter studies the suitability of the dataflow model to implement a computationally demanding, irregular biomedical application for heterogeneous distributed platforms (such as a cluster of CBEs or GPU-equipped PCs). Our use-case application is a computer-aided prognosis system that automates the analysis of digitized neuroblastoma tissue images. Neuroblastoma is a cancer of the sympathetic nervous system that affects children and is often present at birth. Traditionally, the patient's prognosis is determined by a visual examination of tissue samples, where morphological characteristics of the cancerous tissue are evaluated under a microscope by pathologists [31]. This manual analysis is very time consuming and subject to error [31]; hence, using an automated image analysis system will speed the process, and may help reduce error. Beyond the benefit to the biomedical community, this application is a good test-bed for the use of the dataflow model for other computationally demanding, data-intensive applications.

17.2   Dataflow Computing Model and Runtime Support

The dataflow programming model consists of an application task model and an associated middleware runtime system, which sits on top of the operating system(s) of the involved computational nodes and provides the dataflow interface abstraction to the application. In dataflow applications, the processing structure is implemented as a set of components, referred to as *filters*, that exchange data through *logical streams*. A *stream* denotes a uni-directional data flow from one filter (i.e., the producer) to another (i.e., the consumer). Data flows along these *streams* in untyped *buffers* so as to minimize various system overheads. A *layout* is a filter ontology which describes the set of application tasks, streams, and the connections required for the computation. A *placement* is one instance of a *layout* with an actual filter copy to physical processor mappings.

In the filter-stream programming model, all of the computation involved with the application is handled by the filters operating on data flowing in streams. Filters implement specific transformation functions for data items flowing in streams. These transformation functions can take the form of simple one-to-one data buffer transformations, data split functions (where more than one output data buffer is created for each incoming data buffer), and data join functions (where more than one data buffer is required to trigger the execution of a filter's processing function). Thus, when developing a filter-stream application the programmer should only implement such filter transformation functions, and the filter-stream framework should handle all of the communication between these filters.

For certain types of applications, the most natural decomposition into the dataflow model is a directed acyclic graph (DAG), where the execution consists of passing multiple data items through this DAG. For other applications, the computation structure is a cyclic graph where the execution consists of multiple iterations of the processing sequence. In this case, an application starts with data representing an initial set of possible solutions and as these pass through the filters, new candidate solutions are created. In our experience, we have noticed that this type of application decomposition leads to an asynchronous execution, since several candidate solutions (possibly from different iterations) are processed simultaneously at runtime.

The dataflow application decomposition process leads to task, data, and pipelined parallelism. *Task parallelism* is the concurrent execution of independent tasks in a dataflow application on exactly one instance of the application's input data stream. *Pipelined parallelism* is the concurrent execution of dependent tasks on different instances of the input data stream. Dataflow runtime systems also support *data parallelism* at multiple levels. At the application level, multiple copies of the application layout can be instantiated and executed. At the filter level, multiple copies can be transparently created, providing the illusion of a single, higher-bandwidth filter to all upstream and downstream filters. This transparent copy mechanism allows any filter in the application's graph to be replicated over many compute nodes; the data that go through each filter may be partitioned among the copies. Care should be taken when the filter being replicated has state. To ensure application scalability, the filter state should be partitioned. In order to maintain a correct partitioned filter state, incoming messages are sorted with respect to the state variables they update and delivered to the appropriate transparent copy.

The dataflow model, therefore, exploits task, pipelined, and data parallelism as well as communication asynchrony to provide efficient application performance. By dividing the computation into multiple pipeline stages, and replicating each pipeline stage transparently, we can make use of very fine-grained parallelism. Since there is minimal synchronization among parallel tasks, the execution is mostly bottleneck free. By leveraging non-blocking `send` primitives, communication can be overlapped with computation on the sending side of a data transfer. Similarly, dataflow systems can be designed to run in their own threads, allowing the runtime system to prefetch data from remote sources before any `receive` call is made. In this way, with minimal amounts of network bandwidth and filter execution time, most of the communication latency can be overlapped with useful computation. Thus, by decomposing the application into logical functional steps, these runtime overheads are drastically minimized.

The dataflow programming model is, by necessity, supported by a middleware runtime system. The two instances of dataflow runtime systems used in this paper are called DataCutter [3,13] and Anthill [10,34]. The runtime engine performs all steps necessary to instantiate filters on the desired machines and cores, to connect all logical endpoints, and to call the filter's interface functions for processing work. The processing, network, and data copying overheads can be minimized by intelligent placement and scheduling of filters.

17.3   Use Case Application: Neuroblastoma Image Analysis System

Several research studies have been conducted with the aim of developing CAP systems for various types of cancer [12, 18, 19, 23, 28, 32]. While these studies show promising results for CAP applications in pathology, they require significant computational power to analyze the digitized images of cancer tissues. Therefore, we investigate high-performance solutions, making use of multicore microprocessors and accelerator technologies.

Neuroblastoma is a cancer of the sympathetic nervous system which mostly affects children. The prognosis of the disease is currently determined by expert pathologists based on visual examination under a microscope of tissue slides. The slides can be classified into different prognostic groups conditioned to the differentiation grade of the neuroblasts, among other issues [30].

Manual examination by pathologists is an error-prone and very time-consuming process [37]. Therefore, the goal of neuroblastoma image analysis (NBIA) is to assist in the determination of the prognosis of the disease by classifying the digitized tissue samples into different subtypes that have prognostic significance. In this application, we concentrated on the classification of stromal cell development as either stroma rich or stroma poor, which is one of the morphological criterions in NB prognosis that contributes to the categorization of the histology as favorable and unfavorable [31].

Since the slide images can be very high resolution (over 100K × 100K pixels of 24-bit color), the whole-slide NBIA first decomposes the image into smaller image tiles that are processed independently. The analysis can then be performed with a single- or a multi-resolution strategy [25]. In the single-resolution approach, each of the tiles resulting from the decomposition step is processed at the full resolution. The multi-resolution strategy, on the other hand, constructs multiple copies of the image tiles from the decomposition step, each one with a different resolution. For example, a three-layered multi-resolution analysis could be constructed with (32 × 32), (128 × 128), and (512 × 512) image sizes. The analysis proceeds for each of the tiles, starting at the lowest resolution, and stops at the resolution level where the classification satisfies some predetermined criterion.

The multi-resolution analysis strategy is designed to reduce the high levels of computation involved in the analysis of each neuroblastoma image tile, as it mimics the way pathologists examine the tissue slides under the microscope. The image analysis starts from the lowest resolution, which corresponds to the lower magnification levels in a microscope, and uses the higher-resolution representations for the regions where the classification step requires more detailed information.

Tile classification is achieved based on statistical features that characterize the texture of tissue structure. Therefore, NBIA first applies a color space conversion to the La*b* color space, where color and intensity are separated and the difference between two pixel values is perceptually more uniform, enabling the use of Euclidean distance for feature calculation. The texture information is extracted using co-occurrence statistics and local binary patterns (LBPs), which help characterize the color and intensity variations in the tissue structure [25]. Finally, the classification confidence at a particular resolution is computed using hypothesis testing, if the multi-resolution approach has been employed; either the classification decision is accepted or the analysis resumes with a higher resolution if one exists. The result of the image analysis is a classification label assigned to each image tile indicating the underlying tissue subtype, e.g., stroma rich or stroma poor, or background. Figure 17.1 shows the flowchart for the classification of stromal development. More details on the NBIA can be found in [29].

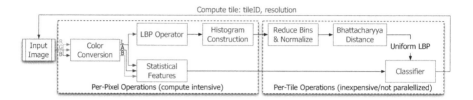

FIGURE 17.1: Neuroblastoma image analysis (NBIA) flow chart.

17.4   Middleware for Multi-Granularity Dataflow

In this section, we detail our approach for designing multi-granularity (fine-grain and coarse-grain) runtime systems to support the dataflow programming model for heterogeneous environments. On the coarse-grained side, we developed mechanisms to efficiently execute dataflow applications on heterogeneous clusters of CPUs and GPUs [6,14,15,36]. The fine-grained framework was designed to allow dataflow applications to be developed for the CBE [13]; this level of hardware granularity matches best with fine-grained application parallelism.

Although they target different hardware and application granularities, both of these coarse-grain and fine-grain types of frameworks utilize the filter-stream programming model as the abstraction to implement applications, distribute and execute tasks and handle communication between the computational elements. The main unifying theme between the two granularities of runtime systems is the heterogeneity of the computational elements: multicore CPUs and GPUs in the coarse-grained framework, and the two CBE processor types in the fine-grained framework. The filter-stream model maps well to these heterogeneous contexts as each application filter can be implemented targeting the most appropriate device, while the runtime system orchestrates the overall execution. This approach is only possible because in filter-stream applications the input and output interfaces of each computing stage are well defined. Provided the interface to the rest of the application's filters is not changed, the specific implementation details of any particular filter are immaterial; that is, computational devices, programming languages, and other internal specifics are all malleable. The next two sections detail each of the frameworks.

17.4.1 Coarse-grained on Distributed GPU Clusters

The Anthill [36], our coarse-grained framework, has been designed to target clusters with computational nodes consisting of multicore CPUs and GPUs. Our approach to exploit these environments allows the programmer to provide multiple implementations of the filters' processing functions targeting different devices, such as CPUs and GPUs. Thus, the application can benefit from all the available computing power, as multiple data buffers arriving in input streams of the application filters can be processed using multiple heterogeneous processors. The next section presents the details of Anthill's architecture and programming framework.

17.4.1.1 Supporting Heterogeneous Resources

The programming abstraction in Anthill is event oriented [34]. The programmer's task is to decompose the application's processing structure into a task graph and provide event-handling functions for the filter tasks. Additionally, the programmer provides event-matching functions that specify data dependency conditions for the execution of these event handlers. Thus, each filter should associate handling functions to each stream arriving at the same filter. The events are asynchronous and independent in the dataflow model, and filters are multi-threaded, meaning that if resources are available, multiple handlers can be executed concurrently on consecutive data buffers in the same stream. The filter runtime environment controls all non-blocking communication, sending and receiving messages through the streams, and chooses the right processing function for each event as defined by the programmer. This event-based approach derives from the message-oriented programming model pioneered by the x-kernel [26] and later extended to include explicit, user-defined events in Coyote [4].

Our solution for coarse-grained filter-stream application development for heterogeneous environments is based on Anthill and makes full use of the event-based dataflow model it exposes. In our approach, the programmer can create multiple versions of each stream handler function, each of which targets a distinct device. At runtime, Anthill automatically chooses among the available alternatives and dispatches events to different event handlers concurrently. In heterogeneous systems, the event handlers may be implemented for either the CPU or the GPU, or both.

It is up to the filter-stream runtime environment to decide, once an event is received, when it may be processed, given the hardware resources available. As long as there are resources available and events pending, execution may continue. Figure 17.2 presents a schematic view of an Anthill filter's environment. The filter shown receives data from multiple input streams (*In1, In2,*

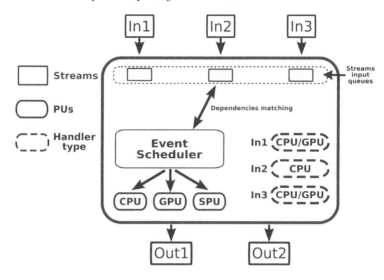

FIGURE 17.2: Filter's environment.

and In3), each of which has an associated handler function. These functions are implemented targeting the most appropriate processing units (PUs). Thus, as depicted in the figure, one input stream can have handler implementations for multiple PUs.

The decision about the mapping of events to the available hardware is made by the "Event Scheduler." The current implementation of Anthill includes two scheduling strategies [35,36]: (i) first-come first-served (FCFS), and (ii) dynamic weighted round robin scheduling (DWRR). The first scheduling decision is to select what stream to execute events from. The two scheduling strategies work similarly for this step; they take events from the next stream in round robin fashion, while observing the constraint that the chosen stream should have an event handler for the available PU.

In the FCFS strategy, the next event to be processed will be the oldest queued for the selected stream. Although very simple, this policy is interesting since the events are dynamically assigned to devices at runtime, according to the processing rate of each available PU. The second policy, DWRR, divides the events to be processed according to a weight that may vary during the execution time, and receives a value for each device. During execution, this weight is then used to order the input queue of each stream according to the available devices, and, when a certain processor is available, the event to be processed is that which has the highest weight. Thus, this policy assumes out-of-order execution of events arriving in each input stream.

17.4.1.2 Experimental Evaluation

The experiments presented in this section were performed using a cluster of 10 PCs connected using a Gigabit Ethernet Switch, where each node has an Intel Core 2 Duo CPU, running at 2.13 GHz, 2 GB of main memory, an NVIDIA GeForce 8800GT GPU, and the Linux operating system. During the experiments when a GPU is used, a single CPU core is reserved to coordinate the GPU's actions. The NBIA input image has 26,742 tiles and two resolution levels: (32×32) and (512×512). For some experiments, we varied the recalculation rate (the percentage of tiles that are reanalyzed at a higher resolution). Thus, as we increased the recalculation rate, each tile has a higher probability of being recomputed at the second resolution level.

The NBIA was implemented in Anthill using four filters: (i) Image reader, which reads the RGB tiles from the disk and sends them to next step; (ii) color conversion: which converts the tiles into the La*b* color space; (iii) computation of the statistical features, which computes the feature vector and LBPs; and (iv) classifier filter, which computes the per-tile operations, receives the feature vector, applies the classification, does the hypothesis testing to decide whether or not the classification is satisfactory, and controls the processing flow. To exploit the target system's heterogeneous resources, we implemented CPU and GPU versions of the "color conversion" and "computation of the statistical features" filters, which were fused into a single filter to avoid unnecessary CPU/GPU data transfers and network communication. When the streamed data arrive at this filter, the Anthill Device scheduler chooses a specific device, and, consequently, a handler function to process the received data.

In the first set of experiments, the performance of the NBIA implementations for the CPU and GPU are analyzed as a function of the input image resolution. During this evaluation, several workloads with varying tile dimensions were generated using a fixed number of 26,742 tiles. In these experiments, the classification is assumed to be successful at the first image resolution. Note that these experiments were run on a single node.

TABLE 17.1: Application speedup variation.

Tile dimensions	32×32	64×64	128×128	256×256	512×512
Speedup	0.62	2.02	6.65	16.97	32.97

The speedup of the GPU versus one CPU core is shown in Table 17.1 for various image tile sizes. The results show high variation of the relative performance between the CPU and the GPU; while the performance of the GPU is worse for 32×32 pixel tiles, the GPU is almost 33 times faster for 512×512 pixel tiles. This is because for small tasks, the overhead of using the

GPU is on the same order as the tile analysis execution time, making using the GPU inefficient.

In real runs of the NBIA, as discussed in Section 17.3, a multi-resolution approach is employed; thus, multiple tiles with different resolutions can be in the processing pipeline simultaneously. In order to improve the execution time in this context we used relative performance speedup, shown in Table 17.1, as the input to our DWRR scheduling algorithm. Thus, when an image tile is received, its parameters are passed to a set of user-provided functions (one for each processor type) which returns the relative speedup. In order to provide the best heuristic ordering of the data buffers, we simply return the number of image rows for the GPU speedup value and the inverse of the number of image rows for the CPU speedup value. This provides an extra, important ordering mechanism for the CPU which means that the CPU will have a preference for processing smaller image tiles when a choice among different resolutions is available.

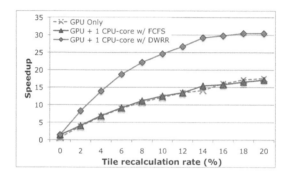

FIGURE 17.3: Speedup of three different versions of the code (GPU only and GPU + CPU with FCFS and GPU + CPU with DWRR) over CPU-only version while varying recalculation rate.

In Figure 17.3, we present the speedup of three versions of the NBIA application as compared to the CPU-only version: the GPU-only version and the collaborative GPU+CPU version with FCFS and DWRR scheduling, respectively. The standard deviation for these experiments was never higher than 4% for FCFS, and 3% for DWRR.

The GPU+CPU results with the FCFS scheduling policy showed significant improvement over the GPU-only version when the recalculation rate was 0% (no recalculation). Since the CPU and GPU have a similar performance for low-resolution tiles, when there is no recalculation and no heterogeneity between the tiles in the processing pipeline, it makes no difference which tile is analyzed by which device. Table 17.2 presents the number of tiles processed by the CPU using 0% recalculation. We can see that FCFS and DWRR have almost the same data partitioning, arriving at a similar performance.

However, when we increased the recalculation rate, we see the DWRR

TABLE 17.2: #tiles processed by CPU at each resolution/scheduling.

Recalc (%)	0		12	
Resolution	Low	High	Low	High
1 CPU core - FCFS	16049	0	263	251
1 CPU core - DWRR	15978	0	21592	4

policy still almost doubling the speedup (compared to a single CPU core) of that of the pure GPU case, while FCFS achieves almost no improvement over the pure GPU case. For instance, for 12% tile recalculation, the GPU-only version of the application is 13.46 times faster than a single CPU core, while using the CPU and GPU together achieved speedups of 13.48 and 26.71 for FCFS and DWRR, respectively.

These experiments are further detailed in Table 17.2, which shows the profile of the tasks processed by the CPU with a 12% tile recalculation rate, under each scheduling policy. As shown, using FCFS, the NBIA processed few low-resolution tiles in the CPU, for which it is faster. When using the DWRR policy, the CPU processed more than 80% of the small tiles, while the GPU processed the majority of the high-resolution ones.

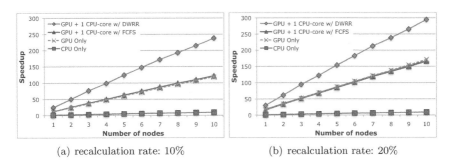

(a) recalculation rate: 10%   (b) recalculation rate: 20%

FIGURE 17.4: Speedup of three different versions of the code (GPU only and GPU + CPU with FCFS and GPU + CPU with DWRR) over CPU-only version while varying the number of compute nodes.

In order to evaluate the performance of our application in a distributed environment, we performed experiments using one copy of each filter per machine, and using 26,742 image tiles, as before. We also varied the recalculation rate from 0% to 20%, but the scalability results did not vary substantially; hence, we only show the results for a 10% and 20% rate here. These speedup results are shown in Figure 17.4. For these experiments, computing the speedups for all versions of the NBIA using the execution time of the respective version on one processor yields near-linear scaling results. Here we plot the speedup using the execution time of the CPU only version on one node as the baseline. Figure 17.4 demonstrates how much speedup multi-node GPU-equipped clusters can achieve with effective task scheduling.

17.4.2   Fine-Grained on Cell

DataCutter-Lite (DCL) [13] is our fine-grain dataflow runtime engine and programming framework for modern multicore microprocessors. As a dataflow framework, it allows developers to write applications as components arranged in a task graph with data dependencies. By registering the filters and their stream interconnections with the application development framework, the runtime engine handles all of the data movement, process invocation, and task execution required to complete the execution. The current implementation of DCL targets only single nodes. However, as we show below, keeping the same dataflow programming model for both intra-node and inter-node programming allows fast and scalable application development.

Many other methods exist for programming multicore and accelerator microprocessors, including block-based methods [1, 7, 9, 16, 20, 33], source-level methods [2, 5, 8, 11, 17, 21], message-passing environments [22, 24, 27], and other streaming frameworks [38]. Unfortunately, none of these programming frameworks provide the capability of scaling from fine-grain tasks and hardware resources to coarse-grain tasks and hardware resources in the same way dataflow frameworks can.

17.4.2.1   DCL for Cell—Design and Architecture

The first implementation of DCL is designed for the CBE. As an instance of a dataflow runtime system, DCL exposes an application programming interface (API) which allows developers to decompose their applications into dependent task graphs, composed of filters and streams. Built on top of IBM's SPE Runtime Management Library Version 2.2 (libspe2), and a two-sided communication library called the CBE Intercore Messaging Library (CIML), DCL's API provides functions to describe the application's filters and stream interconnections, initialize the runtime system, and manage the execution of the application.

The runtime system follows the event-driven model, wherein handler functions are assigned to each stream, such that when data buffers arrive at a filter from a specific stream, the runtime system calls the appropriate processing function. This event-driven model is well-suited to the CBE, since the synergistic processing elements (SPEs) are single-threaded processors and have no capability to switch thread contexts like more traditional processors.

17.4.2.2   NBIA with DCL

With DCL, we have implemented the single-resolution NBIA for the CBE [14]. As discussed above, the input to the overall application is a tissue slide image digitized at high resolution. Each RGB pixel is converted to the LAB color space and some statistics are calculated on a per-tile basis. The luminance channel is then taken from the LAB image and an LBP feature is calculated. The four statistics per image channel and the LBP feature comprise a feature vector which is used in a classification stage to determine the properties of the image tile. To compare with our DCL-based implementation, we have developed a custom CBE implementation of the image analysis application. This implementation uses one main power processor elements (PPE) thread loop which reads the RGB image tiles from a socket and performs the functions on each tile. The operations are slightly decoupled, meaning that the RGB-to-LAB color space transformation and the LBP feature calculation must be scheduled separately on the SPEs. The main loop in the application acts as a task scheduler.

Figures 17.5 and 17.6 show the results of the experiments on the full biomedical image analysis application with various disk I/O and tile decompression times excluded and included, respectively. When end-to-end applications are considered, even the most efficient algorithm implementation is subject to such concerns as disk latency and upstream data processing times. In this case, the upstream data processing is the decompression of the TIFF images to be calculated. When excluding these times, we see near-linear speedup for the DCL implementation. Unfortunately, the baseline version of the application actually begins to suffer when the number of SPEs used rises above 5. This is likely due to the extra scheduling time and memory bandwidth used in the baseline's implementation, since it decouples the two major stages of the operation and schedules them separately. The DCL implementation simply writes one buffer as output from the first stage to the input of the second stage. Since this buffer stays in the SPE's local store memory, it saves main memory bandwidth, and since the second stage is triggered by the runtime system resident on the SPE, the PPE is not involved in the scheduling operation, saving time.

When the disk I/O and TIFF decompression times are included, the application performance decreases, particularly when more SPEs are involved in the computation. The best speedup achieved by DCL and baseline version was 2.7×. The DCL version does not read the decompressed TIFFs from an incoming socket, since this operation would require the use of mutexes in order to share the socket, and we have avoided this type of programming, since it is incompatible with our goal of designing a runtime system devoid of these types of details. As such, each TIFF is decompressed in the same thread which calls the DCL routines to analyze the image.

To solve the problem of insufficient TIFF decompression bandwidth, we have implemented a mixed coarse-grain and fine-grain dataflow version of the image analysis application using DataCutter and DCL. The layout of the application integrates DataCutter for inter-node communications and DCL for intra-node executions. As such, Figure 17.6 shows the results when DataCutter (DC) is used to distribute the TIFF tile decompression stage among several computational nodes, and one CBE processor is used to analyze the tiles with

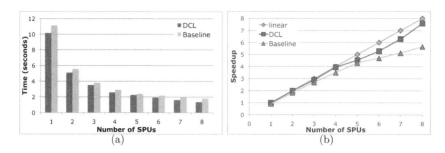

FIGURE 17.5: Execution times and speedups for biomedical image analysis application for 32 image tiles — I/O and decompression time excluded.

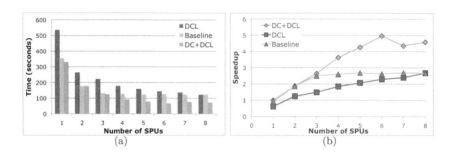

FIGURE 17.6: End-to-end execution times and speedups for biomedical image analysis application for 1024 image tiles.

its 8 SPEs. Unfortunately, OSC's CBE blades are currently only configured to run with Gigabit Ethernet, which limits the amount of help distributed nodes can give.

17.5   Conclusions and Future Work

Biomedical applications are an important focus for high-performance computing (HPC) researchers currently. Due to their large potential benefit to the public good, and their computationally intensive and irregular nature, these applications are as challenging to work with as they are rewarding to produce. In order to enable medical researchers to further progress in their own fields, the HPC community needs to apply effort to give current biomedical applications sufficient performance to remove them from the critical path of medical research. The use of accelerators, with their low cost and high performance, is a natural avenue for investigating methods to provide this high performance.

Unfortunately, due to the irregular nature of the applications, and the difficulties in programming particular accelerator technologies, mapping biomedical applications onto heterogeneous clusters is a challenging task. Here, we have shown that the dataflow programming model and associated runtime systems can, at multiple application and hardware granularities, ease the implementation of challenging biomedical applications for these types of computational resources.

Acknowledgments

This work was supported in parts by CNPq, CAPES, Fapemig, and INWeb; by the DOE grant DE-FC02-06ER2775; by AFRL/DAGSI Ohio Student-Faculty Research Fellowship RY6-OSU-08-3; by the NSF grants CNS-0643969, OCI-0904809, OCI-0904802, and CNS-0403342; and an allocation of computing time from the Ohio Supercomputer Center.

Bibliography

[1] M. Älind, M. V. Eriksson, and C. W. Kessler. BlockLib: a skeleton library for cell broadband engine. In *IWMSE '08: Proceedings of the 1st International Workshop on Multicore Software Engineering*, pages 7–14, New York, NY, USA, 2008. ACM.

[2] P. Bellens, J. M. Pérez, R. M. Badia, and J. Labarta. CellSs: A programming model for the Cell BE architecture. In *SC*, page 86, 2006.

[3] M. Beynon, R. Ferreira, T. M. Kurc, A. Sussman, and J. H. Saltz. Datacutter: Middleware for filtering very large scientific datasets on archival storage systems. In *IEEE Symposium on Mass Storage Systems*, pages 119–134, 2000.

[4] N. T. Bhatti, M. A. Hiltunen, R. D. Schlichting, and W. Chiu. Coyote: A system for constructing fine-grain configurable communication services. *ACM Trans. Comput. Syst.*, 16(4):321–366, 1998.

[5] Brook+. http://ati.amd.com/technology/streamcomputing/AMD-Brookplus.pdf.

[6] U. Catalyurek, T. Hartley, O. Sertel, M. Ujaldon, A. Ruiz, J. Saltz, and M. Gurcan. Chapter: Processing of large-scale biomedical images on a

cluster of multi-core CPUs and GPUs. IOS Press, 2009. Accepted for publication.

[7] CUDA. Home page maintained by Nvidia. http://developer.nvidia.com/object/cuda.html.

[8] L. Dagum and R. Menon. Openmp: An industry standard api for shared-memory programming. *Computational Science and Engineering, IEEE*, 5(1):46–55, Jan-Mar 1998.

[9] K. Fatahalian, D. R. Horn, T. J. Knight, L. Leem, M. Houston, J. Y. Park, M. Erez, M. Ren, A. Aiken, W. J. Dally, and P. Hanrahan. Sequoia: Programming the memory hierarchy. In *SC '06: Proceedings of the 2006 ACM/IEEE conference on Supercomputing*, page 83, New York, NY, USA, 2006. ACM.

[10] R. Ferreira, W. M. Jr., D. Guedes, L. Drummond, B. Coutinho, G. Teodoro, T. Tavares, R. Araujo, and G. Ferreira. Anthill: A scalable run-time environment for data mining applications. In *Symposium on Computer Architecture and High-Performance Computing (SBAC-PAD)*, 2005.

[11] Amd stream computing. http://ati.amd.com/technology/streamcomputing/index.html.

[12] M. N. Gurcan, T. Pan, H. Shimada, and J. Saltz. Image analysis for neuroblastoma classification: Segmentation of cell nuclei. In *Internatioanl Conference of the IEEE Engineering in Medicine and Biology Society*, pages 4844–4847, New York City, New York, USA, 2006.

[13] T. D. R. Hartley and U. V. Catalyurek. A component-based framework for the cell broadband engine. In *Proc. of 23rd Int'l. Parallel and Distributed Processing Symposium, The 18th Heterogeneous Computing Workshop (HCW 2009)*, May 2009.

[14] T. D. R. Hartley, U. V. Catalyurek, A. Ruiz, F. Igual, R. Mayo, and M. Ujaldon. Biomedical image analysis on a cooperative cluster of GPUs and multicores. In *Proceedings of the 22nd Annual International Conference on Supercomputing, ICS 2008*, pages 15–25, 2008.

[15] T. D. R. Hartley, A. R. Rasih, C. A. Berdanier, F. Ozguner, and U. V. Catalyurek. Investigating the use of GPU-accelerated nodes for SAR image formation. In *Proc. of the IEEE International Conference on Cluster Computing, Workshop on Parallel Programming on Accelerator Clusters (PPAC)*, 2009.

[16] IBM. Accelerated Library Framework. http://www-01.ibm.com/chips/techlib/techlib.nsf/techdocs/41838EDB5A15CCCD002573530063D465.

[17] IEEE. Threads extension for portable operating systems (draft 6), February 1992. p1003.4a/d6.

[18] K. Jafari-Khouzani and H. Soltanian-Zadeh. Multiwavelet grading of pathological images of prostate. *Biomedical Engineering, IEEE Transactions on*, 50(6):697–704, June 2003.

[19] J. Kong, H. Shimada, K. Boyer, J. Saltz, and M. Gurcan. Image analysis for automated assessment of grade of neuroblastic differentiation. In *IEEE International Symposium on Biomedical Imaging (IEEE ISBI07)*, pages 61–64, Washington, D.C., USA, 2007.

[20] D. M. Kunzman, G. Zheng, E. J. Bohm, J. C. Phillips, and L. V. Kalé. Poster reception—Charm++ simplifies coding for the cell processor. In *SC*, 2006.

[21] W. I. Lundgren, K. B. Barnes, and J. W. Steed. Gedae: Auto coding to a virtual machine. In *Proc. of the 8th High Performance Embedded Computing Conference*, 2004.

[22] The Message Passing Interface (MPI).

[23] A. Nasser Esgiar, R. Naguib, B. Sharif, M. Bennett, and A. Murray. Microscopic image analysis for quantitative measurement and feature identification of normal and cancerous colonic mucosa. *Information Technology in Biomedicine, IEEE Transactions on*, 2(3):197–203, September 1998.

[24] M. Ohara, H. Inoue, Y. Sohda, H. Komatsu, and T. Nakatani. MPI microtask for programming the Cell Broadband Engine processor. *IBM Syst. J.*, 45(1):85–102, 2006.

[25] T. Ojala, M. Pietikainen, and T. Maenpaa. Multi-resolution gray-scale and rotation invariant texture classification with local binary patterns. *IEEE Trans. on Pattern Analysis and Machine Intelligence (PAMI)*, 24:971–987, 2002.

[26] S. W. O'Malley and L. L. Peterson. A dynamic network architecture. *ACM Trans. Comput. Syst.*, 10(2):110–143, 1992.

[27] S. Pakin. Receiver-initiated message passing over RDMA networks. In *IPDPS*, pages 1–12, 2008.

[28] S. Petushi, C. Katsinis, C. Coward, F. Garcia, and A. Tozeren. Automated identification of microstructures on histology slides. In *IEEE International Symposium on Biomedical Imaging (IEEE ISBI04)*, pages 424–427, Washington, D.C., USA, 2004.

[29] O. Sertel, J. Kong, H. Shimada, U. V. Catalyurek, J. H. Saltz, and M. N. Gurcan. Computer-aided prognosis of neuroblastoma on whole-slide images: Classification of stromal development. *Pattern Recognition, Special*

Issue on Digital Image Processing and Pattern Recognition Techniques for the Detection of Cancer, 42(6):1093–1103, 2009.

[30] H. Shimada, I. M. Ambros, L. P. Dehner, J. Hata, V. V. Joshi, and B. Roald. Terminology and morphologic criteria of neuroblastic tumors: Recommendation by the international neuroblastoma pathology committee. *Cancer*, 86:349–363, 1999.

[31] H. Shimada, I. M. Ambros, L. P. Dehner, J. ichi Hata, V. V. Joshi, and B. Roald. Terminology and morphologic criteria of neuroblastic tumors: Recommendation by the international neuroblastoma pathology committee. *Cancer*, 86(2):349–363, 1999.

[32] M. Tahir and A. Bouridane. Novel round-robin tabu search algorithm for prostate cancer classification and diagnosis using multispectral imagery. *Information Technology in Biomedicine, IEEE Transactions on*, 10(4):782–793, October 2006.

[33] Intel Threading Building Blocks 2.0 for Open Source. http://threadingbuildingblocks.org/.

[34] G. Teodoro, D. Fireman, D. Guedes, W. M. Jr., and R. Ferreira. Achieving multi-level parallelism in the filter-labeled stream programming model. *Parallel Processing, International Conference on Parallel Processing (ICPP)*, 0:287–294, 2008.

[35] G. Teodoro, R. Sachetto, D. Fireman, D. Guedes, and R. Ferreira. Exploiting computational resources in distributed heterogeneous platforms. In *Symposium on Computer Architecture and High-Performance Computing (SBAC-PAD)*, 2009.

[36] G. Teodoro, R. Sachetto, O. Sertel, M. Gurcan, W. M. Jr., U. Catalyurek, and R. Ferreira. Coordinating the use of GPU and CPU for improving performance of compute intensive applications. In *IEEE Cluster*, New Orleans, LA, USA, 2009.

[37] L. Teot, R. Khayat, S. Qualman, G. Reaman, and D. Parham. The problems and promise of central pathology review: Development of a standardized procedure for the children's oncology group. *Pediatric and Developmental Pathology*, 10(3):199–207, 2007.

[38] D. Zhang, Q. J. Li, R. Rabbah, and S. Amarasinghe. A lightweight streaming layer for multicore execution. *SIGARCH Comput. Archit. News*, 36(2):18–27, 2008.

Chapter 18

Accelerator Support in the Charm++ Parallel Programming Model

Laxmikant V. Kalé

University of Illinois at Parallel Programming Lab, Urbana-Champaign

David M. Kunzman

University of Illinois at Parallel Programming Lab, Urbana-Champaign

Lukasz Wesolowski

University of Illinois at Parallel Programming Lab, Urbana-Champaign

18.1 Introduction

The increased popularity of accelerators and heterogeneous systems has naturally lead to their adoption in the realm of high performance computing (HPC). As the name implies, HPC applications are always striving to achieve greater performance, typically to perform scientific simulations, financial predictions, etc. The high peak computation rates (GFlops) of accelerators make them particularly attractive to programmers creating these compute intensive HPC codes. Charm++ is one of several parallel programming models used to develop HPC applications. As we will discuss, we believe that some features of the Charm++ programming model make it well suited for programming heterogeneous clusters as well as clusters with accelerators. We also describe the various extensions that have been introduced into the Charm++ programming model to add support for accelerators and heterogeneous clusters.

18.2   Motivations and Goals of Our Work

Various accelerators have been available for some time now, including field programmable gate arrays (FPGAs), graphical processing units (GPUs), and the Cell processor (Cell). It is clear that the motivation for developers to write code targeting platforms that include accelerators is the great computational performance that accelerators provide some applications. This claim is reinforced by the existence of accelerators in a variety of systems, ranging from desktops to large clusters housed in various government agencies, such as Roadrunner at LLNL [3], Lincoln[1] at NCSA, and MariCel[2] at BSC. The increased use of accelerators naturally creates a need for parallel programming models that assist developers targeting heterogeneous clusters that include accelerators. This need provides the basic motivation of our work. More precisely, the goal of our work is to reduce the programming effort required to program heterogeneous clusters that may or may not include accelerators, while maintaining portability. Towards this end, we have extended the Charm++ programming model. Currently, the development of our extensions is focused on CUDA-based GPUs, Cell, and Larrabee. As such, we will discuss our extensions mainly within the context of these devices, with particular emphasis on the Cell processor.

As we will discuss in Section 18.3.2, we believe that there are several features of the Charm++ programming model that make it well suited for programming accelerators. These features include *virtualization*, *data encapsulation*, and *near future predictability*. Our previous work on the Offload API [19, 20], a general C library for developing code on the Cell and a precursor to this work, illustrates the generality of these features. By taking advantage of these features in our extensions, we enable Charm++ programs to be portable between systems with and without accelerators, with minimal programmer effort. In many cases, programmer effort will be limited to recompiling the program. Programmers may also have to modify data/computation grain sizes or otherwise tune performance parameters, in a manner similar to tuning/optimizing C++ code when porting it between platforms.

There are several aspects of programming models that help reduce programmer effort. One such property, obviously, is portability. Heterogeneous clusters that include accelerators often require the programmer to interleave architecture specific code within the application code. This architecture spe-

---

[1]http://www.ncsa.illinois.edu/UserInfo/Resources/Hardware/Intel64TeslaCluster/
[2]http://www.prace-project.eu/documents/PRACE-Prototypes.pdf

cific code provides access to architectural features that are often the source of the architecture's greatest strengths. For example, to reach peak GFlops rates on most modern processors, a programmer must make use of single instruction multiple data (SIMD) instructions (e.g. streaming SIMD extensions [SSE]). On Cell, to make use of the synergistic processing element (SPE) cores, where most of the functional units reside, the programmer must use direct memory access (DMA) commands to move data to and from programmer managed scratchpad memories. However, if the application is to be portable between platforms, then (1) the non-portable architecture specific code needs to be removed and (2) the required functionality provided by these pieces of code needs to be recreated on the target system(s). A goal of our extensions is that they are portable between various platforms without requiring the programmer to rewrite code, thus easing programmer burden.

Another property of programming models that aids programmers is the ability to write programs in a modular way. Modularity allows programmers to implement significant portions of code independently and then later "glue" them together by using their well defined interfaces. The fact that the independently written pieces of code are going to be linked together into the same application does not affect their internal implementations, allowing developers to more easily develop code and libraries and providing a mechanism for code reuse. However, when developing for accelerators, implementing code in a modular way is often complicated for various reasons. One such complications is that native programming models for accelerators often grant exclusive access to hardware resources. For example, in the Cell, an SPE thread has exclusive use of the SPE and must voluntarily exit or yield before another thread can begin executing. Because programmer code has complete control of the SPE, the SPE thread must *choose* to *play nicely* since nothing will force the thread to yield. Other possible complications include forced groupings of threads, explicit memory management, multiple address spaces, and so on. Towards this end, we have designed our extensions to allow for the development of modular code.

Finally, it would be advantageous if programmers could use the same parallel programming model to parallelize their application across the available host and accelerator cores. First, this means the programmer only has to learn a single programming model and is not otherwise required to add extra code to interface two or more models. Second, the underlying runtime system for the programming model can take advantage of a unified representation of a program to automate various tasks, such as load balancing, that the programmer would otherwise have to explicitly control. By using the same abstraction, *entry methods executing on chare objects*, for both the host cores and the accelerators, the underlying Charm++ runtime system is capable of mapping the *chare objects* as it sees fit for the purpose of load balancing.

18.3   The Charm++ Parallel Programming Model

Before we discuss our extensions to the Charm++ programming model (Charm++), we review the model itself. Charm++ has been developed on top of the C++ programming language. Charm++ applications execute on top of an intelligent and adaptive runtime system, referred to as the *Charm++ runtime system*. The Charm++ runtime system automatically manages many things on behalf of the application, such as mapping chare objects to physical hardware, routing messages between the processors, scheduling entry methods, dynamic load balancing, automatic checkpointing, and fault tolerance. Charm++ is implemented as a library, without a specialized compiler.

18.3.1   General Description of Charm++

Charm++ [18] is an object oriented message driven parallel programming model based on C++. Charm++ programs are composed of a collection of special C++ objects called *chare objects* (or just *chares*). In addition to member functions, chare objects have *entry methods* and can *migrate* between processors. The chare objects in a given application define the algorithm that the application implements. For example, in a matrix-matrix multiplication, each matrix could be divided into a set of tiles, each represented by a single chare object, thereby distributing the matrix data. Further, other chare objects could be responsible for performing the actual multiplication of the individual tiles (i.e. work is also decomposed and parallelized via chare objects).

Chare objects are the basic unit of parallelism in Charm++. This makes Charm++ fairly different from other programming models, such as message passing interface (MPI) [12], where each processing core executes an independent thread. Threads are the main unit of parallelism within MPI, capable of encapsulating local data, communicating with one another as peers, and performing local computation. In Charm++, chare objects encapsulate local data within their member variables. Chare objects are also able to communicate with one another as peers and perform local computations by *invoking entry methods*. Entry methods are "special" member functions that are invoked on chare objects (*chareObject.entryMethod(parameters)*), have passed parameters, and may access the local chare object's member variables. The difference between member functions and entry methods is that entry methods are *asynchronously invoked*. From the point of view of the invoker, the entry method returns immediately without a return value. The runtime system packs the passed parameters into a message, finds the physical processor on which the target chare object resides, and routes the message to it. After the message is received by the target processor, the message is placed in a message queue and eventually scheduled for execution by the Charm++ runtime system. Only a single entry method can be *executing on a chare object* at any given moment, removing race conditions on member variables accessed by multiple entry methods and bounding the amount of parallelism in a Charm++ program to the number of chare objects. With this in mind, Charm++ programs are *over-decomposed* so that there are many more chare objects than there are physical processors. Computation and communication are automatically overlapped since, at any given moment, on a given physical processor there are likely chare objects with entry methods ready to execute (messages have arrived) while other chare objects are waiting for messages to arrive (messages in transit). Over-decomposition also allows the runtime system to automatically migrate the chare objects between the physical processors for the purpose of load balancing. Chare objects, entry methods, and the other programming constructions in Charm++ are identified and prototyped via *interface files*, which are similar to header files in C/C++. That is, interface files are Charm++ specific files that identify which objects are chares, chare arrays, groups, message objects, entry methods, parameters, parameter sizes, and so on.

Charm++ has been used for many years in the development of high performance and highly scalable applications, such as NAMD [5], ChaNGa [16], and OpenAtom [6]. As such, we can only give a brief introduction to the ideas and features that are encompassed in the Charm++ programming model and runtime system. For more information on Charm++ and related material, including documentation and tutorials, please visit the Charm++ website.[3]

18.3.2   Suitability of Charm++ for Exploiting Accelerators

As we have already alluded to, we believe that there are several features of the Charm++ programming model that make it well suited for programming accelerators, including *virtualization, data encapsulation*, and *near future predictability of execution*.

First, in Charm++, programs are over-decomposed into a set of chare objects. Naturally, this leads to an automatic overlap of communication (messages in transit) and computation (execution of entry methods on chare objects as already delivered messages are dequeued from the message queue). We refer to this over-decomposition as *processor virtualization*. From the point of view of an individual chare object, it seems that it has its own *virtual processor*. However, in reality, multiple chare objects are sharing the same physical processor under the control of the Charm++ runtime system.

Second, the application data are *encapsulated* within chare objects and messages. Entry methods typically execute on both the data contained within the associated chare object and the data contained within the message that triggered the entry method's execution (the parameters passed by the caller). Through the programming model, the programmer clearly exposes the possible

---

[3]http://charm.cs.illinois.edu/

working set for an entry method, allowing the runtime system to identify and move application data as required.

Third, the Charm++ runtime system keeps a queue of incoming messages. The runtime system then schedules messages in this message queue based on queue order, entry method type, and message priorities, etc. The result is that the message queue serves as a work list of entry methods that are ready to execute on the specified message and chare objects. We refer to this as *near future predictability* since we can predict with perfect accuracy what portions of application code will need to be executed in the near future.

When combined, these three features make Charm++ quite flexible. With perfect accuracy, the runtime system can predict what pieces of application code need to be executed in the near future. Application code is encapsulated within entry methods and data are encapsulated within message and chare objects. As long as it adheres to one entry method executing per chare object at any given moment, the runtime system is free to schedule entry methods on the available cores without risking data races. As a result of virtualization, there are typically multiple messages in the message queue targeting multiple chare objects on the same physical processor, allowing the runtime system to make use of multiple cores, host and accelerator alike. Virtualization further allows entry methods to be streamed through the available accelerator cores since data movement can be overlapped with code execution. As an example, an SPE's memory controller can move data related to one or more entry methods while the SPE itself executes another entry method.

For these features to be useful for accelerators, we are assuming, of course, that the runtime system is capable of executing entry methods on accelerators. To enable the execution of entry methods on accelerators, we have introduced multiple extensions to Charm++, including the *SIMD instruction abstraction*, *accelerated entry methods*, and *accelerated blocks*. Entry methods come in several varieties, including *threaded*, *sync*, *expedited*, and so on. To this mix, we add *accel*, short for *accelerated*. Section 18.4 will discuss these extensions.

18.4   Support for Cell and Larrabee in Charm++

Two accelerator technologies that have received attention recently are the Cell processor [17] and Larrabee [27]. Both are designed to accommodate general purpose computation, but they differ in various ways. The Cell processor is composed of two types of cores, the power processing element (PPE) and the SPE. In current implementations, there is a single PPE and multiple SPEs, ranging from four to eight. The PPE can serve as a host core with direct access to the network. Larrabee, on the other hand, resembles a commodity multicore processor that serves as a co-processor to the system's host processor.

We have extended the Charm++ programming model to support Cell and

```
// Header file (chare class declaration)
class Tile : public CBase_Tile {  // Chare class "Tile"
  private:
    float* C;   // C[M*N] subtile of overall output matrix
    int M, N;   // Number of rows and columns in local tile
  public:
    Tile(int m, int n) { M = m; N = n; C = new float[M*N];
                         memset(C, 0, sizeof(float)*M*N);   }
    ~Tile() { delete [] C; }
    calcTile(int M, int N, int K, float* A, float* B);
};

// Interface file (entry method declaration)
entry void calcTile(int M, int N, int K,
                    float A[M*K], float B[K*N]);

// Source file (entry method function body)
void Tile::calcTile(int M, int N, int K, float* A, float *B) {
  for (int row = 0; row < M; row++)
    for (int col = 0; col < N; col++) {
      float Cval = 0;
      for (int elem = 0; elem < K; elem++)
        Cval += A[elem+K*row] * B[col+N*elem];
      C[col+N*row] += Cval;
    }
}
```

FIGURE 18.1: This code is written using standard Charm++ constructs. The *calcTile()* entry method of the *Tile* chare class takes two matrix tiles as input, multiplies them together, and adds the corresponding elements of the result to its local matrix tile, *C*.

Larrabee by adding several extensions, including *accelerated entry methods*, *accelerated blocks*, and a *SIMD instruction abstraction*. While we discuss our current work in the context of the Cell processor [21], we have begun adding support for Larrabee [27] and would like to target additional accelerator types in the future. We also target CUDA-based GPUs. However, the large grain size requirement for proper utilization and other details make it difficult to directly map accelerated entry methods to CUDA-based GPUs. This may change as new programming methods become available. Nonetheless, we currently have a separate framework for using CUDA-based GPUs in Charm++ applications, as described in Section 18.5.

Currently, only a limited number of details regarding Larrabee's design are being made publicly available by Intel [27]. Because of this, we are restricted in what we can publicly discuss about our extensions with regards to Larrabee. Instead, we can simply state that we have had early success in applying our Charm++ extensions to Larrabee. Therefore, we will discuss our extensions solely in the context of the Cell processor. The reader, however, should keep in mind that, unless stated otherwise, the discussion of our extensions is applicable to both devices.

Before we introduce our extensions, we first introduce a short piece of example code that will later be modified for illustrative purposes. The example code in Figure 18.1 performs a naive matrix-matrix multiplication in the *calcTile()* entry method of the *Tile* chare class. We present a naive implementation for both brevity and for the sake of keeping the example simple, so our later code changes are clear and straightforward. An optimized version will only complicate the example, without providing a more in-depth illustration of our extensions. The *Tile* chare class has three local variables: C contains the local 2D tile of an overall matrix, M is the number of rows in C, and N is the number of columns in C. These variables are initialized when an instance of the *Tile* chare class is created. In this example, we have a single entry method that receives two input matrix tiles, multiplies them together, adding the corresponding elements of the result into the local C matrix tiles.

18.4.1   SIMD Instruction Abstraction

Many modern processors have SIMD extensions. Only by using these instruction extensions can a program achieve a significant fraction of the peak performance of a processor core that supports them. However, these extensions are architecture specific, often requiring non-portable intrinsics to be interleaved throughout the application code.

The goal of our SIMD instruction abstraction is to create a unified interface for many of the commonly supported SIMD operations that appear in various architectures. By utilizing a common interface, programmers may write their application code once. The operations provided by the SIMD abstraction are then mapped to the specific SIMD instructions provided by the architecture. If the underlying architecture does not provide SIMD instructions, a generic C/C++ implementation is used. It should be noted that we do not wish to prevent programmers from optimizing their code beyond our abstraction. As such, we provide precompiler macros that can be used by programmers to conditionally include architecture specific code as they see fit. We currently provide support for SSE, AltiVec/VMX, and the SPE's SIMD extensions. We will also include support for Larrabee when hardware is made available (work is currently in progress).

The SIMD instruction abstraction presents several data types that represent packed sets of some of the primitive C++ data types, which we refer to as *SIMD registers*. The abstraction also defines several operations that can be performed on SIMD registers in a data parallel manner, such as subtraction, multiplication, and fused-multiply-add. Figure 18.2 demonstrates the application of our SIMD abstraction to the previous code example presented in Figure 18.1. In the modified code, the loop that iterates over the columns in the final matrix is "SIMDized" to operate on *vecf_numElems* columns per iteration. The *vecf* data type is a packed set of *floats*. The *const_vzerof* is a packed set of floats where each element has a value of zero. The *vspreadf()* operation copies a scalar value into all the elements of a SIMD register. The

```
// Header file (chare class declaration)
class Tile : public CBase_Tile {  // Chare class "Tile"
    // Only the Tile() and ~Tile() functions change
    Tile(int m, int n) {
        M = m; N = n;   // Will cast C to 'vecf*' so align
        C = CmiMallocAligned(M*N*sizeof(float), sizeof(vecf));
        memset(C, 0, sizeof(float)*M*N);
    }
    ~Tile() { CmiFreeAligined(C); }
};

// Interface file (entry method declaration)
entry void calcTile(int M, int N, int K,
                    align(sizeof(vecf)) float A[M*K],
                    align(sizeof(vecf)) float B[K*N]);

// Source file (entry method function body)
void Tile::calcTile(int M, int N, int K,
                    float A[M*K], float B[K*N]) {
  vecf* Bv = (vecf*)B;   // Treat B and C as "columns of SIMD...
  vecf* Cv = (vecf*)C;  // vectors," not "columns of scalars"
    int Nv = N / vecf_numElems;  // Number of "vector columns" in C
    for (int row = 0; row < M; row++)
      for (int colv = 0; colv < Nv; colv++) {
        vecf Cval = const_vzerof;  // vecf_numElems columns at once
        for (int elem = 0; elem < K; elem++)
          Cval = vmaddf(vspreadf(A[elem+K*row]), // A*B+C
                        Bv[colv+Nv*elem],
                        Cval);
        Cv[colv+Nv*row] = vaddf(Cval; Cv[colv+Nv*row]);
  }  }
```

FIGURE 18.2: The same chare class and entry method as in Figure 18.1, but using our SIMD abstraction. To keep the example code short, we assume that N is an integer multiple of *vecf_numElems*.

vmaddf() operation performs a fused-multiply-add. Note that the programmer doesn't necessarily care how many elements there are in a SIMD register at compile time, as long as calculations using that value can be performed at runtime.

The SIMD abstraction increases portability between various types of processing cores, including x86 cores, PPEs, and SPEs. However, it does not take care of data movement and other architecture specific code that would still be required to make use of accelerators. To further increase portability, we introduce *accelerated entry methods*.

18.4.2   Accelerated Entry Methods

Accelerated entry methods behave in a manner similar to standard entry methods. A function body operates on the passed parameters along with the chare object's data. However, accelerated entry methods are meant to target computationally intensive calculations within the program (i.e. the function body should perform a relatively large amount of computation with minimal control flow). By marking an entry method as *accelerated*, the programmer is indicating that the runtime system *may* execute it on an accelerator core. This not only allows the runtime system to make use of the available accelerators, but also enables load balancing a portion of the application's overall workload between host and accelerator cores dynamically at runtime.

From the point of view of the programmer, accelerated entry methods, by design, resemble standard entry methods. However, there are some differences. First, accelerated entry methods are completely defined within interface files, giving tools in the Charm++ build process access to the function body code. Second, an entry method is identified as being accelerated by using the *accel* keyword. Third, in addition to passed parameters, the programmer specifies which member variables of the chare class the accelerated entry method will access (the *local parameters*). So far, these three differences are a direct result of Charm++ being implemented as a library. In the future, if the Charm++ build process were to include a custom C++ compiler with Charm++ specific extensions, these differences may be eliminated. Fourth, once the accelerated entry method finishes executing, a programmer specified *callback function* will be executed on the host core. Fifth, the function body of an accelerated entry method is somewhat limited relative to standard entry methods. The most notable difference is that accelerated entry methods cannot invoke other entry methods from within their function bodies. The callback function, however, can invoke other entry methods. Figure 18.3 illustrates how the *Tile::calcTile()* entry method from Figure 18.2 would be implemented as an accelerated entry method. In this particular example, the function body is exactly the same since the original function body in Figure 18.2 did not invoke any other entry methods (had it done so, the invocation would have been moved to the callback function).

As in C++, there may be common functionality shared between two or more entry methods. *Accelerated blocks* are designed to address such cases, by giving programmers a place where they can declare functions and/or macros which can be called by accelerated entry methods. The point is to minimize code duplication by moving the shared functionality into functions declared within the accelerated blocks and then calling these functions within accelerated entry methods. The function bodies in accelerated blocks are limited in the same way accelerated entry method function bodies are limited.

As is the case with standard entry methods, the runtime system automatically handles data movement and scheduling related to accelerated entry methods. Typically, there will be several objects making use of accelerated entry methods and able to make use of the same accelerator (a result of over-decomposition). The runtime system can then stream accelerated entry methods through the accelerator. As one accelerated entry method executes on the accelerator, the input/output data of one or more other accelerated entry methods can be moved to/from the accelerator's memory, respectively.

```
// Header file (chare class declaration)
class Tile : public CBase_Tile {  // Chare class "Tile"
    // Declare CkIndex_Tile (generated code) as a friend class
    friend class CkIndex_Tile;
    void calcTile_callbackFunc() { /* invocations here */ }
};

// Interface file (entry method declaration and function body)
entry [accel] void calcTile(  // Passed parameters...
                    int M, int N, int K,
                    align(sizeof(vecf)) float A[M*K],
                    align(sizeof(vecf)) float B[K*N]
                  ) [  // Local parameters...
                    readWrite : float C[M*N] <impl_obj->C>
                  ] {
    // Function body same as SIMD version (omitted for brevity)
} calcTile_callbackFunc;
```

FIGURE 18.3: The same chare class and entry method as in Figure 18.2, but implemented as an accelerated entry method instead of a basic entry method.

Additionally, because accelerated entry methods are well encapsulated, accelerated entry methods from different chare objects, executing different code, can be interleaved somewhat arbitrarily within the stream. Furthermore, our approach uses the same programming model for both host and accelerator cores (entry methods executing on chare and message objects), allowing for a single underlying and unified runtime system. The Charm++ runtime system already performs various tasks for the programmer, such as load balancing, fault tolerance, and automatic checkpointing. By using the same programming model and runtime system, we also enable these features in relation to accelerators. However, implementing these features will require additional engineering work on the runtime system and remain future work.

18.4.3   Support for Heterogeneous Systems

A natural question to ask at this point is whether or not our abstractions can be applied to heterogeneous systems. In some sense, we have already answered yes in that the Cell is a heterogeneous processor. However, because Charm++ is designed to target clusters, we are also interested in heterogeneous clusters where host cores are architecturally different from one another. Even in the context of heterogeneous clusters, the answer remains yes.

Towards this end, we have modified the Charm++ runtime system to handle architectural differences between host cores, such as endianness. As entry methods, standard or accelerated, are invoked and the passed parameters are packed into a message, the runtime system also includes information describing how the parameters are encoded. Just prior to the entry method being invoked on the target processor, the passed parameters are unpacked from the message and any architectural differences between the sending and receiving processors are reconciled. From the point of view of the programmer, no application code changes are required. In addition to the various tasks the runtime system already takes care of for the programmer, such as message creation and routing, we add automatic modification of application data (passed parameters) to account for architectural differences between host cores.

18.4.4 Performance

We use a simple molecular dynamics (MD) code to demonstrate our extensions. The example MD code is modeled after the popular NAMD [5] application, but greatly simplified. For brevity, we omit a detailed description of the example MD code since it has already been described in previous work [21]. In both NAMD and our example MD code, there are patch objects containing particle data and compute objects that calculate the forces between patches. At each timestep, patches send particle data to computes, computes calculate forces, computes send force data to patches, and patches integrate (updating the particles' positions and velocities). The MD code uses an all-to-all non-cutoff based algorithm $(O(N^2))$, whereas NAMD uses a cutoff based algorithm $(O(N))$. Note that the amount of parallelism in Charm++ is directly related to the number of chare objects (at most one entry method executing per chare object at any moment). To avoid generating excessive parallelism in our executions, we choose a problem size of 92160 particles (144 patches, 10440 compute objects) creating less than a factor of two more compute objects than NAMD simulating the common ApoA1 benchmark (92224 particles, 144 patches, 6017 compute objects). We point this out to provide the reader a context in which to compare our simple code against a production code.

Figure 18.4 presents our performance results when executing the example MD code on a heterogeneous cluster composed of: four IBM QS20 Cell Blades (*blades*: using one Cell each, one PPE & eight SPEs per Cell), four Sony PlayStation 3s (*PS3s*; one Cell each, one PPE & six SPEs per Cell), and one dual-core Intel Xeon 5130 (*Xeons*). A *blade/PS3 pair* is a combination of one blade Cell (using one per node) and one PS3 Cell. Plot A, which we include for reference, presents previously achieved [21] GFlop rates without load balancing. Plot B represents updated results including minor changes we have made to the MD code since our previous work, such as performance tuning and adjusting message priorities. In plot C, we added a simple weight-based static load balancer to the example MD code to account for both chare object differences (patches vs computes) and node differences (blades vs PS3s). Plot D further adds two Xeon cores to the specified number of blade/PS3 pairs. Note that the executions in plot D make use of three core types (Xeons, PPEs, SPEs), three SIMD instructions extensions (SSE, AltiVec/VMX, the SPEs' extensions), and two memory structures (cache hierarchy and scratch-pad memory). Further, only the PPEs and the Xeon cores are peers to one another on the network, while the SPEs are slaves to the PPEs. The example MD code uses our extensions and thus contains no architecture specific

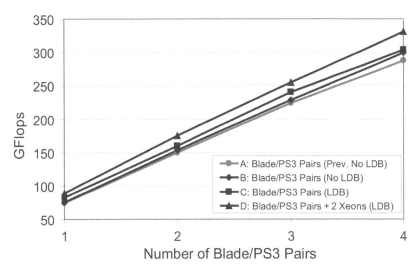

FIGURE 18.4: Performance of the example MD program executing on a heterogeneous cluster, with and without load balancing.

code (i.e. remains portable) even though it makes use of architecture specific features of all core types.

One might expect plot D to only give a constant increase over plot C since two Xeons were added at each point. However, as can be seen, the added Xeons actually cause the MD code to scale better. Upon closer inspection, we noticed that the Xeon cores were better at communication (lower send times) than the Cells (especially the blade Cells which took more than twice the time on average to do a message send, of the same size, compared to the Xeons). The static load balancer is placing most of the patches (communication heavy) on the Xeons and most of the computes (computation heavy) on the Cells. By matching the strengths of each core type to the demands of each object type, better performance is achieved. In plot C, however, the hardware is not as heterogeneous (only Cell processors) and, thus, the static load balancer cannot take advantage of the communication strengths of the Xeons, harming the MD code's overall scaling performance.

18.5   Support for CUDA-Based GPUs

General purpose GPUs are another popular accelerator platform. Designed to exploit the massive parallelism inherent in graphics, GPUs in the past few years have become an attractive solution for accelerating data parallel applications. The key development which made it possible to apply the computational power of GPUs to non-graphics applications was the creation of software tools and a hardware infrastructure for general purpose programming on the GPU. CUDA [24] is the most popular such platform in use today. Currently, we only support CUDA; however, as more portable accelerator programming languages, like OpenCL [15], mature, we may add support for them.

When writing a Charm++ application to run on a cluster of CUDA GPUs, developers can either use CUDA directly or employ the *Charm++ GPU Manager* framework which we have developed. Using CUDA directly within a Charm++ application is possible, but becomes tedious when trying to maximize performance. Since a Charm++ program typically has a number of entry methods waiting to execute, it is important to ensure that CPUs are not blocked when executing GPU kernels or transferring memory between host and device. Secondly, the presence of multiple chare objects per processor, many of which may need to use the GPU, requires that chares be able to execute on the GPU without explicit synchronization with other ongoing GPU work. CUDA allows both of these requirements to be satisfied through the use of CUDA streams and polling functions. Under this model, the user associates a unique CUDA stream parameter to each chare executing on the GPU. Subsequently, in order to check if all GPU events within a stream have finished, the user can make a polling call with the stream value as a parameter.

The above approach is possible but has negative implications in terms of both usability and performance. The approach requires periodic execution of entry methods, on each chare which is using the GPU, in order to check if work for a particular stream has completed. This reduces code clarity and is tedious for the user. In addition, while the number of chares which have work to be executed on the GPU may be large, at any moment, only one will actually be executing on the GPU. A large number of calls to the periodic entry methods which perform the polling will waste CPU cycles. A better-performing scheme is possible, but requires additional programming effort.

Our approach to using CUDA-based GPUs aims to automate the management of GPU kernels and memory transfer, thus simplifying the work of the programmer [28]. Under this model, the user provides the system with a *work request* which indicates the GPU kernel and CPU buffers containing data for the kernel. The user also submits a *callback object*, to be invoked by the system once the GPU work request is complete. The callback object typically specifies the entry method to be executed directly after GPU work is complete. As a result, the user no longer has to poll for completion of GPU operations, and the Charm++ GPU Manager knows which work request is currently executing, allowing it to poll only for currently executing operations.

Although CUDA, at least until the release of the upcoming Fermi architecture, does not allow concurrent execution of multiple kernels, it does allow overlap of kernel execution with memory transfer operations. GPU Manager makes use of this capability to pipeline the execution of work requests. In steady state, the transfer of data from the GPU to the CPU for a finished work request happens concurrently with the kernel execution of a second work request and the transfer of data to the GPU for a work request which will execute next.

18.6   Related Work

Several programming models have been created that target a single processor (perhaps multicore) combined with one or more accelerators in single node (non-cluster) systems, including Sequoia [11], RapidMind [23], Mercury's Multicore Framework [7], CUDA [24], OpenCL [15], Ct [13], CellSs [4], and others. The largest difference these approaches have with our work is that we specifically target heterogeneous clusters that may or may not include accelerators. Further, while these other models could be used to program clusters, they require a second programming model to communicate between host cores, such as TCP/UDP sockets or MPI. We use a single programming model for both host and accelerator cores.

Other approaches extend existing parallel programming models. Open MPI [14] and HeteroMPI [22] extend MPI. Our approach goes further in that it provides support for using SIMD instructions and accelerators in addition to host core heterogeneity. However, unlike Open MPI, our work does not address network heterogeneity. MPI microtasks [25] also extends MPI. The XLC single source compiler [10] makes use of OpenMP [8] to program Cell. Unlike our approach, neither XLC single source compiler nor MPI microtasks address heterogeneous clusters or provide an abstraction for SIMD instructions.

Other programming models specifically target heterogeneous clusters that may include accelerators. In StarSs [2,26] functions are marked as *tasks*. These tasks are then executed on the available cores, accelerated or not. Tasks in StarSs are arranged hierarchically. In our model, chare objects are peers to one another with no implicit synchronization between parent and child *tasks*. StarPU [1] and HMPP [9] both use *codelets* (tasks) to express parallelism. In StarPU, multiple versions of each codelet are required, one for each core type. In our approach, a programmer writes a single version of an accelerated entry method that can execute on multiple core types. HMPP, itself, is limited to a single node, but can be used in conjunction with MPI to program clusters, (i.e. requires a second model unlike our approach). However, StarSs, StarPU, and HMPP also use the same programming constructs for CUDA-based and non-CUDA-based accelerators. In our approach, we use different constructs for CUDA-based and non-CUDA-based accelerators.

18.7   Concluding Remarks

We have extended the Charm++ parallel programming model and runtime system to support accelerators and heterogeneous clusters that include accelerators. We have presented several extensions to the Charm++ programming model, including the *SIMD instruction abstraction, accelerated entry methods*, and *accelerated blocks*. We also briefly discussed our support for CUDA-based GPUs. All of these extensions are continuing to be developed and improved upon, as we increase support for heterogeneous clusters in Charm++.

Bibliography

[1] Cédric Augonnet, Samuel Thibault, Raymond Namyst, and Maik Nijhuis. Exploiting the Cell/BE architecture with the StarPU unified runtime system. In *SAMOS Workshop—International Workshop on Systems, Architectures, Modeling, and Simulation*, Lecture Notes in Computer Science, Samos, Greece, July 2009. To appear.

[2] Eduard Ayguade, Rosa M. Badia, Francisco D. Igual, Jesus Labarta, Rafael Mayo, and Enrique S. Quintana-Orti. An extension of the StarSs programming model for platforms with multiple GPUs. In *Proceedings of the 15th International Euro-Par Conference, Lecture Notes in Computer Science*, Delft, The Netherlands, August 2009. To appear.

[3] Kevin J. Barker, Kei Davis, Adolfy Hoisie, Darren J. Kerbyson, Mike Lang, Scott Pakin, and Jose C. Sancho. Entering the petaflop era: The architecture and performance of Roadrunner. In *SC '08: Proceedings of the 2008 ACM/IEEE Conference on Supercomputing*, pages 1–11, Piscataway, NJ, USA, 2008. IEEE Press.

[4] Pieter Bellens, Josep M. Perez, Rosa M. Badia, and Jesus Labarta. Exploiting locality on the cell/b.e. through bypassing. In *SAMOS '09: Proceedings of the 9th International Workshop on Embedded Computer Systems: Architectures, Modeling, and Simulation*, pages 318–328, Berlin, Heidelberg, 2009. Springer-Verlag.

[5] Abhinav Bhatele, Sameer Kumar, Chao Mei, James C. Phillips, Gengbin Zheng, and Laxmikant V. Kale. Overcoming scaling challenges in biomolecular simulations across multiple platforms. In *Proceedings of IEEE International Parallel and Distributed Processing Symposium 2008*, April 2008.

[6] Eric Bohm, Abhinav Bhatele, Laxmikant V. Kale, Mark E. Tuckerman, Sameer Kumar, John A. Gunnels, and Glenn J. Martyna. Fine grained parallelization of the Car-Parrinello ab initio MD method on Blue Gene/L. *IBM Journal of Research and Development: Applications of Massively Parallel Systems*, 52(1/2):159–174, 2008.

[7] Brian Bouzas, Robert Cooper, Jon Greene, Michael Pepe, and Myra Jean Prelle. MultiCore Framework: An API for Programming Heterogeneous Multicore Processors. Mercury Computer System's Literature Library (http://www.mc.com/mediacenter/litlibrarylist.aspx).

[8] Leonardo Dagum and Ramesh Menon. OpenMP: An industry-standard API for shared-memory programming. *IEEE Computational Science & Engineering*, 5(1), January-March 1998.

[9] Romain Dolbeau, Stephane Bihan, and Francois Bodin. HMPP: A hybrid multi-core parallel programming environment. In *Workshop on General Purpose Processing on Graphics Processing Units*, October 2007.

[10] Alexandre E. Eichenberger, Kathryn O'Brien, Kevin O'Brien, Peng Wu, Tong Chen, Peter H. Oden, Daniel A. Prener, Janice C. Shepherd, Byoungro So, Zehra Sura, Amy Wang, Tao Zhang, Peng Zhao, and Michael Gschwind. Optimizing compiler for the cell processor. In *PACT '05: Proceedings of the 14th International Conference on Parallel Architectures and Compilation Techniques*, pages 161–172, Washington, DC, USA, 2005. IEEE Computer Society.

[11] Kayvon Fatahalian, Timothy J. Knight, Mike Houston, Mattan Erez, Daniel Reiter Horn, Larkhoon Leem, Ji Young Park, Manman Ren, Alex Aiken, William J. Dally, and Pat Hanrahan. Sequoia: Programming the memory hierarchy. In *Proceedings of the 2006 ACM/IEEE Conference on Supercomputing*, 2006.

[12] Message Passing Interface Forum. MPI-2: Extensions to the message-passing interface, 1997. http://www.mpi-forum.org/docs/mpi-20-html/mpi2-report.html.

[13] A. Gholoum, E. Sprangle, J. Fang, G. Wu, and Xin Zhou. CT: A flexible parallel programming model for tera-scale architectures. Technical report, Intel Whitepaper, 2007.

[14] Richard L. Graham, Galen M. Shipman, Brian W. Barrett, Ralph H. Castain, George Bosilca, and Andrew Lumsdaine. Open MPI: A high-performance, heterogeneous MPI. In *Proceedings, Fifth International Workshop on Algorithms, Models and Tools for Parallel Computing on Heterogeneous Networks*, Barcelona, Spain, September 2006.

[15] The Khronos Group. Open computing language, 2008. www.khronos.org.

[16] Pritish Jetley, Filippo Gioachin, Celso Mendes, Laxmikant V. Kalé, and Thomas R. Quinn. Massively parallel cosmological simulations with ChaNGa. In *Proceedings of IEEE International Parallel and Distributed Processing Symposium 2008*, 2008.

[17] J. A. Kahle, M. N. Day, H. P. Hofstee, C. R. Johns, T. R. Maeurer, and D. Shippy. Introduction to the Cell Processor. *IBM Journal of Research and Development: POWER5 and Packaging*, 49(4/5):589, 2005.

[18] Laxmikant V. Kalé. Performance and productivity in parallel programming via processor virtualization. In *Proc. of the First Intl. Workshop on Productivity and Performance in High-End Computing (at HPCA 10)*, Madrid, Spain, February 2004.

[19] David Kunzman. Charm++ on the Cell Processor. Master's thesis, Dept. of Computer Science, University of Illinois, 2006. http://charm.cs.uiuc.edu/papers/KunzmanMSThesis06.shtml.

[20] David Kunzman, Gengbin Zheng, Eric Bohm, and Laxmikant V. Kalé. Charm++, Offload API, and the Cell Processor. In *Proceedings of the Workshop on Programming Models for Ubiquitous Parallelism*, Seattle, WA, USA, September 2006.

[21] David M. Kunzman and Laxmikant V. Kalé. Towards a framework for abstracting accelerators in parallel applications: Experience with cell. In *SC '09: Proceedings of the Conference on High Performance Computing Networking, Storage and Analysis*, pages 1–12, New York, NY, USA, 2009. ACM.

[22] Alexey Lastovetsky and Ravi Reddy. Heterompi: Towards a message-passing library for heterogeneous networks of computers. *J. Parallel Distrib. Comput.*, 66(2):197–220, 2006.

[23] Michael D. McCool. Data-parallel programming on the cell be and the GPU using the rapidmind development platform. In *GSPx Multicore Applications Converence*, 2006.

[24] NVidia. *CUDA Programming Guide, Version 2.2.1*, May 2009. http://developer.download.nvidia.com/.

[25] M. Ohara, H. Inoue, Y. Sohda, H. Komatsu, and T. Nakatani. MPI microtask for programming the Cell Broadband Engine^{TM} processor. *IBM Syst. J.*, 45(1):85–102, 2006.

[26] Judit Planas, Rosa M. Badia, Eduard Ayguad, and Jesus Labarta. Hierarchical tasked-based programming with StarSs. In *International Journal of High Performance Computing Applications*, June 2009.

[27] Larry Seiler, Doug Carmean, Eric Sprangle, Tom Forsyth, Michael Abrash, Pradeep Dubey, Stephen Junkins, Adam Lake, Jeremy Sugerman, Robert Cavin, Roger Espasa, Ed Grochowski, Toni Juan, and Pat Hanrahan. Larrabee: A many-core x86 architecture for visual computing. *ACM Trans. Graph.*, 27(3):1–15, 2008.

[28] Lukasz Wesolowski. An application programming interface for general purpose graphics processing units in an asynchronous runtime system. Master's thesis, Dept. of Computer Science, University of Illinois, 2008. http://charm.cs.uiuc.edu/papers/LukaszMSThesis08.shtml.

Chapter 19

Efficient Parallel Scan Algorithms for Manycore GPUs

Shubhabrata Sengupta

University of California, Davis

Mark Harris

NVIDIA Corporation

Michael Garland

NVIDIA Corporation

John D. Owens

University of California, Davis

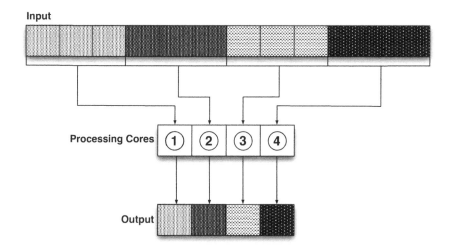

FIGURE 19.1: A simple memory access pattern in which each processor reads a contiguous bounded neighborhood of input (each neighborhood has a different hatching pattern) and produces one output item.

19.1   Introduction

We have witnessed a phenomenal increase in computational resources for graphics processors units (GPU) over the last few years. The highest performing graphics processors from both ATI and NVIDIA already have billions of transistors, resulting in more than a teraflop of peak processing power. This incredible processing power comes from the presence of hundreds of processing cores, all on the same chip.

This computational resource has historically been designed to be best suited for the *stream programming model*, which has many independent threads of execution. In this model a single program operates in parallel on each input, producing one output element for each input element. This model extends nicely to problems in which each output element depends on a small, bounded neighborhood of inputs. This simple streaming memory access pattern is shown in Figure 19.1.

Many interesting problems (sorting, sparse matrix operations) require more general access patterns in which each output may depend on a variable number of inputs. In addition, the ratio between the number of input elements and the number of output elements may not be constant. This implies that processing cores may produce a variable number (including zero)

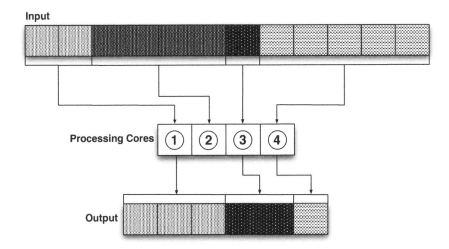

FIGURE 19.2: A more general memory access pattern in which each processor reads input of variable length and produces output of variable length. Processor 2 produces no output.

of elements. Two different scenarios may occur: *expansion*, in which a smaller number of input elements produces a large number of output elements; and *contraction*, in which a larger number of input elements produces a smaller number of output elements. Figure 19.2 shows a variable input/output memory access pattern. Elements being processed by processor core 1 undergo an expansion while those processed by cores 2, 3 and 4 undergo a contraction.

This is a challenging scenario for fine-grained parallelism, since we would like each processor core to work independently on its section of the input, and also to write its output independently into the output array. To allow this parallel execution, the key question is "where in the output array does each processor core write its data?" To answer this question, each processor core needs to know where the processor core to its left will finish writing its data, resulting in an apparent serial dependency.

A recurrence relation is yet another common case in which a serial dependency arises. This is expressed in the following loop.

```
for all i
    B[i] = f(B[i-1], A[i])
```

In this case each result element `B[i]` depends on all elements to its left.

In Section 19.3 we describe a family of algorithmic primitives, the *scan* primitives, that let us solve such seemingly serial problems efficiently in the

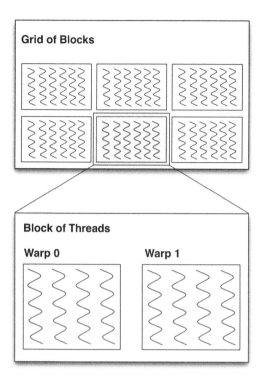

FIGURE 19.3: In the hierarchical architecture of the CUDA programming model, a grid is composed of multiple thread blocks, each of which is composed of multiple warps of threads (squiggly lines). This diagram shows only four threads per warp for simplicity.

data-parallel world. But first it is helpful to look at our programming environment, so that we can present code snippets along the way.

19.2   CUDA—A General-Purpose Parallel Computing Architecture for Graphics Processors

We chose NVIDIA's CUDA GPU Computing environment [14,15] for our implementation. CUDA provides a direct, general-purpose C language interface to the programmable processing cores (also called streaming multiprocessors or SMs) on NVIDIA GPUs, eliminating much of the complexity of writing non-graphics applications using graphics application programming interface (APIs) such as OpenGL.

CUDA executes GPU programs using a grid of *thread blocks* of up to 512 threads each. Each thread executes functions that are either declared `__global__`, which means they can be called from the CPU, or `__device__`, which means they can be called from other `__global__` or `__device__` functions. The host program specifies the number of thread blocks and threads per block, and the hardware and drivers map thread blocks to processing cores on the GPU. Within a thread block, threads can communicate through shared memory and cooperate using simple thread synchronization. Threads are scheduled on processor cores in groups of 32 called warps. All threads in a warp execute one instruction at a time in lockstep. This hierarchy is shown in Figure 19.3. Another popular programming environment for parallel processors, OpenCL [13], has an identical hierarchy, albeit with different terminology.

19.3 Scan: An Algorithmic Primitive for Efficient Data-Parallel Computation

Scan is an algorithmic primitive that solves a wide class of interesting problems on processors that have been designed to provide maximum performance for streaming access [2,3]. These operations are the analogs of parallel prefix circuits [11], which have a long history, and have been widely used in collection-oriented languages dating back to APL [10].

Even though the scan primitive is conceptually quite simple, it forms the basis for a surprisingly rich class of algorithms. These include various sorting algorithms (radix, quick, merge), computational geometry algorithms (closest pair, quickhull, line of sight), graph algorithms (minimum spanning tree, maximum flow, maximal independent set) and numerical algorithms (sparse matrix-dense vector multiply) [2]. They also form the basis for efficiently mapping nested data-parallel languages such as NESL [4] on to flat data-parallel machines. In the following sections we look at scan and its segmented variant.

19.3.1 Scan

Given an input sequence a and an associative[1] binary operator $\oplus$ with identity I, an *inclusive* scan produces an output sequence $b = \text{scan} < inclusive > (a, \oplus)$ where $b_i = a_0 \oplus \cdots \oplus a_i$. Similarly, an *exclusive* scan produces an output sequence $b = \text{scan} < exclusive > (a, \oplus)$ where $b_i = I \oplus \cdots \oplus a_{i-1}$.

---

[1]In practice, this requirement is often relaxed to include pseudo-associative operations, such as addition of floating point numbers.

```
template<class OP, class T>
T scan(T *values, unsigned int n)
{
    for(unsigned int i=1; i<n; ++i)
        values[i] = OP::apply(values[i-1] , values[i]);
}
```

FIGURE 19.4: Serial implementation of inclusive scan for generic operator OP over values of type T.

Some common binary associative operators are add (prefix-sum), min (min-scan), max (max-scan) and multiply (mul-scan).

As a concrete example, consider the input sequence:

$$a = [3 \quad 1 \quad 7 \quad 0 \quad 4 \quad 1 \quad 6 \quad 3]$$

Applying an inclusive scan operation to this array with the usual addition operator produces the result

```
scan<inclusive>(a, +) = [ 3 4 11 11 15 16 22 25 ]
```

and the exclusive scan operation produces the result

```
scan<exclusive>(a, +) = [ 0 3  4 11 11 15 16 22 ]
```

As always, the first element in the result produced by the exclusive scan is the identity element for the operator, which in this case is 0.

19.3.1.1   A Serial Implementation

Implementing scan primitives on a serial processor is trivial, as shown in Figure 19.4. Note that throughout this paper, we use C++ templates to make scan generic over the operator OP and the datatype of values T.

19.3.1.2   A Basic Parallel Implementation

Figure 19.5 shows a simple CUDA C implementation of a well-known parallel scan algorithm [5,9]. This code scans the binary operator OP across an array of $n = 2^k$ values using a single thread block of 2^k threads. We make two assumptions—the number of values is a power of 2 and each thread inputs exactly 1 value—to simplify the presentation of the algorithm.

Analyzing the behavior of this algorithm, we see that it will perform only $\log_2 n$ iterations of the loop, which is optimal. However, this algorithm applies the operator $O(n \log n)$ times, which is asymptotically inefficient compared to the $O(n)$ applications performed by the serial algorithm. It also has practical disadvantages, such as requiring $2 \log_2 n$ barrier synchronizations (__syncthreads()).

```
template<class OP, class T>
__device__ T scan(T *values)
{
    // ID of this thread
    unsigned int i = threadIdx.x;

    // number of threads in block
    unsigned int n = blockDim.x;

    for(unsigned int offset=1; offset<n; offset *= 2)
    {
        T t;

        if(i>=offset)  t = values[i-offset];
        __syncthreads();

        if(i>=offset)  values[i] = OP::apply(t, values[i]);
        __syncthreads();
    }
}
```

FIGURE 19.5: Simple parallel scan of 2^k elements with a single thread block of 2^k threads.

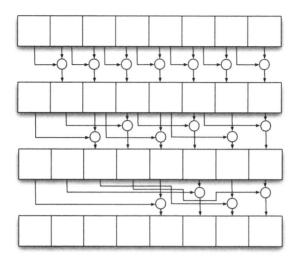

FIGURE 19.6: We show the memory access pattern for the code snippet shown in Figure 19.5. Arrows show the movement of data. Circles show the binary operation. Since our input has 8 elements we need 3 iterations of the `for` loop.

19.3.2   Segmented Scan

Segmented scan generalizes the scan primitive by simultaneously performing separate parallel scans on *arbitrary* contiguous partitions ("segments") of the input vector. For example, an inclusive scan of the + operator over a sequence of integer sequences would give the following result:

$$a = [\ [3 \ 1] \ [7 \ 0 \ \ 4] \ [1 \ 6] \ [3] \ ]$$
$$\mathtt{segscan(a, \ +)} = [\ [3 \ 4] \ [7 \ 7 \ 11] \ [1 \ 7] \ [3] \ ]$$

Segmented scans provide as much parallelism as unsegmented scans, but operate on data-dependent regions. Consequently, they are extremely helpful in mapping irregular computations such as quicksort and sparse matrix-vector multiplication onto regular execution structures, such as CUDA's thread blocks.

Segmented sequences of this kind are typically represented by a combination of (1) a sequence of values and (2) a *segment descriptor* that encodes how the sequence is divided into segments. Of the many possible encodings of the segment descriptor, we focus on using a *head flags* array that stores a 1 for each element that begins a segment and 0 for all others. This representation is convenient for massively parallel machines. All other representations (for example, every element knows the index of the first and/or the last index of its own segment) can naturally be converted to this form.

The head flags representation for the example sequence above is:

```
a.values = [ 3 1 7 0 4 1 6 3 ]
a.flags  = [ 1 0 1 0 0 1 0 1 ]
```

For simplicity of presentation, we will treat the head flags array as a sequence of 32-bit integers; however, it may in practice be preferable to represent flags as bits packed in words.

Schwartz demonstrated that segmented scan can be implemented in terms of (unsegmented) scan by a transformation of the given operator [2,16]. Given the operator $\oplus$ we can construct a new operator $\oplus^s$ that operates on flag–value pairs (f_x, x) as follows:

$$(f_x, x) \oplus^s (f_y, y) = (f_x \vee f_y, \text{ if } f_y \text{ then } y \text{ else } x \oplus y)$$

Segmented scan can also be implemented directly rather than by operator transformation [6,17]. In Section 19.5 we explore both of these implementation strategies.

In the next two sections, we turn to the question of how to efficiently implement the scan primitives on modern GPUs.

19.4   Design of an Efficient Scan Algorithm

19.4.1   Hierarchy of the Scan Algorithm

The most efficient CUDA code respects the block granularity imposed by the CUDA programming model and the warp granularity of the underlying hardware. For example, threads within a block can efficiently cooperate and share data using on-chip shared memory and barrier synchronization. Threads within a warp, however, do not need explicit barrier synchronization in order to share data because they execute in lockstep.

We organize our algorithms to match the natural execution granularities of warps and blocks in order to maximize efficiency. At the lowest level, we design an *intra-warp* primitive to perform a scan across a single warp of threads. We then construct an *intra-block* primitive that composes intra-warp scans together in parallel in order to perform a scan across a block of threads. Finally, we combine grids of intra-block scans into a *global* scan of arbitrary length.

To simplify the discussion, we assume that there is exactly 1 thread per element in the sequence being scanned. The code we present is templated on the input datatype and binary operator used, which is assumed to be associative and to possess an identity value. Input arrays to our scan functions (e.g., `ptr` and `hd`) are assumed to be located in fast on-chip shared memory. The code invoking the scan functions is responsible for loading array data from

```
template<class OP, ScanKind Kind, class T>
__device__ T scan_warp(volatile T *ptr,
                       const uint idx=threadIdx.x)
{
    // index of thread in warp (0..31)
    const uint lane = idx & 31;

    if (lane >=  1)
        ptr[idx] = OP::apply(ptr[idx -  1] , ptr[idx]);
    if (lane >=  2)
        ptr[idx] = OP::apply(ptr[idx -  2] , ptr[idx]);
    if (lane >=  4)
        ptr[idx] = OP::apply(ptr[idx -  4] , ptr[idx]);
    if (lane >=  8)
        ptr[idx] = OP::apply(ptr[idx -  8] , ptr[idx]);
    if (lane >= 16)
        ptr[idx] = OP::apply(ptr[idx - 16] , ptr[idx]);

    if( Kind==inclusive )
        return ptr[idx];
    else
        return (lane>0) ? ptr[idx-1] : OP::identity();
}
```

FIGURE 19.7: Scan routine for a warp of 32 threads with operator `OP` over values of type `T`. The `Kind` parameter is either `inclusive` or `exclusive`.

global (off-chip) device memory into (on-chip) shared memory and storing results back.

19.4.2   Intra-Warp Scan Algorithm

We begin by defining a routine to perform a scan over a warp of 32 threads, shown in Figure 19.7. It uses precisely the same algorithm as shown in Figure 19.5, but with a few basic optimizations. First, we take advantage of the synchronous execution of threads in a warp to eliminate the need for barriers. Second, since we know the size of the sequence is fixed at 32, we unroll the loop. We also add the ability to select either an inclusive or exclusive scan via a `ScanKind` template parameter.

For a warp of size w, this algorithm performs $O(w \log w)$ work rather than the optimal $O(w)$ work performed by a work-efficient algorithm [2]. However, since the threads of a warp execute in lockstep, there is actually no advantage in decreasing work at the expense of increasing the number of steps taken. Each instruction executed by the warp has the same cost, whether executed by a single thread or all threads of the warp. Since the work-efficient reduce/-

downsweep algorithm [2] performs twice as many steps as the algorithm used here, it leads to measurably lower performance in practice.

The removal of explicit barrier synchronization from warp-synchronous code has an unintended consequence. Without synchronization barriers, an optimizing compiler may choose to keep data in registers rather than writing it to shared memory, causing cooperating threads in a warp to read incorrect values from shared memory. Therefore, shared variables used without synchronization by multiple threads in a warp must be declared `volatile` to force the compiler to store the data in shared memory, as shown in Figures 19.7, 19.10, and 19.11.

19.4.3   Intra-Block Scan Algorithm

We now construct an algorithm to scan across all the threads of a block using this intra-warp primitive. For simplicity, we assume that the maximum block size is at most the square of the warp width, which is true for the GPUs we target. Given this assumption, the intra-block scan algorithm is quite simple.

1. Scan all warps in parallel using inclusive `scan_warp()`.

2. Record the last partial result from each warp i.

3. With a single warp, perform `scan_warp()` on the partial results from Step 2.

4. For each thread of warp i, accumulate the partial results from Step 3 into that thread's output element from Step 1.

This organization of the algorithm is only possible because of our assumption that the scan operator is associative. The CUDA implementation of this algorithm is shown in Figure 19.8. The individual steps are labeled and correspond to the algorithm outline.

19.4.4   Global Scan Algorithm

The `scan_block()` routine performs a scan of fixed size, corresponding to the size of the thread blocks. We use this routine to construct a "global" scan routine for sequences of any length as follows.

1. Scan all blocks in parallel using `scan_block()`.

2. Store the partial result (last element) from each block i to `block_results` [i].

3. Perform a scan of `block_results`.

```
template<class OP, ScanKind Kind, class T>
__device__ T scan_block(T *ptr,
                           const uint idx=threadIdx.x)
{
    const uint lane   = idx & 31;
    const uint warpid = idx >> 5;

    // Step 1: Intra-warp scan in each warp
    T val = scan_warp<OP,Kind>(ptr, idx);
    __syncthreads();

    // Step 2: Collect per-warp partial results
    if( lane==31 )  ptr[warpid] = ptr[idx];
    __syncthreads();

    // Step 3: Use 1st warp to scan per-warp results
    if( warpid==0 ) scan_warp<OP,inclusive>(ptr, idx);
    __syncthreads();

    // Step 4: Accumulate results from Steps 1 and 3
    if (warpid > 0) val = OP::apply(ptr[warpid-1], val);
    __syncthreads();

    // Step 5: Write and return the final result
    ptr[idx] = val;
    __syncthreads();

    return val;
}
```

FIGURE 19.8: Intra-block scan routine composed from `scan_warp()` primitives.

4. Each thread of block i combines element i from Step 3 to its output element from Step 1. The combination simply uses the binary operator for the scan operation. For example, if the operation is add, the element i from Step 3 is added to all output elements from Step 1.

Because they require global synchronization, Steps 1 and 2 are performed by the same CUDA kernel, while Steps 3 and 4 are performed in their own separate kernels; thus we require three separate CUDA kernel invocations. Indeed, Step 3 may require recursive application of the global scan algorithm if the number of blocks in Step 1 is greater than the block size.

Aside from the decomposition into kernels, the structure of this global algorithm is strikingly similar to the intra-block algorithm. Indeed, they are nearly identical except for Step 3, where the fixed width of blocks guarantees that the intra-block routine can scan per-warp partial results using a single warp, while the variable block count necessary in the global scan does not provide an analogous guarantee.

19.5   Design of an Efficient Segmented Scan Algorithm

To implement efficient segmented scan routines, we follow the same design strategy already outlined in Section 19.4 for scan. We begin by defining an intra-warp primitive, from which we can build an intra-block primitive, and ultimately a global segmented scan algorithm. The implementations are also quite similar, with the added complications of dealing with arrays of head flags.

19.5.1   Operator Transformation

As described in Section 19.3.2, segmented scan can be implemented by transforming the operator $\oplus$ into a segmented operator $\oplus^s$ that operates on flag–value pairs [16]. This leads to a particularly simple strategy of defining a `segmented<>` template such that `scan<segmented<OP>>` applied to an array of flag–value pairs accomplishes the desired segmented scan. Sample code for such a transformer is shown in Figure 19.9. This trivially converts the inclusive `scan_warp()` and `scan_block()` routines given in Section 19.4 into segmented scans. Achieving a correct exclusive segmented scan via operator transformation requires additional changes to the inclusive/exclusive logic in these routines.

Although a reasonable approach, one downside of relying purely on operator transformation is that it alters the external interface of the scan routines. It accepts a sequence of flag–value pairs rather than corresponding sequences of values and flags. We can restore our desired interface, and si-

```
template<class OP>
struct segmented
{
    template<class T>
    static __host__ __device__
    inline T apply(const T a, const T b)
    {
        T c;
        c.flag  = b.flag | a.flag;
        c.value =
            b.flag ? b.value : OP::apply(a.value, b.value);
        return c;
    }
};
```

FIGURE 19.9: Code for transforming operator OP on values of type T into an operator segmented<OP> on flag–value pairs.

multaneously accommodate correct handling of exclusive scans, by explicitly expanding scan_warp<segmented<OP>>() into the segscan_warp() routine shown in Figure 19.10. The structure of this procedure is exactly the same as the one shown in Figure 19.7 except that (1) it does roughly twice as many operations and (2) it requires slightly different logic for determining the final inclusive/exclusive result.

19.5.2   Direct Intra-Warp Segmented Scan

We have also explored an alternative technique for adapting our basic scan_warp procedure into an intra-warp segmented scan. This routine, which is shown in Figure 19.11, operates by augmenting the conditionals used in the indexing of the successive steps of the algorithm. Each thread computes the index of the head of its segment, or 0 if the head is not within its warp. This is the *minimum index* of the segment, and is recorded in the variable mindex. We compute mindex by writing the index of each segment head to the hd array and propagating it within the warp to other elements of its segment via a max-scan operation (as defined in Section 19.3.1). We take advantage of the unpacked format of the head flags and use them as temporary scratch space.

We use the minimum segment indices to guarantee that elements from different segments are never accumulated. The unsegmented routine shown in Figure 19.7 is essentially just the special case where mindex=0. Figure 19.12 illustrates an example of the resulting data movement for a warp of size 8.

```
template<class OP, ScanKind Kind, class T>
__device__ T segscan_warp(volatile T *ptr,
                          volatile flag_type *hd,
                          const uint idx = threadIdx.x)
{
   const uint lane = idx & 31;

   if (lane >=  1) {
       ptr[idx] =
           hd[idx] ?
           ptr[idx] : OP::apply(ptr[idx - 1] , ptr[idx]);
       hd[idx]  = hd[idx -  1] | hd[idx];
   }
   if (lane >=  2) {
       ptr[idx] =
           hd[idx] ?
           ptr[idx] : OP::apply(ptr[idx - 2] , ptr[idx]);
       hd[idx]  = hd[idx -  2] | hd[idx];
   }
   if (lane >=  4) {
       ptr[idx] =
           hd[idx] ?
           ptr[idx] : OP::apply(ptr[idx - 4] , ptr[idx]);
       hd[idx]  = hd[idx -  4] | hd[idx];
   }
   if (lane >=  8) {
       ptr[idx] =
           hd[idx] ?
           ptr[idx] : OP::apply(ptr[idx - 8] , ptr[idx]);
       hd[idx]  = hd[idx -  8] | hd[idx];
   }
   if (lane >= 16) {
       ptr[idx] =
           hd[idx] ?
           ptr[idx] : OP::apply(ptr[idx - 16] , ptr[idx]);
       hd[idx]  = hd[idx - 16] | hd[idx];
   }

   if( Kind==inclusive )
       return ptr[idx];
   else
       return (lane>0  && !hd[idx]) ?
               ptr[idx-1] : OP::identity();
}
```

FIGURE 19.10: Intra-warp segmented scan derived by expanding `scan_warp` `<segmented<OP>>`.

```
template<class OP, ScanKind Kind, class T>
__device__ T segscan_warp(volatile T *ptr,
                          volatile flag_type *hd,
                          const uint idx = threadIdx.x)
{
    const uint lane = idx & 31;

    // Step 1: Convert head flags to minimum-index form
    if( hd[idx] ) hd[idx] = lane;
    flag_type mindex = scan_warp<op_max, inclusive>(hd);

    // Step 2: Perform segmented scan across warp
    //         of size 32
    if( lane >= mindex + 1 )
        ptr[idx] = OP::apply(ptr[idx - 1] , ptr[idx]);
    if( lane >= mindex + 2 )
        ptr[idx] = OP::apply(ptr[idx - 2] , ptr[idx]);
    if( lane >= mindex + 4 )
        ptr[idx] = OP::apply(ptr[idx - 4] , ptr[idx]);
    if( lane >= mindex + 8 )
        ptr[idx] = OP::apply(ptr[idx - 8] , ptr[idx]);
    if( lane >= mindex +16 )
        ptr[idx] = OP::apply(ptr[idx -16] , ptr[idx]);

    // Step 3: Return correct value for
    //         inclusive/exclusive kinds
    if( Kind==inclusive )
        return ptr[idx];
    else
        return (lane>0 && mindex!=lane) ?
               ptr[idx-1] : OP::identity();
}
```

FIGURE 19.11: Intra-warp segmented scan using conditional indexing.

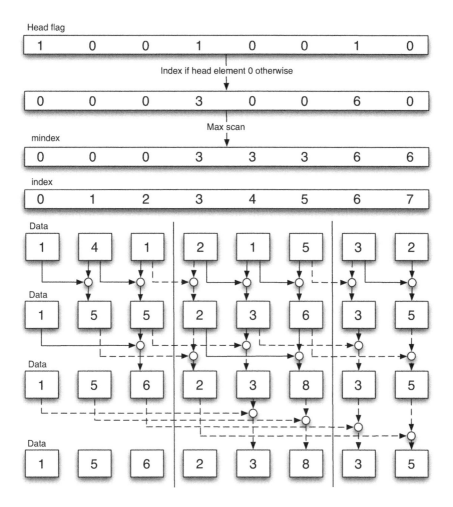

FIGURE 19.12: Data movement in intra-warp segmented scan code shown in Figure 19.11 for threads 0–7 of an 8-thread warp. Data movement in the unsegmented case (dotted arrows) crossing segment boundaries (vertical lines) is not allowed.

19.5.3   Block and Global Segmented Scan Algorithms

As we have done in Section 19.4.1, to maximize efficiency, we need to organize our algorithms to match the natural execution granularities of warp and block. At the lowest level, we have shown two ways to design an *intra-warp* primitive to perform a segmented scan across a single warp of threads. We can then use either of these two ways to construct an *intra-block* primitive that composes intra-warp segmented scans together to perform a segmented scan across a block of threads. Finally, we combine grids of intra-block segmented scans into a *global* segmented scan of arbitrary length.

The methodology used for constructing an intra-block segmented scan is essentially identical to our construction of the intra-block scan routine in Section 19.4.3, with two additional complications. First, when writing the partial result produced by the last thread of each warp in Step 2, we also write an aggregate segment flag for the entire warp. This flag indicates whether there is a segment boundary within the warp, and is simply the (implicit) OR-reduction[2] of the flags of the warp. Second, we accumulate the per-warp offsets in Step 4 only to elements of the first segment of each warp. This process is illustrated in Figure 19.13.

The full CUDA implementation of this algorithm is shown in Figure 19.14. We first record if the warp starts with a new segment, because the flags array is converted to `mindex`-form by the first `segscan_warp()` call. Step 2b determines whether any flag in a warp was set. This step also determines if the thread belongs to the first segment in its warp by checking if (1) the first element of the warp is not the first element of a segment (the warp is *open*) and (2) the index of the head of the segment is 0. The remaining steps compute warp offsets using a segmented scan of per-warp partial results and accumulate them to the per-thread results computed in Step 1.

The global segmented scan algorithm can be built in the same way. We observe that another way to check if a thread belongs to the first segment of a warp (or block) is to do a min-reduction of `hd`. This gives the index of the first element of the second segment in each warp. Each thread then checks if its thread index is less than this index before adding the result in Step 4. This is a typical time vs. space tradeoff. The method used in Figure 19.14 must carry one value per thread but no reduction is necessary; this alternative must carry only one value per warp but requires a reduction. We use the former when propagating data between warps because shared memory loads and stores are cheap. However, when we are doing a global segmented scan, Steps 1 and 3 happen in different kernel invocations, so data must be written to global memory, which is much slower. Therefore, for global scans we do a min-reduction and write only one index per block instead of one per thread.

---

[2]A reduction operation used with the logical OR operator. See Section 19.7.1 for an explanation of reduction.

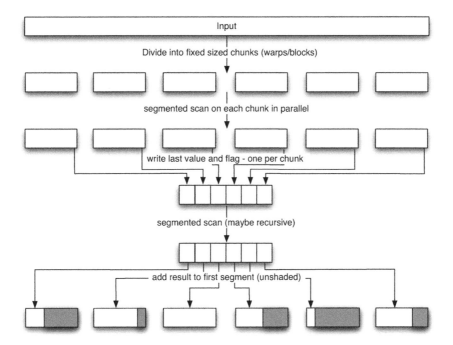

FIGURE 19.13: Constructing block/global segmented scans from warp/block segmented scans. The unshaded part in the last row shows the extent of the first segment.

```
template<class OP, ScanKind Kind, class T>
__device__ T segscan_block(T *ptr, flag_type *hd,
                           const uint idx = threadIdx.x)
{
    // Right shift by 5 as warp size is 32
    uint warpid     = idx >> 5;

    uint warp_first = warpid << 5;
    uint warp_last  = warp_first + 31;

    // Step 1a:
    // Before overwriting the input head flags, record whether
    // this warp begins with an "open" segment.
    bool warp_is_open = (hd[warp_first] == 0);
    __syncthreads();

    // Step 1b:
    // Intra-warp segmented scan in each warp.
    T val = segscan_warp<OP,Kind>(ptr, hd, idx);

    // Step 2a:
    // Since ptr[] contains *inclusive* results, irrespective of Kind,
    // the last value is the correct partial result.
    T warp_total = ptr[warp_last];

    // Step 2b:
    // warp_flag is the OR-reduction of the flags in a warp and is
    // computed indirectly from the mindex values in hd[].
    // will_accumulate indicates that a thread will only accumulate a
    // partial result in Step 4 if there is no segment boundary to
    // its left.
    flag_type warp_flag = hd[warp_last]!=0 || !warp_is_open;
    bool will_accumulate = warp_is_open && hd[idx]==0;

    __syncthreads();

    // Step 2c: The last thread in each warp writes partial results
    if( idx == warp_last )
    {
        ptr[warpid] = warp_total;
        hd[warpid]  = warp_flag;
    }
    __syncthreads();

    // Step 3: One warp scans the per-warp results
    if( warpid == 0 )
        segscan_warp<OP,inclusive>(ptr, hd, idx);

    __syncthreads();

    // Step 4: Accumulate results from Step 3, as appropriate.
    if( warpid != 0 && will_accumulate )
        val = OP::apply(ptr[warpid-1], val);
    __syncthreads();

    ptr[idx] = val;
    __syncthreads();

    return val;
}
```

FIGURE 19.14: Intra-block segmented scan routine built using intra-warp segmented scans.

19.6   Algorithmic Complexity

Given the hierarchy of blocks and warps described above, we can calculate the algorithmic complexity to scan n elements. Let B and w represent the block and warp size in threads, respectively. We assume that each block scans $O(B)$ elements (i.e., $O(1)$ elements per thread). To scan n elements we use n/B blocks. Let S and W denote step and work complexity, respectively. The *work complexity* of an algorithm is defined as the total number of operations the algorithm performs. Multiple operations may be performed in parallel in each step. The *step complexity* of an algorithm is the number of steps required to complete the algorithm given an infinite number of threads/processors; this is essentially equivalent to the critical path length through the data dependency graph of the computation. A subscript of n, b or w indicates the complexity of the whole array, a block, or a warp, respectively.

As already discussed, our `scan_warp()` routine has step complexity $S_w = O(\log_2 w)$ and work complexity $W_w = O(w \log_2 w)$. For a block containing B/w warps, we arrive at the following step and work complexity for a block scan.

$$S_b = O(S_w \log_w B) = O\left((\log_2 w)\left(\frac{\log_2 B}{\log_2 w}\right)\right) = O(\log_2 B) \qquad (19.1)$$

$$W_b = O\left(\sum_{i=1}^{\log_w B} \left\lceil \frac{B}{w^i} \right\rceil W_w\right) = w \log_2 w \left(\sum_{i=1}^{\log_w B} \left\lceil \frac{B}{w^i} \right\rceil\right) = O\left(B \log_2 w\right) \qquad (19.2)$$

The same pattern extends to the array (global) level. The step and work complexity for an array comprising an arbitrary number of blocks, n/B, is given by the following expressions.

$$S_n = O(S_b \log_B n) = O((\log_2 w)(\log_B n)) = O(\log_2 n) \qquad (19.3)$$

$$W_n = O\left(\sum_{i=1}^{\log_B n} \left\lceil \frac{n}{B^i} \right\rceil W_b\right) = w \log_2 w \left(\sum_{i=1}^{\log_B n} \left\lceil \frac{n}{B^i} \right\rceil \sum_{j=1}^{\log_w B} \left\lceil \frac{B}{w^j} \right\rceil\right) = O\left(n \log_2 w\right) \qquad (19.4)$$

For any given machine, it is safe to assume that the warp size w is in fact some constant number. Under this assumption, the step complexity of our algorithm is $S_n = O(\log n)$ and the work complexity is $W_n = O(n)$, both of which are asymptotically optimal.

19.7   Some Alternative Designs for Scan Algorithms

In this section we show two different ways to structure the data flow among the various levels of a scan implementation. Both work at the intra-block and the inter-block level and they can be combined. They also work for both segmented and unsegmented scans, though for the sake of simplicity we restrict the discussion to the unsegmented version. We also refer readers to the recent technical report by Merrill and Grimshaw [12] with their memory-bandwidth-bound scan implementation and extensive performance analysis.

19.7.1   Saving Bandwidth by Performing a Reduction

Dotsenko et al. [7] have showed an alternative strategy for an efficient scan implementation. They too follow a bottom-up strategy where the intra-block scan is built from smaller scans (which could be warp-wide or smaller, but rarely larger for efficiency reasons) and the scan of an arbitrary number of elements is based on intra-block scans.

The key contribution of this method is the use of a reduction operation as the first stage. A reduction operation on an array of elements is a binary operator applied to all elements in a "telescopic" fashion. This means the binary operator is applied to the first two elements, the result of which is combined with the third element by the same operator. A simple example is performing a reduce on an array with an add operator that produces the sum of all elements in the array.

To illustrate, we show how a scan of an arbitrary number of elements can be performed using a block-wide reduce (`reduce_block()`) and scan (`scan_block ()`) primitive.

1. Do a reduction operation on all blocks in parallel using `reduce_block()` and store the result from each block i to `block_results[`i`]`.

2. Perform a scan of `block_results`.

3. One thread of each block i combines element i from Step 3 to its first element from Step 1. The combination just uses the binary operator for the scan operation. For example, if the operation is add, the element i from Step 3 is added to the first output element from block i in Step 1.

4. Scan all blocks in parallel using `scan_block()`.

The same strategy can be used to construct an intra-block scan (`scan_block ()`) by doing a reduction and subsequent scan on warp-wide or even smaller set of elements. The reduction and scan at this level is done sequentially: one thread does the reduction and scan operation on all elements of this smallest set.

An interesting thing to note is that this method has to save less state between Step 1 and Step 4. While the method shown in Section 19.4.4 has to write out N elements at the end of Step 1, this one only writes out N/B elements where B is the block size. On the other hand this method does more computation since the simple add operation in Step 4 in Section 19.4.4 is replaced by a reduction operation in Step 1 here. This is a commonly used optimization strategy on GPUs where off-chip bandwidth is limited while computational resources are rarely so.

On the flip side, a possible disadvantage is that such a combination of reduction then scan may assume that the operator is commutative. This would restrict the operators that can be used for scan to those which are now not only associative but also commutative.

19.7.2 Eliminating Recursion by Performing More Work per Block

Step 3 in the method described in Section 19.4.4 and Step 2 in the method described in Section 19.7.1 require a recursive call to scan. If each block is performing a scan on B elements, then `block_results` ends up with N/B elements after the first step. If $N/B > TC$, where TC is the number of threads in a block, we have to do a global scan on `block_results` recursively. Otherwise we can use our intra-block scan algorithm (Section 19.4.3) to scan `block_results`, eliminating all further recursive kernel calls and associated data transfer to and from off-chip memory.

Thus given an N we choose B such that $N/B \leq TC$. Given that all our intra-block algorithms assume that each thread processes one element, and given that each block can only have TC threads (TC can take a maximum value of 512 in current architectures), for large enough N, TC will be smaller than B.

Figure 19.15 shows that small modifications to the existing intra-block algorithm (Section 19.4.3) allow us to perform a scan on a block of B elements when $TC \leq B$. As we can see, the function `scan_block_anylength` divides the elements into blocks of TC and iterates over the block serially. The key difference is the presence of the variable `reduceValue` that carries the result of the reduce operator on one block of TC elements to the next.

The same kind of modification can be done to the approach shown in Section 19.7.1 to make the intra-block scan function operate on inputs of arbitrary length.

```
template<class OP, ScanKind Kind, class T>
__device__ void scan_block_anylength(T *ptr,
                                     const T *in,
                                     T *out,
                                     const uint B,
                                     const uint idx=
                                          threadIdx.x,
                                     const uint bidx=
                                          blockIdx.x,
                                     const uint TC=
                                          blockDim.x)
{
    const uint nPasses = float(ceil(B/float(TC)));

    T reduceValue = OP::identity();

    for (uint i = 0; i < nPasses, ++i)
    {
        const uint offset = i * TC + (bidx * B);

        // Step 1: Read TC elements from global (off-chip)
        // memory to shared memory (on-chip)
        T input = ptr[idx] = in[offset + idx];
        __syncthreads();

        // Step 2: Perform scan on TC elements
        T val = scan_block<OP, Kind, T>(ptr);

        // Step 3: Propagate reduced result from previous
            block
        // of TC elements
        val = OP::apply(val, reduceValue);

        // Step 4: Write out data to global memory
        out[offset + idx] = val;

        // Step 5: Choose reduced value for next iteration
        if (idx == (TC-1))
        {
            ptr[idx] =
                (Kind == exclusive) ?
                OP::apply(input, val) : val;
        }
        __syncthreads();

        reduceValue = ptr[TC-1];
        __syncthreads();
    }
}
```

FIGURE 19.15: A block level scan that scans a block of B elements with the help of scan_block, which does a scan of TC elements at a time, where $TC \leq B$. TC is the number of threads in a block.

19.8   Optimizations in CUDPP

The code given in Sections 19.4 and 19.5 illustrate the core parallel scan algorithms we use. We make our implementation available as open-source components of CUDPP, the "CUDA Data Parallel Primitives" library (available at http://www.gpgpu.org/developer/cudpp/). To achieve peak performance for scan kernels, our CUDPP library combines the basic algorithms described above with an orthogonal set of further optimizations.

The biggest efficiency gain comes from optimizing the amount of work performed by each thread. We find that processing one element per thread does not generate enough computation to hide the I/O latency to off-chip memory, and so we assign eight input elements to each thread. In our implementations this is handled when data are loaded from global device memory into shared memory. Each thread reads two groups of four elements from global memory and scans both groups of four sequentially. The rightmost result in each group of four (i.e., the reduction of the four inputs) is fed as input to a routine very similar to the block level routines shown in Figure 19.8 and Figure 19.14. The key difference is that the block level routines are slightly modified to handle two inputs per thread instead of one as shown here. When the block level routine terminates, the result is accumulated back to the groups of four elements that were scanned serially.

The next most important set of optimizations focuses on minimizing the number of registers used. The GPU architecture relies on multithreading to hide memory access latency, and the number of threads that can be co-resident at one time is often limited by their register requirements. Therefore, it is important to maximize the number of available co-resident threads by minimizing register usage. Optmizing register usage is particularly important for segmented scan since many more registers are needed to store and manipulate head flags. The CUDPP code uses a number of low-level code optimizations designed to limit register requirements, including packing multiple head flags into the bits of registers after they are loaded from off-chip memory.

19.9   Performance Analysis

In this section, we analyze the performance of the scan and segmented scan routines that we have described and compare them to some alternative approaches. All running times were collected on an NVIDIA GeForce GTX 280 GPU with 30 SMs. These measurements do not include any transfers of data between CPU and GPU memory, under the assumption that scan and seg-

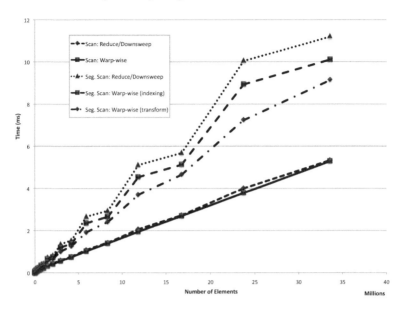

FIGURE 19.16: Scan and segmented scan performance.

mented scan will be used as building blocks in more complicated applications on GPUs.

Our first test consists of running both scan and segmented scan routines over sequences of varying length using the CUDPP test apparatus. These tests scan the addition operator over sequences of single-precision floating point values.

Figure 19.16 compares two scan implementations: our warp-wise scan and the reduce/downsweep algorithm used by Sengupta et al. [17]. Our warp-wise approach is 1 to 20% faster, improving most on scans of non-power-of-two arrays and for arrays smaller than 65K elements. Figure 19.16 compares the reduce/downsweep segmented scan [17] with both of our warp-wise segmented scan kernels: one based on conditional indexing (Figure 19.11) and one based on operator transformation (Figure 19.10). We see the same trend as in the unsegmented case. Our direct warp-based algorithm using conditional indexing is up to 29% faster than the reduce/downsweep algorithm. The kernel derived from operator transformation improves on it by a further 6 to 10%. Compared to the results reported by Sengupta et al. [17] for sequences of 1,048,576 elements running on an older NVIDIA GeForce 8800 GTX GPU, our scan implementation is 2.8 times faster and our segmented scan is 4.2 times faster on the same hardware.

Multiple factors contribute to the performance increase in scan and segmented scan. First, using warp-wise execution minimizes the need for bar-

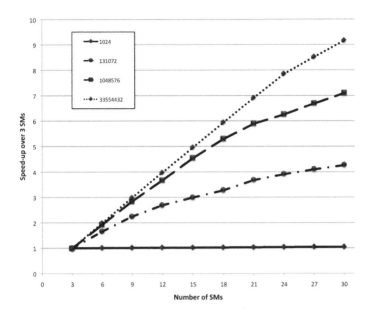

FIGURE 19.17: Parallel scaling of segmented scan on sequences of varying length.

rier synchronization, because most thread communication is within warps. In contrast, the reduce/downsweep algorithm requires barrier synchronizations between each step in both the reduce and downsweep stages. Second, we halve the number of parallel steps required, as compared to the reduce/downsweep algorithm. This comes at the "expense" of additional work, which is actually not a cost at all since threads in a warp execute in lockstep. The intra-warp step complexity is in fact optimal. Third, our segmented scan based on operator transformation interleaves scans of flags and data. Like software pipelining, this increases the distance between dependent instructions. Consequently, the performance of this kernel is higher than the indexing kernel, which performs the scan of the flags before the scan of the values. Finally, while our reduce/downsweep implementations expend much effort to avoid shared memory bank conflicts [8], the intra-warp scan algorithm is inherently conflict free.

The main efficiency advantage of segmented scan is that its performance is largely invariant to the way in which the sequence is decomposed into subsequences. Thus, it implicitly load-balances work across potentially quite unbalanced subsequences. Bell and Garland [1] explore this phenomenon in detail using the very important case of sparse-matrix vector multiplication.

Finally, Figure 19.17 illustrates the scaling of our segmented scan algorithm on sequences of various sizes. We use the running time on 3 SMs of a GeForce

GTX 280 as a baseline and show the speed-up achieved over this baseline for 6–30 SMs. Small sequences show relatively little scaling, as they are small enough to be processed efficiently by a small number of SMs. For sequence sizes above 512K elements, we see strong linear scaling. Scaling results for the scan algorithm are similar. This demonstrates the scalability of both the GPU architecture itself and our algorithmic design.

19.10   Conclusions

The modern many-core GPU is a massively parallel processor and the CUDA programming model provides a straightforward way of writing scalable parallel programs to execute on the GPU. Because of its deeply multithreaded design, a program must expose substantial amounts of fine-grained parallelism to efficiently utilize the GPU. Data-parallel techniques provide a convenient way of expressing such parallelism. Furthermore, the GPU is designed to deliver maximum performance for regular execution paths—via its single instruction multiple data (SIMD) architecture—and regular data access patterns—via memory coalescing—and data-parallel algorithms generally fit these expectations quite well.

We have described the design of efficient scan and segmented scan routines, which are essential primitives in a broad range of data-parallel algorithms. By tailoring our algorithms to the natural granularities of the machine and minimizing synchronization, we have produced one of the fastest scan and segmented scan algorithms yet designed for the GPU.

Acknowledgments

Thanks to Dominik Göddeke for his thoughtful comments and suggestions on this chapter. The UC Davis authors thank our research supporters: the DOE/SciDAC Institute for Ultrascale Visualization, the National Science Foundation (award 0541448) and NVIDIA for both a Graduate Fellowship and equipment donations.

Bibliography

[1] Nathan Bell and Michael Garland. Implementing sparse matrix-vector multiplication on throughput-oriented processors. In *SC '09: Proceedings of the 2009 ACM/IEEE Conference on Supercomputing*, pages 18:1–18:11, November 2009.

[2] Guy Blelloch. *Vector Models for Data-Parallel Computing*. MIT Press, 1990.

[3] Guy E. Blelloch. Scans as primitive parallel operations. *IEEE Transactions on Computers*, 38(11):1526–1538, November 1989.

[4] Guy E. Blelloch, Siddhartha Chatterjee, Jonathan C. Hardwick, Jay Sipelstein, and Marco Zagha. Implementation of a portable nested data-parallel language. *Journal of Parallel and Distributed Computing*, 21(1):4–14, April 1994.

[5] W. J. Bouknight, Stewart A. Denenberg, David E. McIntyre, J. M. Randall, Amed H. Sameh, and Daniel L. Slotnick. The Illiac IV system. *Proceedings of the IEEE*, 60(4):369–388, April 1972.

[6] Siddhartha Chatterjee, Guy E. Blelloch, and Marco Zagha. Scan primitives for vector computers. In *Supercomputing '90: Proceedings of the 1990 ACM/IEEE Conference on Supercomputing*, pages 666–675, November 1990.

[7] Yuri Dotsenko, Naga K. Govindaraju, Peter-Pike Sloan, Charles Boyd, and John Manferdelli. Fast scan algorithms on graphics processors. In *Proceedings of the 22nd Annual International Conference on Supercomputing*, pages 205–213. ACM, June 2008.

[8] Mark Harris, Shubhabrata Sengupta, and John D. Owens. Parallel prefix sum (scan) with CUDA. In Herbert Nguyen, editor, *GPU Gems 3*, chapter 39, pages 851–876. Addison Wesley, August 2007.

[9] W. Daniel Hillis and Guy L. Steele Jr. Data parallel algorithms. *Communications of the ACM*, 29(12):1170–1183, December 1986.

[10] Kenneth E. Iverson. *A Programming Language*. Wiley, New York, 1962.

[11] Richard E. Ladner and Michael J. Fischer. Parallel prefix computation. *Journal of the ACM*, 27(4):831–838, October 1980.

[12] Duane Merrill and Andrew Grimshaw. Parallel scan for stream architectures. Technical Report CS2009-14, Department of Computer Science, University of Virginia, December 2009.

[13] Aaftab Munshi. *The OpenCL Specification (Version 1.0, Document Revision 48)*, 6 October 2009.

[14] John Nickolls, Ian Buck, Michael Garland, and Kevin Skadron. Scalable parallel programming with CUDA. *ACM Queue*, pages 40–53, March/ April 2008.

[15] NVIDIA Corporation. NVIDIA CUDA compute unified device architecture programming guide. `http://developer.nvidia.com/cuda`, January 2007.

[16] Jacob T. Schwartz. Ultracomputers. *ACM Transactions on Programming Languages and Systems*, 2(4):484–521, October 1980.

[17] Shubhabrata Sengupta, Mark Harris, Yao Zhang, and John D. Owens. Scan primitives for GPU computing. In *Graphics Hardware 2007*, pages 97–106, August 2007.

Chapter 20

High Performance Topology-Aware Communication in Multicore Processors

Hari Subramoni, Fabrizio Petrini and Virat Agarwal

IBM T.J. Watson Research Center

Davide Pasetto

IBM Computational Science Center Dublin

Abstract The increasing computational and communication demands of the scientific and industrial communities require a clear understanding of the performance trade-offs involved in multicore computing platforms. Such analysis can help application and toolkit developers in designing better, topology aware, communication primitives intended to suit the needs of various high end computing applications. In this chapter, we take on the challenge of designing and implementing a portable intra-core communication framework for streaming computing and evaluate its performance on some popular multicore architectures developed by Intel, AMD and Sun. Our experimental results, obtained on the Intel Nehalem, AMD Opteron and Sun Niagara 2 platforms, show that we are able to achieve an intra-socket small message latency between 120 and 271 nanoseconds, while the inter-socket small message latency is between 218 and 320 nanoseconds. The maximum intra-socket communication bandwidth ranges from 0.179 (Sun Niagara 2) to 6.5 (Intel Nehalem) GB/s. We were also able to obtain an inter-socket communication performance of 1.2 and 6.6 GB/s on the AMD Opteron and Intel Nehalem, respectively.

20.1   Introduction

In the earlier days of high performance computing, communication performance almost always implied inter-node data exchange happening through some kind of network interconnect, such as Gigabit Ethernet [1], 10 Gigabit Ethernet [2], Myrinet [3], Quadrics [4] and InfiniBand [5]. The advent of modern multicore processors has changed the scenario of high performance computing, with more and more users attempting to consolidate their distributed jobs within a small set of nodes, if not a single node. In this context intra-node communication and the hardware support for accelerating it have become critical to obtain optimal application performance.

To improve performance, instruction set designers have introduced innovative methods to transfer data from one memory location to another. Vector instructions and Streaming SIMD Extensions [6] are examples of the mechanisms present in the latest generations of processors. These instructions offer many useful features such as no cache pollution and higher performance. Comparing the latencies and bandwidths offered by these new communication mechanisms can only give us limited insight into what is actually happening inside a processing node. In order to gain proper understanding of the underlying mechanisms, we can monitor various system hardware parameters, such as cache hits and misses at various levels in the cache hierarchy, contention for resources in the system and the like. In this context, the enhanced set of performance counters available in many multicore architectures, like the Nehalem series of processors from Intel, allow us to delve deeper into the various issues involved.

In this chapter, we describe the design rationale and implementation details of a portable intra-node communication library for multi-processor architectures. We study the performance trade-offs involved in the intra-node communication mechanisms and evaluate how they scale on different architectures, including the impact of inter- and intra-socket scenarios. Our experimental results show that we are able to achieve an intra-socket small message latency of 120, 222 and 271 *nanoseconds* on Intel Nehalem, AMD Opteron and Sun Niagara 2 respectively. The inter-socket small message latencies for Intel Nehalem and AMD Opteron are 320 and 218 *nanoseconds*, respectively. The maximum intra-socket communication bandwidth obtained were 6.5, 1.37 and

0.179 *GB/s* for Intel Nehalem, AMD Opteron and Sun Niagara 2, respectively. We were also able to obtain an inter-socket communication performance of 6.6 and 1.2 *GB/s* for Intel Nehalem and AMD Opteron, respectively.

Our analysis shows that a single communication strategy cannot achieve the best performance for all message sizes. While the basic memory-based copy scheme (*memcpy*) gives good performance for small messages, hand-coded vector loads and stores provide better performance for medium sized messages. But since both these schemes involve two copy operations to move the data from the source to the destination, this can adversely affect the communication performance for large messages. In this scenario we achieve better performance by using a kernel-based approach to copy the data from the source to the destination. An interesting effect we observed is the apparent poor performance of the streaming instructions for intra-socket as well as inter-socket communication. Deeper analysis using the various performance counters showed that the performance degradation was related to the high number of resource stalls caused by the streaming instructions.

More specifically, the contributions of this chapter can be summarized as follows.

- Design and implementation of a novel communication framework for streaming architectures.

- Performance evaluation of this communication framework on multiple processor architectures.

- In-depth analysis of the performance results using hardware performance counters.

The rest of the chapter is organized as follows. Section 20.2 gives a brief overview of the related technologies. Section 20.3 explains the methodology followed in this work. Our experimental results and analysis are detailed in Section 20.4. We explore the related scholarly work that has already been published in the field in Section 20.5 and we provide some concluding remarks in Section 20.6.

20.2   Background

In this section we present a brief overview of the multicore architectures evaluated in this chapter, Intel Nehalem, Sun Niagara, AMD Opteron, as well as the message passing interface (MPI) communication library that we use as a baseline performance comparison.

20.2.1   Intel Nehalem

The Intel Xeon 5500 processor cluster is the latest server implementation of a new 64-bit micro architecture [7]. The *Nehalem* machines are based on the non-uniform memory access (NUMA) architecture [8] with each socket containing four processing cores. Each core has exclusive L1 and L2 caches and all the cores in a socket share an 8 MB L3 cache. Each core is clocked at 2.93 GHz, with a theoretical peak performance of 44.8 Gflop/s per chip and each compute node is capable of reaching a peak performance of 89.6 Gflop/s. The compute cores use the fast Quick Path Interconnect (QPI) [9] links to communicate with other cores, and the memory controller. Each QPI link offers a bandwidth of 12.8 GB/s in each direction. The Intel Nehalem architecture is shown in Figure 20.1.

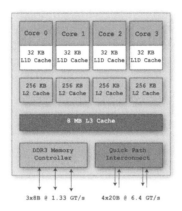

FIGURE 20.1: Intel Nehalem processor architecture.

The Xeon 5560 includes several new mechanisms that help improve application performance. Turbo mode provides frequency-stepping that enables the processor frequency to be increased in increments of 133 MHz. The amount of Turbo boost available varies with the processor. Intel's hyper-threading technology enables two threads to execute on each core to hide data access latencies: the two threads can execute simultaneously, filling each other's unused stages in the functional unit pipelines.

20.2.2   Sun Niagara

The Sun Niagara 2 is a multi-threaded multicore CPU. The general block diagram of the Sun Niagara 2 processor is shown in Figure 20.2. It contains eight cores, each capable of running four simultaneous threads. It has a speed bump feature by which it can increase the clock rate of each thread from 1.2 GHz to 1.6 GHz. The processor can interact with the external world through a PCI Express port and two 10 GigE ports. The 4 MB of L2 cache memory are divided into eight banks with 16-way associativity. The Sun Niagara 2 also contains a number of hardware encryption engines as well as four dual-channel FBDIMM [10] memory controllers.

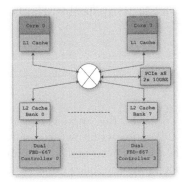

FIGURE 20.2: Sun niagara 2 processor architecture.

20.2.3   AMD Opteron

The architecture of the AMD Opteron processor is shown in Figure 20.3. The AMD Opteron can execute 32-bit as well as 64-bit applications without any performance penalty. The CPU sockets communicate using the Direct Connect Architecture [11] over high speed Hyper Transport links. Opteron also is one of the first X86 architecture processors to use a NUMA architecture [8].

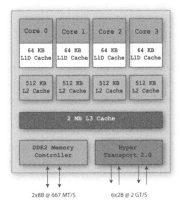

FIGURE 20.3: AMD opteron processor architecture.

20.2.4   MPI

The MPI [12] is a programming model used for writing parallel applications in the cluster computing area. MPI libraries provide basic communication support for a parallel computing job. In particular, several convenient point-to-point and collective communication operations are provided. High performance MPI implementations are closely tied to the underlying network dynamics and try to leverage the best communication performance on the given interconnect. In this chapter we utilize the MPICH2 1.0.8p1 [13] MPI library for our baseline evaluation; however, our observations in this context are quite general and should be applicable to other high performance MPI libraries as well.

20.3   Methodology

In this section, we describe the approach used to evaluate the various multicore architectures. We create our own communication infrastructure to analyze and evaluate the various intra-node communication mechanisms available. In particular, we compare the following message exchange mechanisms.

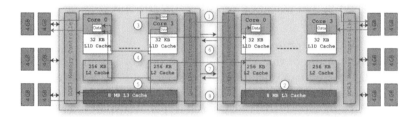

FIGURE 20.4: Data paths in modern multicore architectures.

20.3.1   Basic Memory-Based Copy

We use the basic *memcpy* function call to copy the data from the user buffer to the shared communication buffer and back to the user buffer on the other side. This form of data transfer will, in general, take path number 2 as shown in Figure 20.4: data start from a socket attached memory bank and cross all the cache hierarchy (L3, L2, L1) inside the chip, data proceed through the inter-socket communication link and then traverse the second socket memory hierarchy to its destination. Modern compilers contain highly optimized versions of memcpy, specifically tuned for individual CPUs, but these implementations normally do not consider the fact that the originating thread is "sending" the data to another thread and will not access them any more, so this implementation may not be optimal for a communication library.

20.3.2   Vector Instructions

The vector operations available in the Intel Streaming SIMD Extensions (SSE2) [6] set use the *movdqa* and *movdqu* instructions to transfer 16 bytes of data from the source to the destination buffer. While the *movdqa* instruction is meant for use only with 16 bytes aligned memory locations, the *movdqu* instruction can be used with either aligned or un-aligned memory locations. Such alignment issues do not affect the performance of applications in the latest Nehalem processor due to the presence of a hardware unit which makes appropriate corrections for this. The data transferred using this mechanism will also take path 2 shown in Figure 20.4. These instructions are not available on the Sun Niagara CPU.

20.3.3 Streaming Instructions

The SSE2 and SSE 4.1 [6] instruction sets contain a number of operations that should in theory achieve better performance for this workload. The streaming instructions will follow path 1 shown in Figure 20.4, completely bypassing the cache hierarchy inside the CPU and avoiding cache pollution. There are basically two streaming instructions available in the SSE instruction set:

- **movntdt** is a non-temporal store which copies the data from the source address to the destination address without polluting the cache lines. If the data are already in cache, then the cache lines are updated. This instruction is capable of copying 16 bytes of data at a time and the data must be 16 byte aligned. This assembly level instruction is wrapped up inside the intrinsic [6] *_mm_stream_si128*.

- **movnti** is a non-temporal instruction similar to *movntdq* except for the fact that we only copy 4 bytes of data here and the data need not be aligned to any byte boundary. This assembly level instruction is wrapped up inside the intrinsic *_mm_stream_si32*.

20.3.4 Kernel-Based Direct Copy

All the memory-copy based approaches exploit a two-copy method to transfer the data from the source buffer to the destination buffer. Although, the overhead of this two-copy method is not an issue for small messages, it becomes a critical bottleneck as the size of the message increases. In this context, we opt for a single copy approach with kernel-based memory copies using the LiMIC2 library [14]. This library abstracts kernel-based memory transfers into a few user level functions. This approach has some costs associated with it as well, like the CPU privilege switching between user and kernel space. Since this overhead can be high, this method is not ideal to transfer small or medium sized messages as the overhead will overshadow the benefits obtained from removing one memory copy. This means that we can use the kernel-based copy method only for large message sizes and must utilize another approach for smaller messages.

20.4 Experimental Results

Our experimental setup consists of three different platforms.

- **Intel Nehalem**: This host is equipped with the Intel Nehalem series of processors with two quad-core processor nodes operating at 2.93 GHz with 48 GB RAM and a PCIe 2.0 interface. Red Hat Enterprise Linux Server release 5.2 with Linux Kernel version 2.6.28 was used on the host.

- **AMD Opteron**: This host is equipped with the AMD Opteron series of processors with four quad-core nodes operating at 1.95 GHz with 16 GB RAM and a PCIe 1.0 interface. Red Hat Enterprise Linux Server release 5.4 with Linux Kernel version 2.6.18 was used on the host.

- **Sun Niagara**: This host is equipped with a Sun Niagara 2 processors with 16 GB of memory.

The legends in the various graphs are as given below. The first half of the legend refers to the communication mechanism used (Memcpy, Stream, Vector and LiMIC) and the second half refers to the type of compiler used—GCC or ICC. GCC compiler version *4.4.0* and ICC Compiler version *11.0* were used in all the experiments. The only compiler optimization used was the *-O3* flag. As only the latest Intel-64 architecture gives support for the various new instruction formats and performance counters, we do most of our in-depth evaluation on this system.

20.4.1   Intra-Socket Performance Results

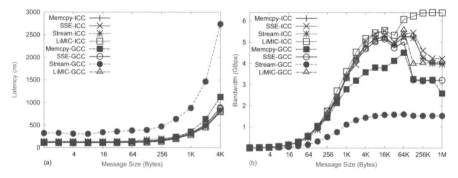

FIGURE 20.5: Intra-socket performance (a) Small message latency and (b) Bandwidth.

We analyze the intra-socket performance of various communication modes in this section. Figures 20.5 (a) and (b) show the intra-socket latency and bandwidth, respectively, for the various communication mechanisms under consideration. As we are interested in the latencies for small messages, we only detail the latency numbers for messages smaller than 4 KB. The communication performance of the entire message range is shown by the bandwidth graphs. As we can see, the kernel-based memory copy approach with the ICC compiler (LiMIC-ICC) gives the best bandwidth, while the standard memory copy scheme with the ICC compiler (Memcpy-ICC) gives the best performance for very small messages. The reason behind this performance difference is that vector-based instructions can only transfer data in chunks of 16 bytes and will incur that overhead even for messages of less than 16 bytes. Overall, we see that *memcpy*-based approaches give the best performance for very small messages, the hand-coded *SSE*-based approach gives the best performance for small to medium sized messages while the kernel-based copy approach using *LiMIC* gives the best performance for large messages.

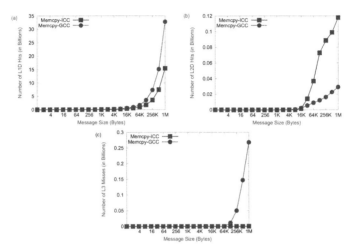

FIGURE 20.6: Intra-socket *memcpy* cache performance for Intel Nehalem architecture: (a) L1D hits (b) L2D hits and, (c) L3 misses.

It is worth noting the performance difference shown by the standard *Memcpy* scheme as implemented by the GCC and the ICC compilers. The performance delivered by the ICC compiler is much higher than that offered by the GCC compiler. In order to understand the reasons behind this performance difference, we analyze the performance using the hardware performance counters provided by the latest Intel x86-64 architecture. In particular, we look at how effectively the compilers are able to use the cache hierarchy to achieve the best performance. Figures 20.6 (a), (b) and (c) show the number of L1 data hits, L2 data hits and L3 cache misses, respectively. As we can see, the cache performance of the memcpy code generated by the ICC compiler is far better than that obtained by the GCC compiler for the same set of optimization flags.

We also observe similar patterns in the performance of SSE instructions implemented by the GCC and ICC compilers for large message sizes alone. Figures 20.7 (a), (b) and (c) show the cache performance of the SSE instructions for large message sizes. Though the code generated by the GCC compiler is able to achieve a slightly higher number of L1 data hits as opposed to the code generated by the ICC compiler, the large number of L3 misses overshadows any performance gains seen in this regard. Similar trends are observed for the *LiMIC*-based approach as well, but we don't show them there due to lack of space.

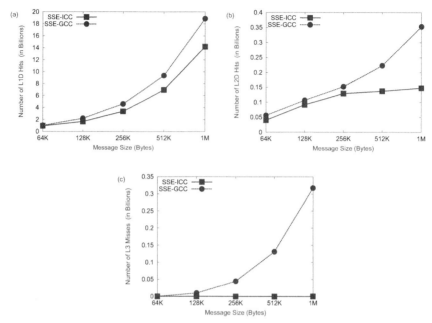

FIGURE 20.7: Intra-socket *SSE* cache performance for Intel Nehalem architecture: (a) L1D hits (b) L2D hits and, (c) L3 misses.

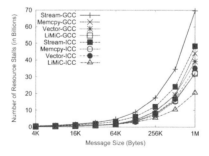

FIGURE 20.8: All intra socket resource stalls for streaming instructions for Intel Nehalem architecture.

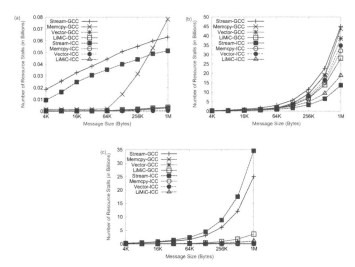

FIGURE 20.9: Intra-socket resource stalls for streaming instructions for Intel Nehalem architecture: (a) Load stalls, (b) Store stalls and (c) Re-order buffer stalls.

Analyzing the behavior of streaming instructions yields some interesting trends as seen in Figures 20.5 (a) and (b). The first is the difference in performance of the streaming instructions generated by the GCC and the ICC compilers. The performance of *Stream-ICC* is comparable to *Memcpy-ICC*, while there is a substantial difference between the GCC version of the two mechanisms. Analyzing the assembly level code generated by the two compilers shows that the ICC compiler is replacing non-temporal (streaming) instructions, with a performance optimized version of *memcpy* implemented by Intel called *intel_fast_memcpy*. Due to the architecture specific performance optimizations done by the Intel compiler, the code generated is identical to the normal *memcpy* operation. We are currently exploring multiple options through which we can disable just this code replacement while retaining the other performance enhancements done by the compiler.

We can also see that the streaming instructions, in general, exhibit very poor performance. This is the case for both the GCC and ICC. To further explore the causes for this difference in performance, we can look at the possibility of resource contention in the system. We again rely on the performance counters to obtain insights into the performance. Figure 20.8 shows the overall number of resource stalls observed in the system for the various communication mechanisms. As we can see, the number of stalls is highest for the GCC generated streaming instructions followed closely by the streaming instructions generated by ICC. Resource stalls can be present for various reasons, like load stalls, store stalls, pipeline buffer stalls etc. Figures 20.9 (a), (b) and (c) depict the stalls caused due to Load, Store and Re-Order Buffer (ROB) full scenario that we faced in the system. We chose these three metrics over

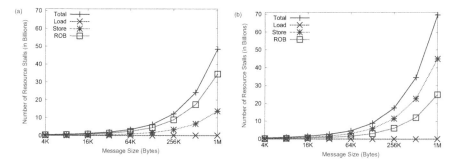

FIGURE 20.10: Intra-socket resource stalls split up for streaming instructions for Intel Nehalem architecture with: (a) ICC compiler and, (b) GCC compiler.

the others as they formed a significant percentage of the total resource stalls. As we can see, the stalls due to store instructions and ROB full scenario account for a significant percentage of all the resource stalls for the streaming instructions. Figures 20.10 (a) and (b) show the resource stall split up for the ICC and GCC generated versions of the streaming instructions, respectively.

20.4.2   Inter-Socket Performance Results

We now analyze the inter-socket performance of the various communication modes under investigation in this section. Figures 20.11 (a) and (b) show the latency and bandwidth performance of inter-socket communication. As we can see, they follow the trends seen for intra-socket communication seen in Section 20.4.1. Due to the nature of data transfer, we might expect that the performance of streaming instructions would be better than that for the intra-socket operations, but in reality, we do not see any such benefits. We believe that this is due to the high number of resource stalls encountered in the system while attempting to perform streaming instructions. Whatever performance benefits that would have been obtained by using the streaming instructions seem to get negated by the high number of resource stalls.

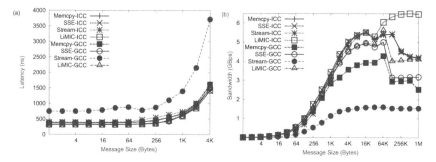

FIGURE 20.11: Inter-socket performance of Intel Nehalem architecture (a) Small message latency (b) Bandwidth.

20.4.3 Comparison with MPI

In this section, we compare the performance of a standard MPI implementation with our communication mechanism. As we saw from the performance evaluation of intra-socket and inter-socket communication, the best performance is obtained by using a combination of *Memcpy*, *Vector* and kernel-based memory copy using *LiMIC*. We refer to this scheme as *Hybrid* and compare it against an MPI implementation. Figures 20.12 and 20.13 show the intra-socket and inter-socket performance comparison of our hybrid scheme against MPI, respectively. One point to note here is that, while the MPI-based tests keep the pipeline full by performing a window-based send/recv operation, we only perform one send and recv at a time. Pipelining the transfers should get us better bandwidth than what is being shown in the graphs. We also need to keep in mind that MPI has the extra overheads of tag matching and additional MPI level packet headers. This will have some impact on the latency of small messages.

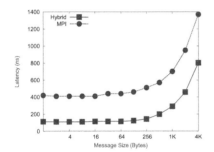 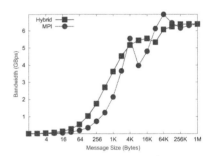

FIGURE 20.12: Intra-socket performance comparison with MPI for Intel Nehalem architecture: (a) Small message latency and (b) Bandwidth.

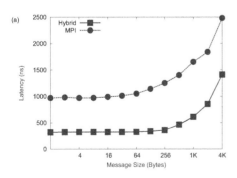 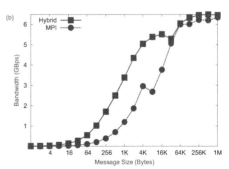

FIGURE 20.13: Inter-socket performance comparison with MPI for Intel Nehalem architecture: (a) Small message latency and (b) Bandwidth.

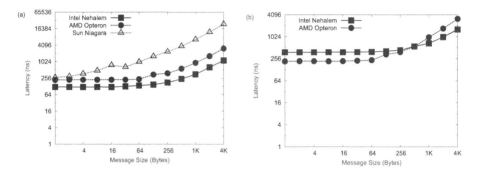

FIGURE 20.14: Latency comparison of various architectures: (a) Intra-socket and (b) Inter socket.

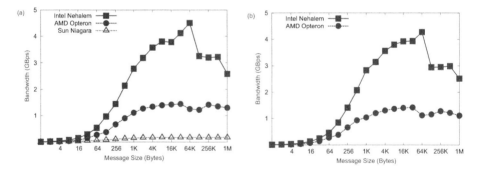

FIGURE 20.15: Bandwidth comparison of various architectures: (a) Intra-socket and (b) Inter-socket.

20.4.4 Performance Comparison of Different Multicore Architectures

In this section we compare the intra-node and inter-node communication performance of the various multicore architectures used. As *GCC* is the only common compiler available on all the machines used for our experiments, we compare the performance of the various architectures under consideration using the basic memory-based copy approach (memcpy). A comparison of inter-node and intra-node performance for various architectures is shown in Tables 20.1 and 20.2. We see that Nehalem gives the best performance for intra-socket communication followed by AMD. Inter-socket performance, on the other hand, shows an interesting trend where the performance of Nehalem machines for small messages is lower than that of the AMD machines. The AMD numbers were taken on sockets that were one hop apart, which is comparable to the Nehalem architecture. Due to technical difficulties, we were not able to gather the inter socket numbers on the Sun Niagara machines.

TABLE 20.1: Communication performance comparison of various architectures

	Intel Nehalem	AMD Opteron	Sun Niagara 2
Latency (*nanoseconds*)			
Intra-Sock	110	222	271
Inter-Sock	321	218	-
Bandwidth (*GB/s*)			
Intra-Sock	6.5	1.4	0.18
Inter-Sock	6.6	1.4	-

TABLE 20.2: Communication performance of Intel Nehalem

	Memcpy ICC	SSE ICC	Stream ICC	LiMIC ICC	Memcpy GCC	SSE GCC	Stream GCC	LiMIC GCC
Latency (*nanoseconds*)								
Intra-Sock	123	113	139	110	121	112	308	114
Inter-Sock	382	321	397	325	382	327	751	331
Bandwidth (*GB/s*)								
Intra-Sock	5.4	5.5	5.3	6.5	4.6	5.1	1.6	5.7
Inter-Sock	5.5	5.5	5.5	6.6	4.3	5.1	1.6	5.7

20.5   Related Work

In this section we report work done by others on performance evaluation of interprocess communication mechanism for modern multicore CPUs. In [15] the authors look at the performance and power consumption of the Intel Nehalem machines using the SPEC CPU benchmarks [16]. In [17] the authors compare the performance of the Intel Nehalem processor with the Barcelona and Tigerton processors from Intel; they also look at bandwidth and latency performance of intra-socket and inter-socket communication on these architectures. In [18] the authors do a performance comparison of Intel Nehalem and the Intel Harpertown architectures; they use the SPEC MPI 2007 [16] benchmark for the evaluation. In [19] the authors compare and contrast the computation and communication capability of the Intel Nehalem architecture and see how the improvements in processor and interconnect technology have affected the balance of computation to communication performance.

20.6   Conclusion and Future Work

In this chapter we have evaluated the communication performance of three families of multicore architectures.We have analyzed the results and gathered valuable insight into several performance aspects. We have discovered that, though the streaming instructions are expected to deliver good performance, the current implementation generates a high number of resource stalls and consequently yields poor performance. We have also found out that the intra-node communication performance is highly dependent on the memory and cache architecture as we saw from the performance analysis of the basic memory-based copy as well as SSE2 vector instructions. The results of the experimental analysis done with our intra-node communication infrastructure showed that we can achieve an intra-socket small message latency of 120, 222 and 371 nanoseconds on Intel Nehalem, AMD Opteron and Sun Niagara 2, respectively. The inter-socket small message latencies for Intel Nehalem and AMD Opteron are 320 and 218 nanoseconds. The maximum intra-socket communication bandwidth obtained were 6.5, 1.37 and 0.179 GB/s for Intel Nehalem, AMD Opteron and Sun Niagara 2. We were also able to obtain inter-socket communication performance of 1.2 and 6.6 GB/s for the AMD Opteron and Intel Nehalem.

Bibliography

[1] H. Frazier and H. Johnson, "Gigabit Ethernet: From 100 to 1000Mbps."

[2] "Ethernet Working Group," http://ieee802.org/3/.

[3] N. J. Boden, D. Cohen, R. E. Felderman, A. E. Kulawik, C. L. Seitz, J. N. Seizovic, and W. K. Su, "Myrinet: A Gigabit-per-Second Local Area Network," http://www.myricom.com.

[4] F. Petrini, W. C. Feng, A. Hoisie, S. Coll, and E. Frachtenberg, "The Quadrics Network (QsNet): High-Performance Clustering Technology," in *Hot Interconnects*, 2001.

[5] Infiniband Trade Association, http://www.infinibandta.org.

[6] "Streaming SIMD Extensions 2," http://en.wikipedia.org/wiki/SSE2.

[7] "Intel 64 Architecture," http://www.intel.com/technology/intel64/.

[8] "NUMA," http://en.wikipedia.org/wiki/Non-Uniform_Memory_Access.

[9] Intel Corporation, "Quick Path Interconnect," http://www.intel.com/technology/quickpath/index.htm.

[10] "Fully Buffered DIMM," http://en.wikipedia.org/wiki/FBDIMM.

[11] "Direct Connect Architecture," http://www.amd.com/us/products/technologies/direct-connect-architecture/Pages/direct-connect-architecture.aspx.

[12] MPI Forum, "MPI: A Message Passing Interface," in *Proceedings of Supercomputing*, 1993.

[13] "MPICH2: High Performance Portable MPI Implementation," http://www.mcs.anl.gov/research/projects/mpich2.

[14] Hyun-Wook Jin, Sayantan Sur, Lei Chai, and Dhabaleswar K. Panda, "Lightweight Kernel-Level Primitives for High-Performance MPI Intra-Node Communication over Multi-Core Systems," 2007.

[15] Axel Busch and Julien Leduc, "Evaluation of Energy Consumption and Performance of Intel's Nehalem Architecture," http://openlab-mu-internal.web.cern.ch/openlab-mu-internal/Documents/2_Technical_Documents/Technical_Reports/2009/CERN_openlab_report-Eval-of-energy-consumption-and-perf-of-Intellem-achitecture.pdf, Tech. Rep., 2009.

[16] SPEC Benchmark, http://www.spec.org/.

[17] Kevin J. Barker, Kei Davis, Adolfy Hoisie, Darren J. Kerbyson, Mike Lang, Scott Pakin, and Jose Carlo Sancho, "A Performance Evaluation of the Nehalem Quad-Core Processor for Scientific Computing," vol. 18, 2008.

[18] H. O. Bugge, "An Evaluation of Intel's Core i7 Architecture Using a Comparative Approach," Tech. Rep., 2009.

[19] H. Subramoni, M. Koop, and D. K. Panda, " Designing Next Generation Clusters: Evaluation of InfiniBand DDR/QDR on Intel Computing Platforms," in *17th Annual Symposium on High-Performance Interconnects (HotI'09)*, New York City, NY, August 2009.

Index